E. Hörbst · C. Müller-Schloer · H. Schwärtzel

Design of VLSI Circuits

Based on VENUS

With 192 Figures

Springer-Verlag Berlin Heidelberg New York
London Paris Tokyo 1987

Dr. phil. Egon Hörbst
Managing Director of the CAD Project "VENUS", Corporate Applied Computer Sciences, Siemens AG, Munich

Dr. Ing. Christian Müller-Schloer
Group Leader "Systems Architectures", Corporate Applied Computer Sciences, Siemens AG, Munich

Prof. Dr. techn. Heinz Schwärtzel
Managing Director of Corporate Applied Computer Sciences, Siemens AG, Munich

ISBN-13: 978-3-642-95527-3 e-ISBN-13: 978-3-642-95525-9
DOI: 10.1007/978-3-642-95525-9

Library of Congress Catalog Card Number.
Hörbst, Egon. VENUS, design methods for integrated circuits.
Bibliography: p. Includes index.
1. Integrated circuits – Design and construction – Data processing. I. Müller-Schloer, C. (Christian), 1950– .
II. Schwärtzel, H. III. Title. TK7874.H673 1987 621.381'73 87–9624
ISBN-13: 978-3-642-95527-3

Softcover reprint of the hardcover 1st edition 1987

Typesetting: Triltsch, Würzburg;
2362/3020-543210

Preface

Microelectronics are certainly one of the key-technologies of our time. They are a key factor of technological and economic progress. They effect the fields of automation, information and communication, leading to the development of new applications and markets.

Attention should be focused on three areas of development:

- process and production technology,
- test technology,
- design technology.

Clearly, because of the development of new application fields, the skill of designing integrated circuits should not be limited to a few, highly specialized experts Rather, this ability should be made available to all system and design engineers as a new application technology – just like nrogramming technology for software.

For this reason, design procedures have to be developed which, supported by appropriate CAD systems, provide the design engineer with tools for representation effective instruments for design and reliable tools for verification, ensuring simpre, proper and easily controllable interfaces for the manufacturing and test processes. Such CAD systems are called standard design systems. They open the way to fast and safe design of integrated circuits.

First, this book demonstrates basic principles with an example of the Siemens design system VENUS, gives a general introduction to the method of designing integrated circuits, familiarizes the reader with basic semiconductor and circuit technologies, shows the various methods of layout design, and presents necessary concepts and strategies of test technology.

Second, the book imparts all information necessary for designing integrated circuits with a standard design system, including essential user instructions for VENUS in a simplified form. In addition, the entire design process for an integrated circuit is described with an example of a simple circuit serving as a learning tool. Moreover some of the cell libraries which are offered as part of VENUS are listed, and a choice of cells sufficient for one's own designs are described in detail as a cell catalogue. Methods of cell library development and quality assurance are also presented. Finally, this book shows developing trends and the possibility of placing design systems for integrated circuits within the framework of CAD systems used for developing larger types of electronic systems.

The drive to develop standard design systems comes from the field of systems technology which has also helped to shape their current capabilities. The development of the design system VENUS is based upon the entire range of electronic systems technology such as computer technology, whose requirements were pointed out by Dr. Braeckelmann, communications technology – whose requirements were pointed out by Dr. Pfrenger, Dr. von Sichart, Dr. Stegmeier –, and automations technology – whose requirements were pointed out by Mr. Schaeff and Mr. Dittmann.

It is particularly important for standard design procedures that the manufacturer guarantees reliable operation and correctness of cell libraries. This is possible only if the libraries and processes are developed with careful quality control. This was done by Mr. Saehn, Dr. Schrader, and Dr. Zibert from the Component Division.

Further contributions to single chapters of the book were provided by: M. Gonauser, Dr. A. Gilg, M. Hernandez and K. Forster. The authors owe special gratitude to Ms. E. Pfeuffer, Mr. M. Spaventa and Mr. M. Müller who were responsible for the English version of the book. Ms. B. Fruehauf, Ms. K. Brosseder and Ms. C. Zeiser rewrote the text several times with great patience.

Munich, April 1987

E. Hörbst
C. Müller-Schloer
H. Schwärtzel

List of Co-Authors

Paul Birzele	Chapter 3	Karlheinz Horninger	Chapter 2
Manfred Gerner	Chapter 4	Gerd Sandweg	Chapter 7
Ernst Göttler	Chapter 5	Thomas Wecker	Chapter 3
Otto Grüter	Chapter 6		

All co-authors are members of Corporate Applied Computer Sciences, Siemens AG, Munich.

Contents

1 Introduction to the Design of Integrated Circuits

This book has three main goals:

First, we want to show that the development of modern design procedures for integrated circuits which have simple and formal interfaces with the design engineer as well as with the production process (standard design procedures) have resulted in new ways of solving user-specific problems: A new kind of application technology has thus been formed. It might find as wide an application as the programming technology for software which, in the beginning of the sixties, became a universal application technology. Second, we want to give general introduction to the methods of designing integrated circuits. We will discuss basic technologies as well as various methods of layout design and required testing concepts.

Third, we want to familiarize the user with the simplified methodology of designing integrated circuits with a standard design procedure. VENUS [1] serves as a learning tool for explaining the current level of desgin automation. The book is, however, not meant as a substitute for a detailed VENUS user manual, nor is it just a catalogue of the functions and cells that any user manual contains.

Chapters 1 to 4 are meant for the reader who wishes to obtain a summary of design methods for integrated circuits, technologies for microelectronics, various layout design methods and the aspects of testing.

Chapters 5 and 6 impart the practical know-how required for designing integrated circuits with VENUS. They also include basic instructions for using VENUS. All steps, starting with the preparatory considerations necessary for each part of the design up to the test, are described and illustrated with an example used as pedagogic tool. Furthermore, Chap. 5 describes the full range of cell libraries and masters offered in VENUS 1; it deals with the problems of developing them and assuring their quality, and it offers a choice of cells arranged in a cell catalogue adequate for user designs.

Finally, in Chap. 7, VENUS is placed in the wider scope of designing large electronic systems.

1.1 User Technology for Digital Electronics

During the last 30 years digital technology has, to an incredible extent, been successful in solving problems arising within applied electrical engineering. The most outstanding and common example in the field of digital electrical engineering is the processor

---

[1] Registered trademark of Siemens AG, Munich. The name is the acronym for "VLSI-Entwicklung und Simulation" (VLSI development and simulation).

which forms the heart of the computer. Together with the computer, a new user technology was developed in the early sixties, namely, programming technology for software. Packages of methods and tools were developed: The *software (SW) programming systems.*

Their main components are: the programming language, the compiler and a package of tools for specification, editing, test and documentation. More recent developments combine all these components, thus creating programming environments.

The scope of the computer, supplemented by other digital circuits in order to reach specific system solutions, is constantly expanding into other fields of technology, such as testing technology, automatic control technology, communications technology, office technology and many others. In order to explain this phenomenon, one needs only look at the advantages of digital technology in terms of reliability, adaptability and efficiency in problem solving, due to progress in the field of circuit integration.

The standardizing effect digital technology has had on the methods of design has probably been a decisive factor for the spread of digital circuits. The components of SSI [2] circuit families in TTL [3] or CMOS [4] technologies has influenced a whole generation of circuit designers. It was no longer the level of distinct individual transistors (the smallest analog units) that formed the base of the design, but rather the complex, digital level of logic gates, flipflops, counters and decoders. With that, a building block system has been created which allows the solving of customer-specific problems by means of free configurations formed quickly from standard components.

For this purpose, the circuit designer had to design a circuit on gate level, based on a specification. It was possible for him to make use of the standard circuits offered by the manufacturers of integrated circuits (ICs). The subsystems built on printed circuit boards (pc boards) were finally combined in a customer-specific system solution (see Fig. 1.1). With about ten equivalent circuit functions per IC and thirty ICs per board, a circuit with 1500 gates thus required approximately five boards!

With increasing complexity of integration, complete subsystems with several thousands of transistors could be placed on one component or chip. For IC manufacturers the problem arose that the specialization, owing to increased complexity, was reflected in smaller production volumes per type. This specialization had to be reconciled with the demand for a high number of pieces, raised by the production lines. The solution was to develop generally applicable standard components, the most prominent type of which finally became the microprocessor. With the microprocessor, that part of a circuit which is application-, or customer-specified could be shifted back to the software.

From the very beginning, of course, the microprocessor was developed for a group of users different from that of computers. The group which uses the microprocessor is that of system engineers. Together with the microprocessor, a second, new application technology for electronics was developed in the second half of the seventies, whose tools were combined in microprocessor (μP) development systems. Their major

---

[2] SSI: Small Scale Integration.

[3] TTL: Transistor Transistor Logic.

[4] CMOS: Complementary Metal Oxide Semiconductor.

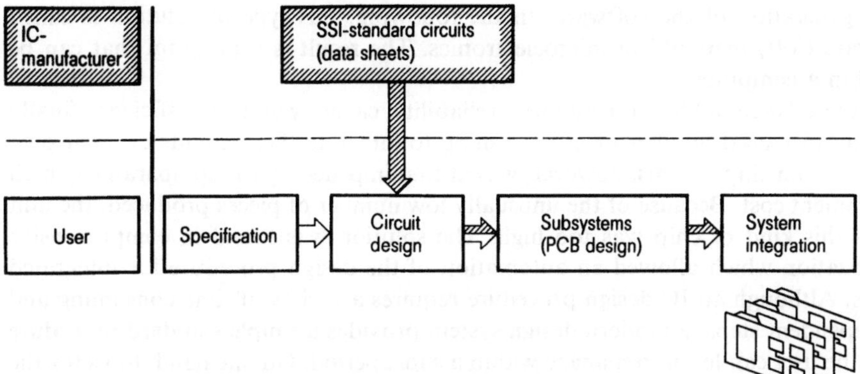

Fig. 1.1. Problem solving with SSI components

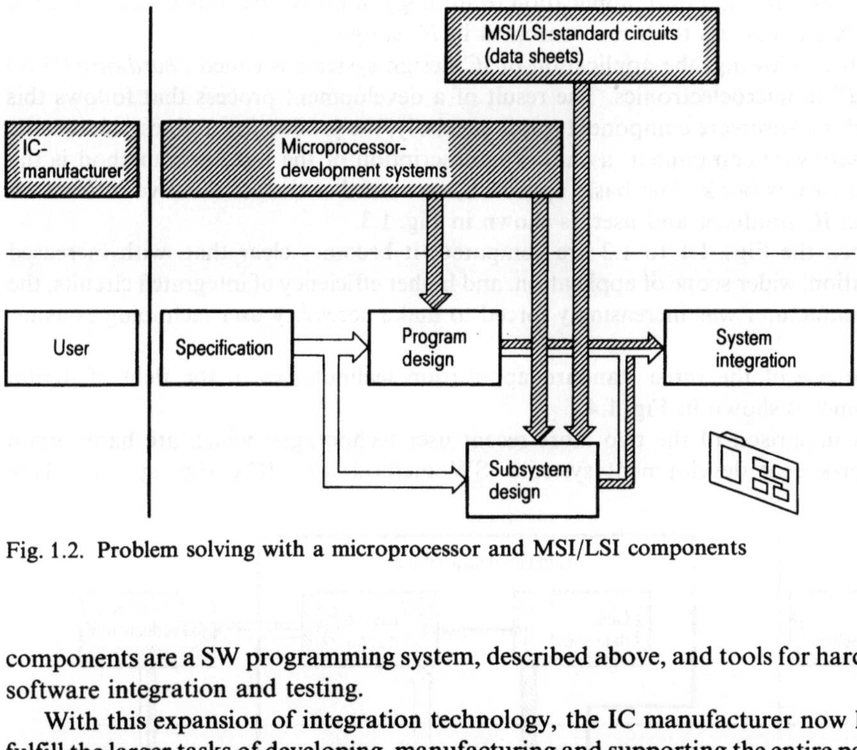

Fig. 1.2. Problem solving with a microprocessor and MSI/LSI components

components are a SW programming system, described above, and tools for hardware/software integration and testing.

With this expansion of integration technology, the IC manufacturer now had to fulfill the larger tasks of developing, manufacturing and supporting the entire package of software and hardware development as well as system integration and test. This resulted in the microprocessor development system described above.

Thus, the development process can be carried out as follows:

From the variety of processors, peripherals and memories, the user chooses the appropriate components, combines them on a board in a suitable way, and finally programs his computer in order to obtain the customer-specific system solution (Fig. 1.2). If standard single-board computers are used, the user's work is limited to the generation of the software. In the following, this type of solution is called "software (SW) method" in microelectronics. The result is a program that can be loaded in a computer.

Increased demand for compactness, reliability, capacity and cost efficiency finally led to an increased number of users wishing to integrate their circuits on their own chip ("custom chip"). This, however, was at first impeded by the comparatively high development cost. Because of the unusually low number of pieces produced, the unit cost of this kind of chip was too high. The solution was found in comprehensive formalization which allowed an automation of the design procedure for integrated circuits. Although an IC design procedure requires a variety of time-consuming and interdependent steps, a modern design system provides a simple standard procedure which the user can learn to manage within a short period. On one hand, tools for the individual steps of design are available for use in such design systems. On the other hand, the organizational process which has become more complicated, is supervised and automatically adhered to. For developing chips which are specifed by their users ("personal chips"), a third application technology of microelectronics was formed in the early eighties. Its tools are combined in *IC design systems*.

In the following, the application of IC design systems is called "*hardware* (HW) *method*" in microelectronics. The result of a development process that follows this method is a hardware component: more precisely, the production and test documents for a hardware component, a chip. The description of the hardware method is the content of this book. The basic approach, combined with the intensive interaction between IC producer and user, is shown in Fig. 1.3.

When the Figs. 1.1 to 1.3 are compared, it becomes clear that, with increased integration, wider scope of application, and higher efficiency of integrated circuits, the IC manufacturer was increasingly forced to make necessary user technologies available.

The rise of the three standard application technologies in the field of digital electronics is shown in Fig. 1.4.

A comparison of the two more recent user technologies which are based upon microprocessor development systems (SW method) and IC-design systems (HW

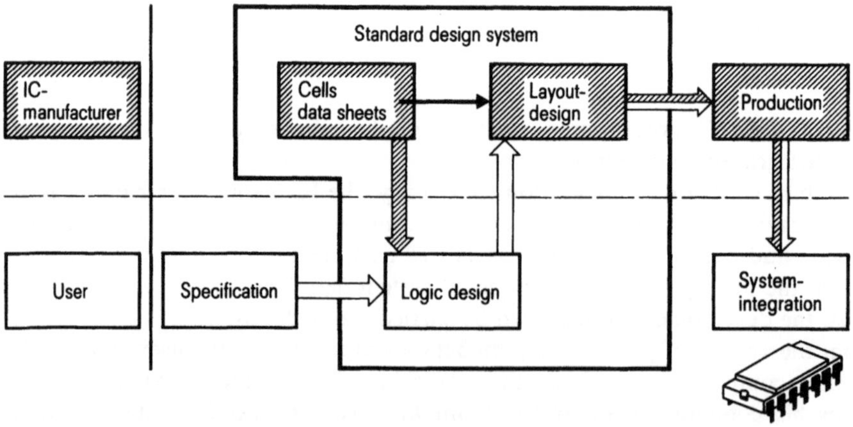

Fig. 1.3. Problem solving with application specific VLSI components

<table>
<tr><th colspan="3">Technological medium</th><th>System support for application technology</th><th>Date of origin</th></tr>
<tr><td rowspan="3">Digital circuit</td><td colspan="2">Processor</td><td>SW programming system</td><td>Since 1963</td></tr>
<tr><td rowspan="2">Integrated circuit</td><td>Microprocessor</td><td>μP development system</td><td>Since 1973</td></tr>
<tr><td>Application specific IC</td><td>IC design system</td><td>Since 1983</td></tr>
</table>

Fig. 1.4. Application technologies of digital electronics

method), shows interesting analogies. For this reason, the development processes for one problem, utilizing the different technologies, are compared in the following two paragraphs.

This comparison shows some important similarities, but also a number of distinctions, resulting in a better understanding of the special features of the respective development process.

Both alternatives have the basic methods of problem solving in common. Those methods will be discussed in more detail in the following paragraph. Thereafter, SW method and HW method will be outlined, showing the individual steps, and compared by means of an example.

1.2 The Process of Problem Solving

In connection with microelectronics development, two new user technologies have arisen for systems development. For problem solving with microprocessors, the microprocessor development systems are available. This kind of problem approach is called "SW method". For problem solving with problem-specified integrated circuits, IC design systems have been developed. This way of problem solving is called "HW method". A comparison of the two approaches makes clear that, at least in this rough sketch, there are no differences in principle between SW method and HW method.

1.2.1 Phase Model of the Problem Solving Process

The description of every complicated process requires the framework of a basic model. For this reason, we first want to outline a basic model of the problem solving process on an abstract level.

Today, it is common to use a phase model as the basic model for complex system development. For organizational reasons, this is done mainly in order to divide the process of problem solving into controllable sections, where results can simply be stated, and tasks can be distributed.

Because of computer aided automation, this is mainly done to obtain exact interfaces for the flow of information, and to ensure the consistency of the data. The phase model is divided into six phases.

The starting point of the problem solving process is always an exact verbal description of the requirements (requirement catalogue). This is obtained as the result of a *study phase* (1).

Based upon this study, the specification or performance description is developed by applying available techniques for problem solving. The description consists of two components, object specification and test specification. The object specification includes the solution concept, the latter consisting of the functional concept, the performance concept and the structure concept.

For the development of subsystems, their arrangement in the main system plays a crucial role. The interface between subsystem and main system is defined by criteria such as type, format and range of data and command flow. For object specification, the performance requirements are also defined. These are, in most cases, the determination of critical time conditions, or of maximum values, such as the maximum current supply or the size of memory.

Test specification, a tool which becomes more and more important as problem solving in electronics becomes increasingly complex, describes test methods, test tools, cases and processes.

The specification is the result of the specification phase (2).

Now the realization or actual development begins with the design phase or, as in this case where complex problems have to be solved, the *system design phase* (3). Starting with the specification, a more formalized description or representation, is begun. Usually a top-town method is applied. The system function is divided into hierarchical levels from top to bottom. It is divided into single functions which are placed in a static connection structure, and whose interactions are controlled by means of a dynamic cooperation structure. Connection and cooperation structures are called function structures. In autonomous processing systems, however, the cooperation structure often consists in turn of two components: one for the information flow, and one for the control flow.

The subsystems are designed in more detail in a step-by-step resolution process. At each level of this process, the target function of the subsystem is divided into a number of subfunctions which are represented by function blocks (modules), and are connected to each other through a functional structure. The flow of control commands which integrates these subfunctions according to the functional structure of the subsystem, can be described with procedural diagrams, for example, in the form of Nassi-Schneidermann diagrams. The hierarchically layered resolution process is ended when the elementary functions have been reached.

In this phase of the development process, the corner-stones are set for an organization which, often necessarily, divides the task of succeeding design steps, that is, it allocates the individual subfunctions of the subsystems to parallel working groups. In this phase, the complete interface architecture is also developed. The test system is

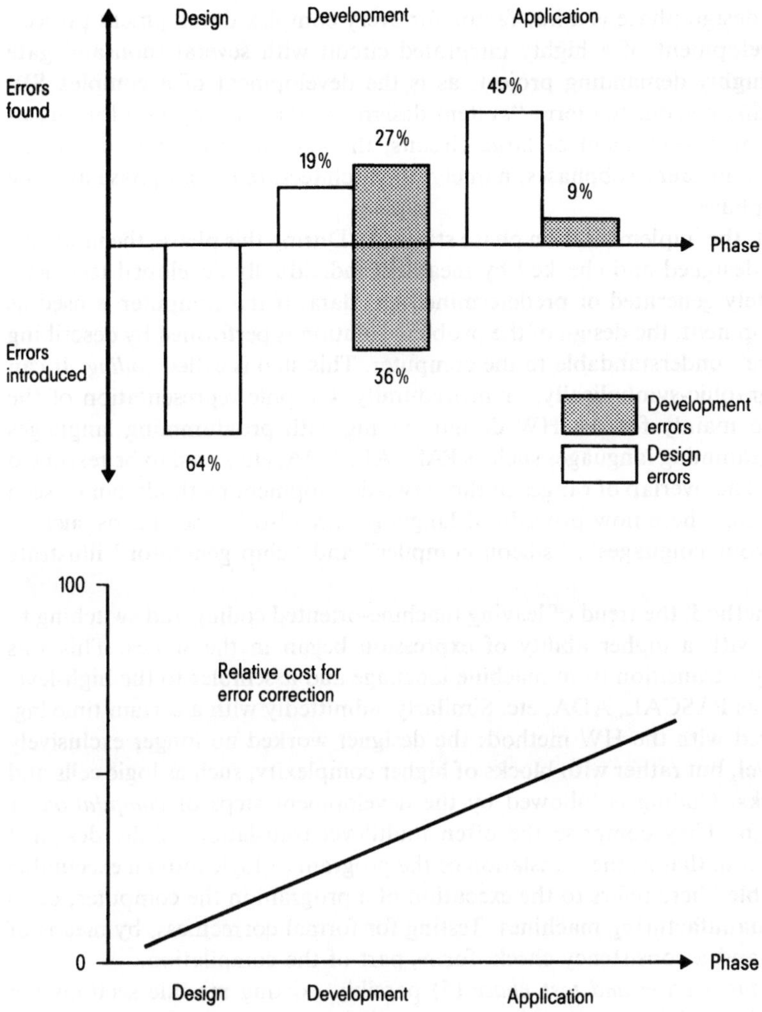

Fig. 1.5. Program errors and debugging expenditure

also prepared which must allow for the individual test (module test) as well as the system test (integration test).

In the system design phase, far-reaching assessments are made, such as the formal representation of system functions (target functions), division into subfunctions by hierarchically layered resolution, design of interface architecture, connection structure and functional structure. For complex problems such decisions cannot, in general, be made with certainty. Wrong decisions and design errors in this phase usually cause especially undesirable effects: they are often discovered very late and then require a disproportionately high amount of repair – if repair is possible at all. Systematic observations during SW development led to recognition of the correlation between error frequency and debugging effort within each phase of the problem solving process (Fig. 1.5).

The system design phase is a key factor for every complex development project. Surely, the development of a highly integrated circuit with several thousand gate functions is a highly demanding project, as is the development of a complex SW program. For this reason, the term "system design" is also appropriate for circuit development. For development of large circuits, the system design phase is often divided into two different subphases, namely, the architecture design phase and the module design phase.

At this point, the implementation phase starts (4). During this phase, the individual modules are designed and checked by means of individually developed test cases with appropriately generated or predetermined test data. If the computer is used as a tool for development, the design of the problem solution is performed by describing it in a "language" understandable to the computer. This step is called *coding*. It can be performed graphic-symbolically, or procedurally. Graphic representation of the design was used mainly for the HW design, coding with programming languages (high-level programming languages such as PASCAL, ADA, etc.) used to be restricted to SW systems. The overlap of ranges of those two development methods can be seen in the HW design, where now procedural languages are also in use. Terms such as "procedural layout languages", "silicon compiler" and "chip generator" illustrate this fact.

In the SW method, the trend of leaving machine-oriented coding and switching to language tools with a higher ability of expression began in the sixties. This was characterized by a transition from machine language and assembler to the high-level languages such as PASCAL, ADA, etc. Similarly, admittedly with a certain time lag, this also occurred with the HW method: the designer worked no longer exclusively at the layout level, but rather with blocks of higher complexity, such as logic cells and functional blocks. Coding is followed by the development steps of *compilation* or automatic design. They comprise the often multilevel translation of the designed object's description, that is, the translation of the program or logic into an executable form. "Executable" here refers to the execution of a program in the computer, or to the control of manufacturing machines. Testing for formal correctness, by means of a syntax check and a consistency check, forms part of the compilation.

During the *integration and test phase* (5) possibly existing module sections are combined and checked for functional correctness and compatibility. It is guaranteed (that is: verified), that the prototype thus created really complies with the specification. If, as is often the case in HW design (but also more and more often in SW design), the production of a prototype is tied to high cost, then the verification [5] is performed during the development by simulations with increasing depth of detail. In this case, the result of the integration and test phase is not an implemented prototype, but one which is completely and accurately described in a database.

The *production phase* (6) is initiated through the transfer of development results from the development environment to the production environment. Product management receives the results for hardware as well as software insuring that operational maturity has been achieved.

Here the paths of SW and HW method evidently branch off.

[5] For the evaluation of a design (showing correctness) various terms are used in the pertinent literature [1.2, 1.3]. In the industry, the term "verification" is used in most cases.

Phase			Content	Result
1	Study		Definition of the requirements as target	Requirement catalogue
2	Specification		Development of a solution concept and the test method	Specification - object specification - test specification
3	System design		Architecture development, functional definition, module structure, test environment	Object structure - architecture - structure of functions - module structure Test conditions - test system - test environment
4	Implementation		Construction of modules and test profiles (coding, compilation)	Tested module
5	Integration and test		Module integration and design verification of the complete object	Prototype
6	Production	HW	Transfer to product management and production	Released product
		SW	Tansfer to product management	

Fig. 1.6. Phase model of problem solution process

In software development, usually distribution is released in combination with the final test. The transfer to the application environment has been achieved. In hardware development, manufacturing or production is initiated. The transfer to the prodution environment follows. Manufacturing consists of producing the verified prototype in quantity. Frequently, this goes together with integration into the final system environment. The release consists of the production output test and the test qualification by the product managers. At this point, the transfer to the application environment has been completed.

The total process described is summarized in Fig. 1.6.

This sketch of the general problem solution process follows, for reasons of better understanding, the ideal *top-down method.* In practice, however, one often has to deviate from the pure top-down method, because in many cases, individual (key)

1	2	3	4	5	6
Study	Specification	System design	Implemen-tation	Integration + test	Production

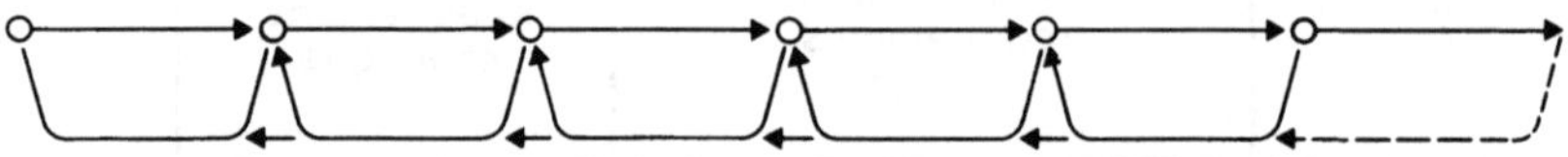

Fig. 1.7. Iteration cycles upon the phase model of the problem solving process

functions require extremely detailed design. Only then the development process can proceed. In such cases, one partially follows the *bottom-up method*, i.e. the design of the target function composed layer by layer of elementary functions which are, in case of the SW method, the program language statements, in the HW method the logic elements or even the basic components of transistors.

The combination of both methods, i.e. a systematic refinement starting at the top, sometimes descending to the most detailed level, and, while going up again, becoming more and more consolidated, is called the *yo yo method* which is supposed to be the best-known method. It shows that problem solving consists of more or less far reaching cycles which are iteratively placed upon the phase model. Figure 1.7 demonstrates the iteration cycles placed on the linear process. In the following, the linear process is also referred to as the "ideal process", as it raises particularly simple demands on data flow and data consistency.

In Fig. 1.7 the characteristic dynamics of the problem solution are highlighted. This kind of presentation will be used repeatedly in later chapters, particularly when the required attributes and performance features of modern IC design systems are derived, but also when the design procedure's placement in the environment of operational organization is considered. Primarily in computer-based processes, certain iteration cycles are normally excluded; otherwise the variety of interfaces and information flow will sharply increase. The remaining processes are considered "permissible" which means that they are compatible with the computer-based operation (see Sect. 1.5.5).

1.2.2 CAD Tools Used for the Process of Problem Solving

The entire process of problem solving is supported by a system of computer aided design (CAD) tools. This support applies equally to problem solving in SW and HW.

For HW development, there is a close connection between CAD and computer aided manufacturing (CAM). The unbroken and consistent flow of information from the product's specification to its reproduction which frequently has to be provided for years and with considerable effort, call for a close connection between development and production.

But the goal of modern CAD systems is to also provide the engineer with an interface at the highest possible, most productive description level and to avoid technical information which is not relevant. Performance features of a modern CAD system for the IC design are:

- interactive-gaphics method with hierarchic detailing and abstraction,
- modeling, verification and simulation tools on several levels,
- testability and automatic support for test preparation,
- automatic design of the production data with the possibility of interactive, system-controlled corrections,
- automatic generation of test profiles and test programs with the possibility of interactive, system-controlled modifications,
- consistency of information and data base.

Avoiding technical detail information means first of all concealing information which is only needed for production and for the organization of the production process. At present, this goal has not quite been reached yet. At least, all technical information which is handed over to the design engineer from lower design levels, must be formulated in the abstract form of compulsory design rules and design recommendations.

1.3 Procedure Example ADUS

In the case of hardware method applications, the steps of the problem solving process have to be interpreted differently than with software applications, i.e. when microprocessor development systems are used.

In order to accentuate the differences between the two methods, we will, by applying both methods, run through a specific problem solution process, namely the development of the analog-to-digital-converter-controller (*A*nalog-*D*igital-*U*msetzer-*S*teuerung) ADUS (standard example and pedagogic tool of this book).

There is, however, another reason for comparing these two problem solving techniques. According to the authors' observations, the SW method is more commonly used in industrial laboratories and universities than the HW method, probably because the former was developed earlier. One may therefore expect that this comparison makes it easier for an engineer who is used to work with the SW method to familiarize himself with the HW method.

The following demonstration assumes that the result of the study phase has already been submitted. The results of the individual design steps and the corresponding explanations have been listed in the charts 1.1 to 1.7 in such a way that, at the various levels, the respective presentational views and their analogies can be compared and illustrated. In the simplified procedure, however, the check and verification steps (syntax check, test run, rule check, simulation) have not been considered. It is true that they, combined with the necessary iterative corrections, improve the design's accuracy, but they do not effect the visible representations at the various levels. For reasons of simplification, the test preparation is not considered. The example of the analog-to-digital-converter-controller ADUS has been developed by means of the VENUS design system.

The SW solution (SW-ADUS) assumes that a one-chip microcomputer with the required number of input and output pins and a development system which allows programming in PASCAL is available. Furthermore, it assumes that the input and output variables represent the corresponding I/O signals in the memory-mapped

Design step	Result	
	HW method	SW method
Specification		

Specification of functions:

The ADUS component is used in the 16-channel analog input unit of the SIMATIC process control system [1.1]. The ADUS controller executes the following functions:

- control of the analog/digital (A/D) converter
- coordination of the data flow between A/D converter and the central processing unit (Zentral Prozessor ZP)
- cyclical scan of the 16 measurement channels by means of relay activation (transient time: 6 ms)
- storage of the digitalized values in a random access memory (RAM) (2 bytes/value), where they are accessed through the central processing unit.
- indication of a line break
- prevention of a collision (simultaneous access of the A/D converter and the central processing unit to RAM)
- timing of clock generation (not considered here)

Performance requirements:

- time of a A/D conversion: 60 ms
- control clock: 1.2 MHz (clock for additional functions: 9.8304 MHz)
- supply current $I < 100$ mA, supply voltage 5 V

System environment:

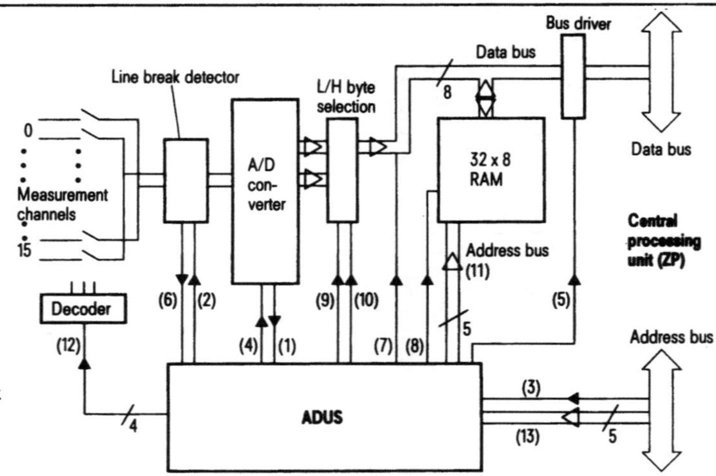

Figure 1.8 Simplified system environment of the ADUS component. The figures in brackets refer to the corresponding variables in the interface definition.

Chart 1.1 ADUS example: specification.

<table>
<tr><th rowspan="2">Design step</th><th colspan="2">Result</th></tr>
<tr><th>HW method</th><th>SW method</th></tr>
<tr><td rowspan="2">Specification
(continued)</td><td colspan="2">Functional description:
Through the ACTIVE.CHANNEL address (12), ADUS selects one of the 16 measurement channels. Connection to a metering channel is carried out through a relay which has a transient time of 6 ms. If a line break is detected (2) the otherwise undefined input voltage is set at 0V (6).

Then, the actual conversion is triggered (4), the end of which is indicated with the A.D.ACTIVE signal (1). Through LOW/HIGH byte selection ((9), (10)), the lowbyte and the highbyte are sequentially switched to the 8-bit wide, local data bus. The 32 x 8 bit RAM has been arranged in form of a cyclic memory where the measured values which are controlled by the write signal (8) and the address bus (11), are stored. In case of a line break, an otherwise unused bit is set (7). The central processing unit (ZP) indicates its read requirement through a decoded address bit (3) and accesses the requested data through (13).

In case of a collision, read-out is prevented by the READ.LOCK (5).</td></tr>
<tr><td colspan="2">Interface definition:

control variables (input):
(1) A.D.ACTIVE — true, if the A/D converter is active, otherwise false.
(2) LINE.BREAK — true, if line break has been detected, otherwise false.
(3) ZP.READ.REQUIREMENT — true, if ZP wants to read data from the RAM.

control variables (output):
(4) A.D.START — if true, an A/D conversion is initiated
(5) READ.LOCK — if true, ZP access to RAM is prevented.
(6) ZERO.INPUT — if true, input voltage of 0V is forced on the A/D converter
(7) LINE.BREAK.INDICATOR — if true, a line break is indicated to ZP.
(8) RAM.WRITE — if true, the data is transferred to RAM
(9) LOBYTE — if true, the low-order-byte of the A/D converter is switched to the data bus
(10) HIBYTE — if true, the high-order-byte of the A/D converter is switched to the data bus

adresses (output):
(11) RAM.ADR — 5 bit wide address for selecting one of 32 RAM cells
(12) ACTIVE.CHANNEL — 4 bit wide address for selecting one of 16 measurement channels

adresses (input):
(13) ZP.ADDRESS — 5 bit wide address of ZP</td></tr>
</table>

Chart 1.2 ADUS example: specification (continued).

Result

Design step	HW method	SW method

System design (Block representation)

HW method

In the block diagram (fig. 1.9), the A/D converter control, the clock rate divider and the digital timer have been omitted for reasons of simplification. At this level only the large functional blocks and their connections with each other are indicated.

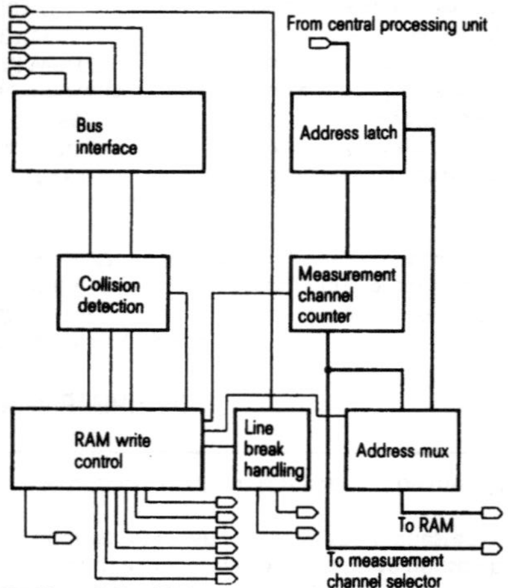

Figure 1.9 Simplified block representation

For the following steps we want to limit our considerations to the measurement channel counter. It can be loaded parallel to the ZP through the address latch.
The RAM write controller causes the measurement channel counter to be cyclically incremented from 0 to 15. The counter output is used first for selecting the active measurement channel, and second for accessing RAM when the measured values are to be stored. If ZP wants to read data from RAM the MUX address connects the address stored in the address latch to the address bus.

SW method

Figure 1.10 shows the structure of the ADUS main program. For large program systems, sections of this main structure are in turn constructed of more and more detailed structures until sufficient detail for coding has been reached.

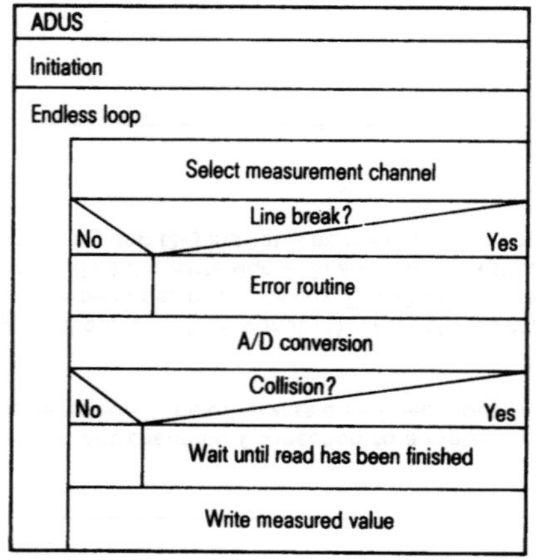

Figure 1.10 ADUS structure

Chart 1.3 ADUS example: system design.

<table>
<tr><th rowspan="2">Design step</th><th colspan="2">Result</th></tr>
<tr><th>HW method</th><th>SW method</th></tr>
<tr>
<td>Implemen-
tation
(coding)</td>
<td>Figure 1.11 shows the address latch, address multiplexer system and measuring channel counter, coded on the cell level. Cells are predetermined and verified basic components, such as NAND gate, counter or multiplexer.
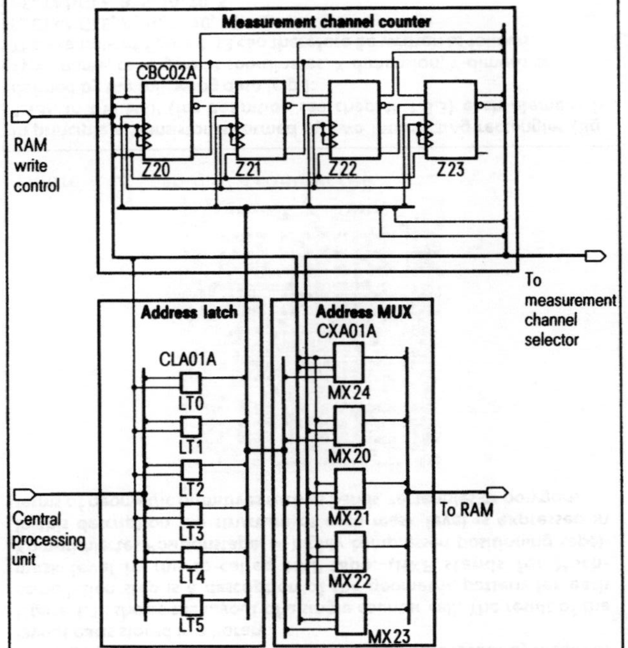

Figure 1.11 ADUS section on cell level (logic plan)</td>
<td>PASCAL-program of ADUS control:

program ADUS;
var
  (* control variables and addresses (input) *)
  A.D.active: boolean;
  line.break: boolean;
  ZP.read. requirement: boolean;
  ZP.address: 0..31;
  (* control variables and addresses (output) *)
  A.D.start: boolean;
  read.lock: boolean;
  zero.input: boolean;
  line.break.indicator: boolean;
  RAM.write: boolean;
  LoByte: boolean
  HiByte: boolean;
  RAM.Adr: 0..31;
  active.channel: 0..15;
  (* internal memory *)
  I: 0..15;
  (* wait procedure *)
  procedure wait (time: integer); external;

begin (* algorithm*)
  (* initialization*)
  I := 0;
  A.D.start: = false
  repeat (* endless loop *)
    (* select measurement channel *)
    I := (I + 1) mod 16;
    active.channel: = I;
    wait (6); (* transient time of relay: 6 ms *)
    (* line break error routine *)
    line.break.indicator := line.break;
    zero.input := line.break;
    (* A.D.conversion*)
    A.D.start := true;
    repeat until A.D.active;
    A.D.start := false;
    repeat until not A.D.active;
    (*collision treatment: .... *)
  until false; (* endless loop*)
  (* write data ... *)
end.</td>
</tr>
</table>

Chart 1.4 ADUS example: implementation (coding).

Design step	Result: HW method	Result: SW method
Implementation (compilation)	The logic plan (fig. 1.11) is compiled into geometric layout by means of layout parts stored in a library. Figure 1.12 shows the layout of a single counter cell. The result of the compilation step is a description of the geometric pattern for each mask level in the so-called HKP-tape. (HKP stands for Hoch-Komprimiertes Positionstape = highly compressed positioning tape). In this description the structure of each mask level is expressed in terms of geometric primitives such as bands, rectangles or polygons.	A compilation of the program shown above into assembly language results in the following:

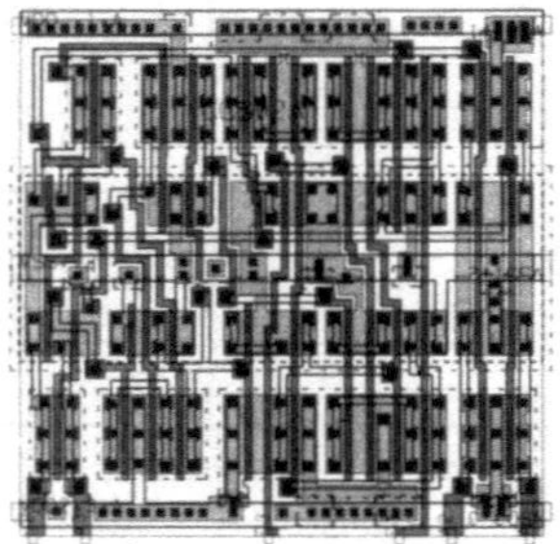

Figure 1.12 Layout of a counter cell

36	1	00000000	0	L 15,88(0,13)
36	1	00000004	4	BALR 10,15
36	1	00000006	BLOCK	://ADUS 64 100 -1
36	2	00000010	22	LA 1,0(0,0)
36	2	00000020	26	STC 1,61(0,11)
37	3	00000024	30	STC 1,52(0,11)
43	5	00000028	34	LA 1,0(0,0)
43	5	00000020	38	IC 1,61(0,11)
43	5	00000030	42	AR 1,0
43	5	00000032	44	LR 2,1
43	5	00000034	46	SRA 2,0,4(0)
43	5	00000038	50	SLA2,0,4(0)
43	5	00000030	54	SR 1,2
43	5	0000003E	56	STC 1,61(0,11)
44	6	00000042	60	STC 1,60(0,11)
......	...	....	...	....
END ADUS		L =	170	

In principle a transistor is formed by two intersecting rectangles (fig. 1.13). In the HKP (for definition see chapter 1.5.3) each element is defined by the following data (e.g.):
Type, name, calling point coordinates, X-dimension, Y-dimension
The example of figure 1.13 can therefore be written as follows:
RECTANGLE, A, 10, 5, 10, 15
RECTANGLE, B, 5, 10, 20, 5

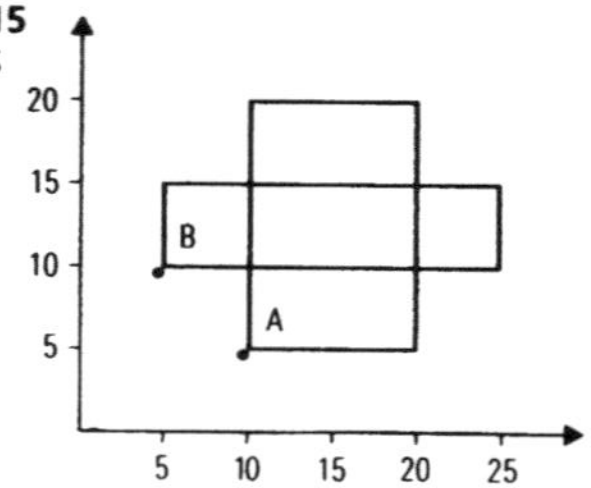

Figure 1.13
Layout of a transistor structure

Chart 1.5 ADUS example: implementation (compilation).

<table>
<tr><th rowspan="2">Design step</th><th colspan="2">Result</th></tr>
<tr><th>HW method</th><th>SW method</th></tr>
<tr><td>Implemen-
tation
(compilation)
continued</td><td>In the HKP format mask tapes are generated by means of compilation programs.
Similarly, from the logic plan test bit patterns are generated.</td><td>The loadable machine program with absolute addresses in hexadecimal code does not display a structure which is recognizable to the designer:

0A0B0A 07FE50ED 02581814 0AB112FF 077E42F4 000607FE50ED0258
182F5836 0000D501 0A0B2A 20A73000 47802024 D241D90205AD235
0A0B4A 0004D207 D9013000 411D0900 0A6E12FF 4780205212444740
0A0BAA 40404040 40404040 40404040 40000000 000023C C2C34040
0A0BCA 40404040 40404040 40404040 40404040 4040404040404040
0A0BEA 40404040 40404040 40404040 40404040 4000000050ED0258
....</td></tr>
<tr><td>Production</td><td>The mask tape bearing the machine layout is used directly for controlling a mask making machine. Figure 1.14 displays some mask layers.
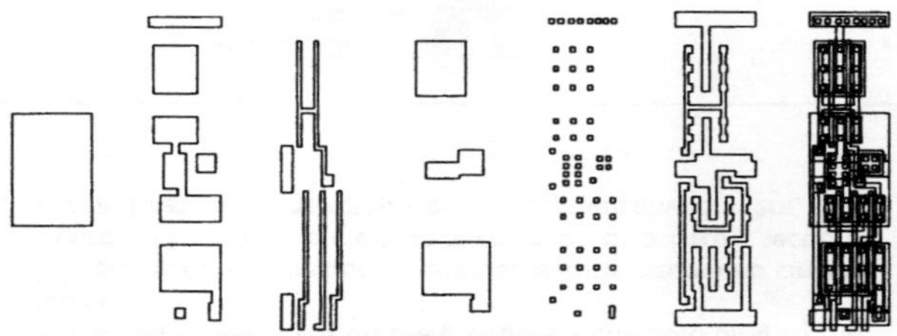
Figure 1.14 Various mask levels and their overlay</td><td></td></tr>
</table>

Chart 1.6 ADUS example: implementation (continued) and production.

Design step	Result: HW method	Result: SW method
Production (continued)	The circuit is generated on the production line according to the masks. This produces the functional customer-specific integrated circuit. Figure 1.15 shows the electron-micrograph of a circuit section, and, as final result, the component in its system environment. 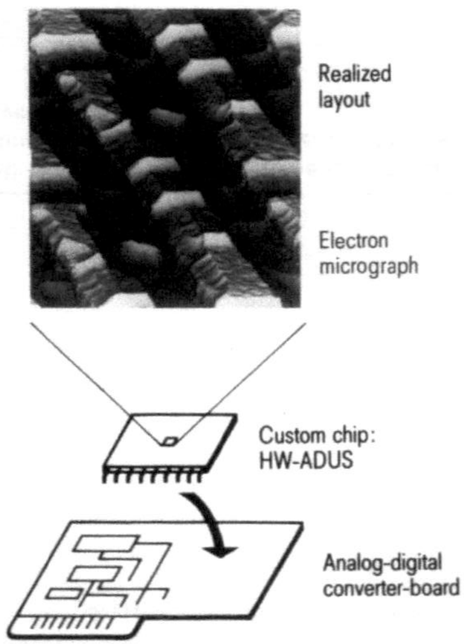 Figure 1.15 Final result: the custom chip in its system environment	Instead of production, one may assume that, in this case, following the appropriate verification, the machine program is "burnt in" the PROM of a one-chip-microcomputer, and is thus introduced into its system environment (Figure 1.16). 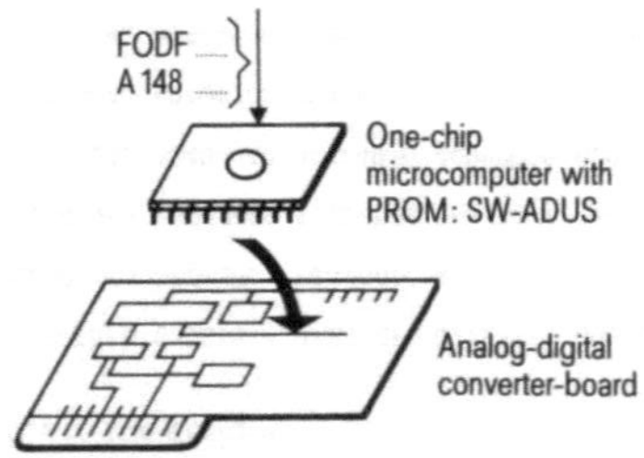 Figure 1.16 Final result: the software in its system environment

Chart 1.7 ADUS example: production (continued).

mode. A logical "one" means true, a logical "zero" means false. The standard test system of the microcomputer development system is used for SW test.

For the HW solution (HW-ADUS) the following are selected:

- the design system VENUS,
- a standard package with a sufficient number of pins,
- and CMOS technology.

The programs and test data are produced by VENUS. The pin-out corresponds to that of the SW-ADUS. All input and output signals are CMOS-compatible.

1.4 HW Method Versus SW Method

Up to and including the specification phase, one may, as is shown in the ADUS example, make use of the same representations for both methods. The specification phase includes the first steps of design realization, following the study phase. First, the function and performance scope of the total system is determined in textual and tabular form, the placement in the system environment is displayed graphically, the functional conception is described in textual form, and the interfaces are defined in tabular form. Here, differences between the two methods are object-specific. For the SW method the specification includes the determination of time conditions (if real time requirements are set) and available memory size, as well as the definition of system interfaces. In addition to time conditions, for the HW method one has to determine other data, such as supply voltage, maximum current consumption and environmental conditions. The specification of system interfaces includes data and instruction formats for input and output, as well as electrical parameters: for example, parameters for voltage levels, driver capability and overload resistance. In both cases, the result of the specification phase is the object specification.

Differences appear with system design regarding representation and design methods. For the SW design one generally prefers displaying the algorithms and the data and control flows on flow charts and tables. For the HW method, the function blocks are predominantly displayed graphically, including the input and output signals and supplies. As graphic design representations are increasingly used also for the SW method, the differences begin to decrease. In the following paragraphs we assume that the first sketch of the development object is drafted during system design. With the design of a first solution, we have completed the highest level of design. It is called system design or system plan.

At the beginning of the next phase, the similarities of HW and SW method are still prevalent: During coding, the next lower level is designed in both cases by means of components taken from a given set of elementary functions. Here one has to follow certain rules for their combination (syntax).

In the given context, the difference between the descriptional methods (procedural languages for the SW method – graphic symbols and their combinations for the HW method) is irrelevant.

As for the differences between the HW method and the SW, we have to discuss four major topics:

- number of design levels,
- verification procedure,
- test preparation,
- production.

1.4.1 Design Levels

Design levels are representation levels of various degrees of abstraction. A design step consists of transformation of a higher-level description to a lower-level representation. For this transformation procedure, automatic tools already exist to a certain extent. In the SW method, compilers automatically transform the higher-level program language (in our example: PASCAL) into an efficient machine code. Intermediate results (e.g. assembly language), generated during the compilation, are usually hidden, and are thus not regarded as representations on an independent design level.

Considering the *visible* design levels we recognize *three* types of description for the SW method:

- system design (module design, block description),
- HLL[6] program (PASCAL program),
- machine code.

In contrast, there are *five* visible types of description and design for the HW method:

- system plan,
- logic plan,
- circuit diagram,
- layout design,
- production data.

As is shown in detail below, the HW method is also increasingly subject to automation of design steps. Therefore, some levels also become "invisible"; thus, if the standard VENUS design system is applied the design engineer does not "see" the circuit design any more (transistor level), and he only has a restricted view of the layout design.

The design on each level passes through the "implementation" and "test" (also "construction" and "verification"[7]) phases. So, this double-step procedure is repeated one or two times more often for the HW method than for the SW method, depending on the procedure chosen.

---

[6] HLL: High Level Language.

[7] For developing electronic systems, even four design steps are distinguised: partitioning, design, verification and integration. In the context of this book the alternating steps "construction" and "verification" are sufficient. See also Chap. 7!

1.4.2 Verification procedure

We will deal with verification in detail in Chap. 1.5.4. Here, for the purpose of comparison between HW and SW method, we will refer only to the verification object, the functional sample: SW is immaterial during the entire design process. The cost for producing a loadable functional sample is restricted to a compiler run.

The proof of accuracy is done by the compiler carrying out a semantic and syntactic analysis. The HLL program is automatically transormed into a code executable by the target machine. This automatic code production is possible, because the microprocessor represents an abstract machine that can be formally described. In the SW method the "visible" representation of the solution is a loadable code. At this point, the lowest level of design has been reached. The functional sample is automatically generated by the compiler. The final verification is executed by means of testing the functional sample. Usually, the design verification for the SW development consists of an iteration of the following steps: coding, compilation, validation; the latter is often carried out in the HW environment where the system is to be run.

In contrast, the production of a functional sample (chip) in the HW method requires considerable effort. For this reason, accuracy has to be proven, to a great extent, by means of simulation. The CAD system assumes the functions of the compiler. It replaces the cells called up as elementary functions in the logic plan with appropriately prepared and verified circuit parts, and thus designs the circuit diagram and the layout design according to given rules of placement and wiring. This is followed by a compilation into the mask control patterns. In order to prove accuracy, formal verifications of compliance with design rules and comparisons between the various levels of representation are performed in addition to simulation. At present, the design of the layout from the logic plan is only "semi-automatic". This is because the translation process can only be described to some extent by means of a complete rule control system. It still requires heuristic knowledge which can only partially be described formally. Because of this, the so-called "vertical verification" (see below) is much more important for the HW method. It guarantees correctness and compliance with the specifications on all design levels.

1.4.3 Test Preparation

During the production of integrated circuits random errors may cause defects. For this reason, the proper functioning of the manufactured chips has to be checked by means of testers. In addition to mask tapes, the test programs are the second essential result of the HW design process. Experience shows that the design of test programs should be started before completing the design process; the test preparations should even be initiated at an early stage. In addition to the object representations of the different levels which are found in the SW method, here the test tools are generated, too. We can identify a correspondance between the results of design and test preparation with the following correlations:

- object specification: test specification
- system and module design: test frame
- logic design: test plan
- mask control data: test program.

1.4.4 Production

The most striking differences between HW method and SW method are found in the production phase.

The effort of SW production is limited to the transfer to product management. It includes the organizational measures for version planning, delivery, error reporting and debugging. The "manufacturing" of SW consists in multiplying the programs. Random production errors do not occur. With reference to the programming of microcomputer systems, the term "manufacturing" could at best refer to burning the program into the PROMs (*p*rogrammable *r*ead *o*nly *m*emories).

In addition to transfer to the prouct management, the production of integrated circuits requires, as opposed to copying programs, considerable procedural and organizational effort. In particular random errors lead to occasional failures of manufactured ICs. For selecting the functional circuits, one has to take considerable test preparation measures – as has been mentioned – during the development process and later at the end of the production phase.

Finally the result of both approaches is a released product, delivered in production quantities.

Figure 1.17 compares the design process of HW and SW method. The results for the HW method are listed explicitly only on the most important design levels.

Some distinctions between the two methods have to be emphasized once more:

- The high cost of realization for functional samples makes it necessary in the HW method to carry out verifications as early as possible *before* compiling and producing the samples. For the SW method the final verification may be carried out on the manufactured functional sample *after* compilation, because the functional sample can be comparatively easily produced automatically.
- The result of the SW design process is a correct, loadable binary code, but with symbols that are easy to decode, symbols which can be used as control information for a processing unit. The result of the HW development is a three-dimensional complex silicon pattern for the design of transistors and their connections in silicon (layout, mask tape). The circuit that can be produced in this way allows the simulation of logic processes by means of controlling charge flows and potentials. This is made possible through the transformation of logic conditions (yes/no) into physical terms (high/low potential).

The result of the two approaches is, although functionally identical, actually different. When deciding in favor of one approach or the other, the following criteria should be checked and evaluated:

- development effort, development time, unit cost,
- maintainability, prevention of duplication, know-how protection.

In particular, the last three criteria are often the decisive factors in choosing the HW method. The individual, personalized chip can be copied only – if this is possible at all – with considerable effort; the SW program, however, can generally be copied with little cost.

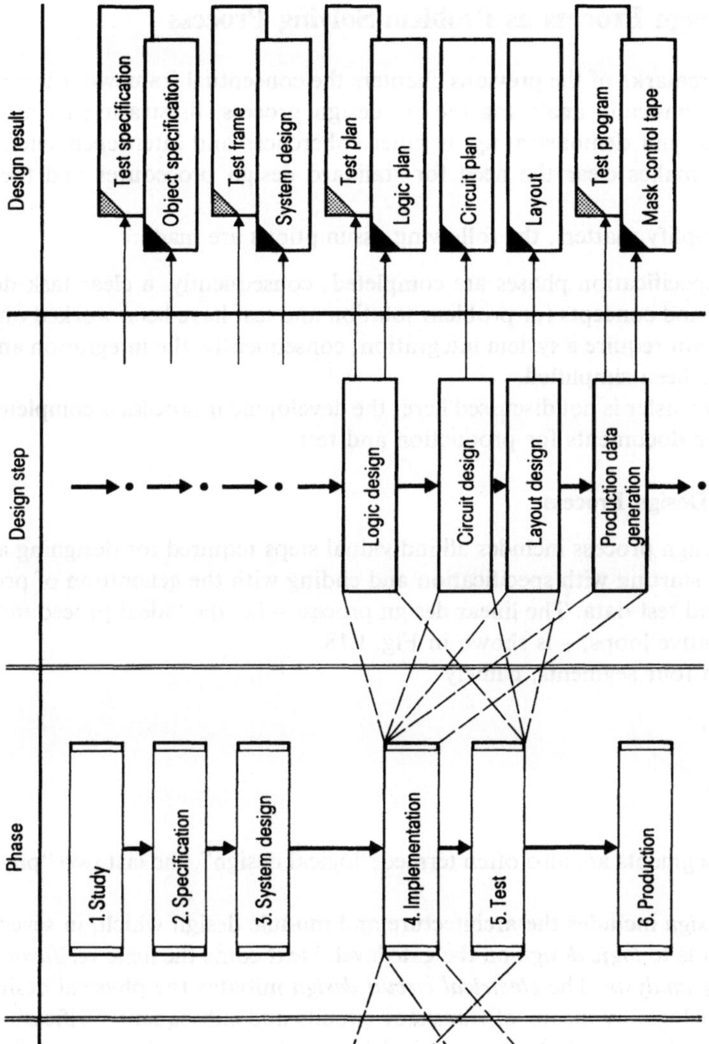

Fig. 1.17. Comparison between HW and SW method

1.5 The IC Design Process as Problem Solving Process

From the general remarks of the previous chapters the conceptual framework is now available for systematically analyzing the IC design process, illustrating its most important features, and demonstrating its inner coherences and interdependencies. The analysis also makes clear the need for standard design procedures and their development.

In order to simplify matters, the following assumptions are made:

- the study and specification phases are completed; consequently, a clear task description exists, and concepts for problem solution and test have been worked out,
- the circuit does not require a system integration; consequently, the integration and test phases have been simplified,
- the production transfer is not discussed here; the development process is completed by providing the documents for production and test.

1.5.1 General IC Design Process

The general IC design process includes all individual steps required for designing an integrated circuit, starting with specification and ending with the generation of production control and test data. The linear design process – i.e. the "ideal procedure", not including iterative loops, – is shown in Fig. 1.18.

It is divided in four segments, namely

A: system design,
B: logic design,
C: circuit design,
D: layout design.

The first two segments are also often termed "logical design", the last two "physical design".

The *system design* includes the architecture and module design which, in several detailing steps, leads to *logic design* on the gate level. Next come the *logic verification* and the testability *analysis*. The *electrical circuit design* initiates the physical design with realization of logic by means of transistor circuits and subsequent verification. Finally, the *layout design* includes geometrical layout design (i.e. the design of all masks required for production on all levels), corresponding steps of verification and, in order to establish a path to production, *generation of the mask tapes*. The *generation of test data* is based upon the logic design which results in executable test programs.

This segmentation in various levels is characteristic for IC development and shows the process from the abstract representation on a conceptual level in step-by-step realization to the mask tapes which describe in detail the manufacturing process. Each higher level constitutes an abstract representation of the next lower level. Vice versa, moving from a higher to a lower level requires special design know-how corresponding to the design method and technology chosen. Thus, the circuit designer has to know, for example, a way to realize the logic function of a D-flipflop by means of transistors and their interconnection utilizing CMOS technology.

However, it is obvious that if certain rules (design rules) have to be followed, errors can be made. As has already been shown in the ADUS example, a sequence of

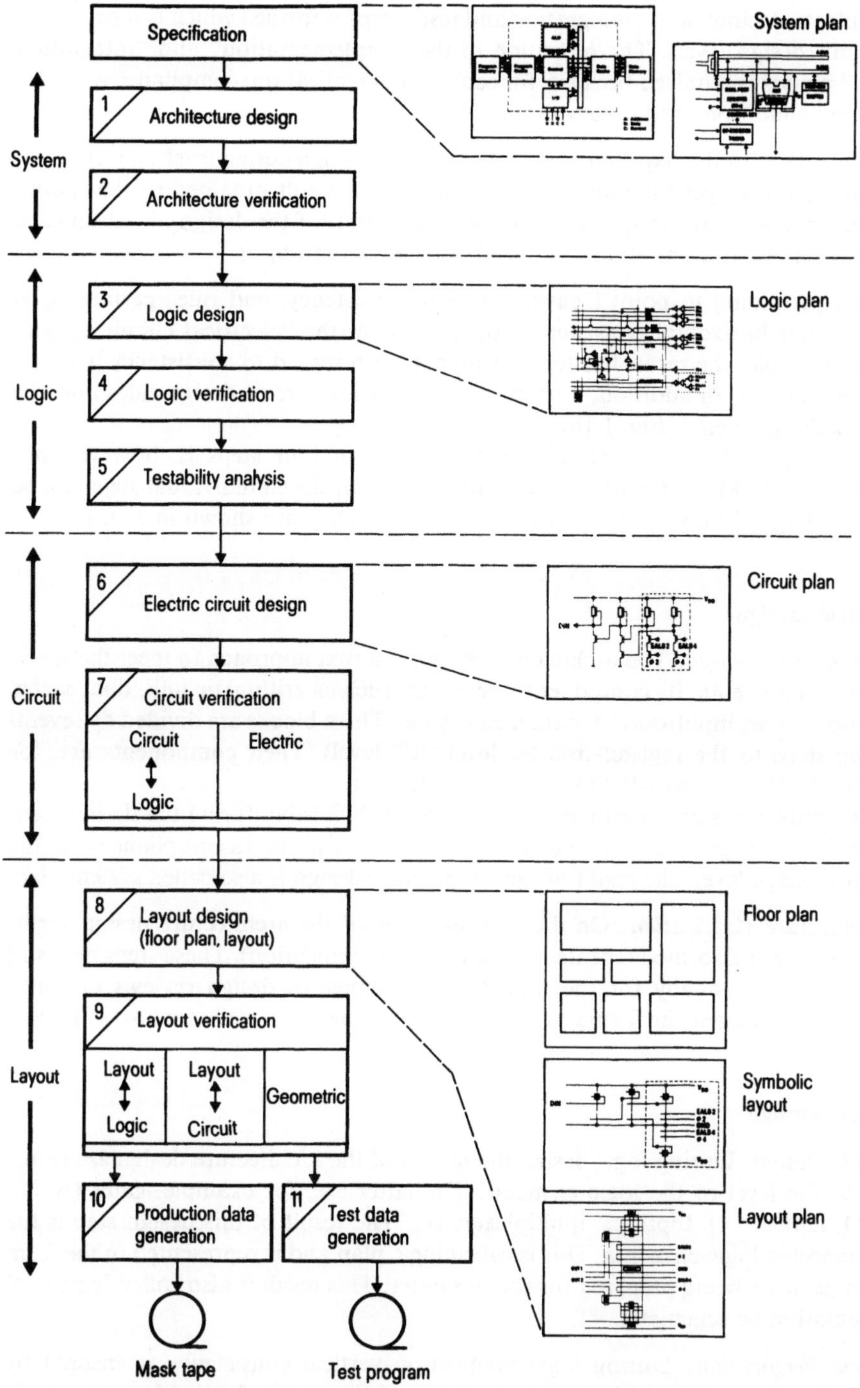

Fig. 1.18. General IC design process

the "implementation" and "integration and test" steps is formed which is repeated on the various design levels. The repetition of the "implementation" and "integration and test" phases is required because, for each design realization, compliance with two correctness conditions has to be examined:

- Compliance with the requirements of the previous design design levels, i.e. ensuring that the given target function is met by the design result obtained to that point.
- Compliance with the design rules for the actual draft of the design object on each level of representation, across the whole of the design object.

Rules pertaining to point 1 ensure vertical consistency, and rules pertaining to point 2 ensure horizontal consistency. In the case of the "electrical circuit design" step, for example, the result has to be examined with regard to consistency with the given logic design. In addition, a check of the electrical circuit design rules must be done (see design step 7, Fig. 1.18).

The analysis of the general IC design process in four steps is, however, not accurate enough. More details are necessary. Basically, eleven individual steps can be identified. We will now take a closer look at them. They are shown in Fig. 1.18.

A. System Design

1. *Architecture Design:* The architecture design is a first approach to meet the specification requirements. Its components are blocks such as arithmetic unit, data cache, instruction cache, input/output, data memory etc. These blocks are divided by several detailing steps to the register-transfer level (RT level). Their components are, for example, ALU, dual-port-RAM, buses, decoders, etc.

For complex system developments, a further subdivision during the design steps of the module design is recommended. We will forgo this here. In compliance with the following design levels the result of the architecture design is also called system plan.

2. *Architecture Verification:* On the various levels of the architecture design verification steps are performed repeatedly (architecture verification). These steps may still be carried out manually for the first design outlines, as design reviews or walkthroughs. With increasing formalization, however, simulation tools are also available.

B. Logic Design

3. *Logic Design:* During logic design the results of the architecture design are transposed to the level of the logic elements. The latter are, for example, gates (AND, NAND, EXOR ...), flipflops, multiplexers etc. The result of this design step is the interconnected logic elements. This is called logic plan and is represented in the form of a net list after being prepared for the computer. This result is also called logic level representation or schematic.

4. *Logic Verification:* During logic verification vertical consistency is ensured by means of a comparison with the requirements of the system level. Moreover, logic simulators or, if information on switching times is available, timing simulators are used to check the proper function. By carrying out simulations, one may also discover

some of the design errors in circuit segments or in the interaction of these circuit segments on the logic level (horizontal design errors).

As an example for ensuring the horizontal consistency, the test design rules should also be mentioned. They include, for example, feedback prohibition for outputs of purely combinational blocks to their own inputs, or the requirement for an externally definable initial circuit state.

5. *Testability Analysis:* The testability analysis gives a measure of observability and controllability for all nodes, thus allowing estimates of the degree of testability and the testing cost expected. If the test design rules have been adhered to the testability analysis will in most cases result in an acceptable degree of error detection.

C. Circuit Design

6. *Electric Circuit Design:* For electric circuit design, the logic design is transformed into transistor circuits. This is also known as transistor level representation. In this case, technology-specific decisions, such as the use of CMOS or ECL technology, have already been postulated. Figure 1.19 shows the transformation of a 2-input-NAND-gate into a circuit, using CMOS technology. The design rules which are to be followed on the transistor level are the basic rules of electric circuit design. They include, for example: "p-channel transistor with source at V_{DD}" or, for CMOS technology: "no open input permissible".

7. *Circuit Verification:* The vertical consistency is checked by a comparison of representation levels (transistor level with logic level), and by circuit simulation. For checking the electrical design rules, special checker programs (electrical rule checkers) are available.

D. Layout

8. *Layout Design:* The layout design constitutes the beginning of the layout phase. It has an especially close connection with the technology characteristics. The representation of each mask level has to be geometrically exact. There are at least ten for the CMOS technology. The criteria for optimization are area and/or speed. As the overall

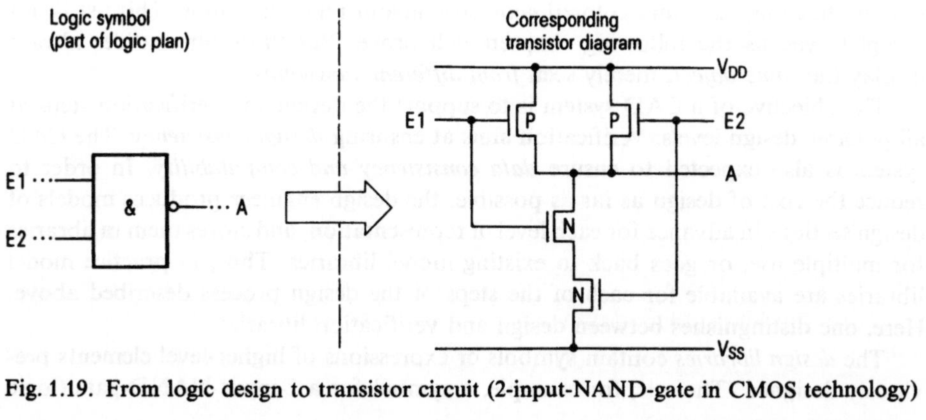

Fig. 1.19. From logic design to transistor circuit (2-input-NAND-gate in CMOS technology)

layout has to fit into a rectangle, one starts with the global design of a rectangular area which will then be specified step-by-step. Stick diagrams support this layout design, as they allow designing the layout topology before the geometry is chosen. Design rules at this level are technology design rules describing geometric restrictions due to production which the layout must satisfy. These design rules state, for example, the minimum distance between two structures at the aluminum level, or the extent to which polysilicon may overlap a contact hole. These technology-related rules usually change when the process technology is modified (see Chap. 3).

9. *Layout Verification:* The layout verification is carried out after extraction of the electrical values by comparing them with the requirements of the circuit level, or (sometimes also) with those of the logic level. This level's compliance with the design rules which ensure the horizontal consistency, i.e. the technology design rules, is checked by means of special programs (design rule checkers).

10. *Production Data Generation:* The layout representation of the integrated circuit, thus designed and verified, are transformed into control data for mask making machines (production data generation). This step is carried out fully automatically, verification is not required. The result is the mask tape.

11. *Test Data Generation:* During logic verification test patterns are already generated as stimuli for the simulator. Those patterns are being supplemented with other test patterns until the required test coverage is reached (test data generation). The test patterns can be generated automatically. Next, the test programs have to be written, in order to test certain parameters of the manufactured circuit and to carry out functional tests by means of the test patterns. The process of test program generation together with the previous stimuli generation is called test data generation. Hence, the second result of the design process is a test program.

1.5.2 Models and Libraries

For the observer of the general IC design process there is a striking feature: several representations, quite different from each other, are designed for the integrated circuit. The architecture design provides block representations (system plans), the logic design provides logic plans built of logic symbols, the circuit design provides transistor circuit diagrams. During the layout design phase floor plans are drawn, as well as stick diagrams and colored geometric layout representations. This list is not complete yet, as the following chapters will prove. But those alternatives always display the *same object*, merely seen *from different viewpoints*.

The objective of a CAD system is to support the design and verification steps at all possible design levels. Verification aims at ensuring *design consistency*. The CAD system is also expected to ensure *data consistency and compatability*. In order to reduce the cost of design as far as possible, the design engineer produces models of design sections in advance for each level of representation, and stores them in libraries for multiple use, or goes back to existing model libraries. Thus, in practice model libraries are available for each of the steps of the design process described above. Here, one distinguishes between design and verification libraries.

The *design libraries* contain symbols or expressions of higher-level elements previously designed. There is, for example, a symbol for a 2-input-NAND-gate (logic

Design step		Model libraries	
		Design	Verification
1	Architecture	Block symbols, RT symbols	
2	Architecture verification		Register transfer models
3	Logic design	Circuit symbols	
4	Logic verification		Logic design rules, logic models
5	Testability analysis		Test design rules, test models
6	Electric circuit design	Circuit element symbols	
7	Circuit verification		Electrical design rules, circuit models, switch level models
8	Layout design	Stick elements layout contours	
9	Layout verification		Geometric layout design rules, electrical layout design rules, propagation delay models
10	Production data generation	Layout segments, technology requirements	
11	Test data generation	Model test programs, electric data	

Fig. 1.20. Model libraries for the IC design

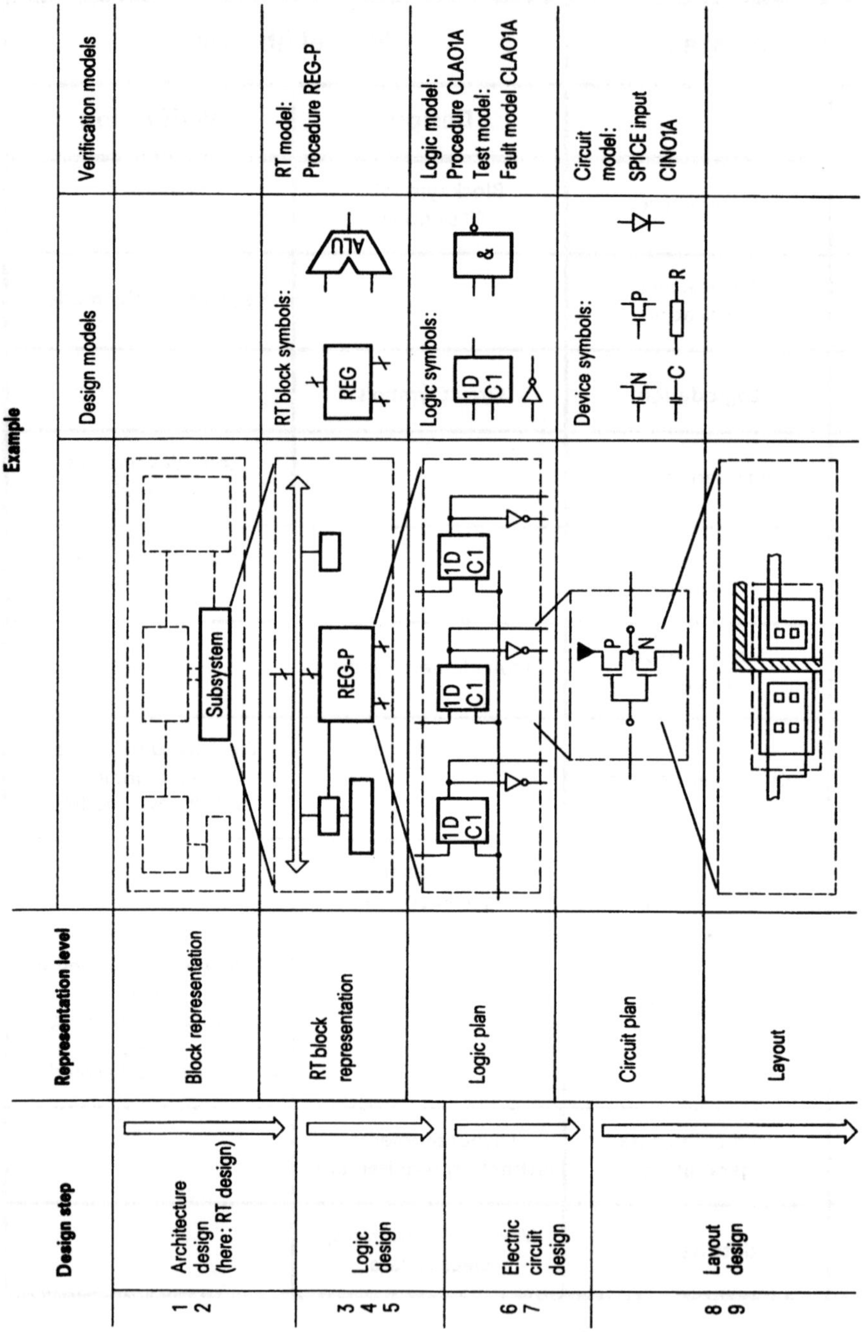

Fig. 1.21. Illustrative design procedure

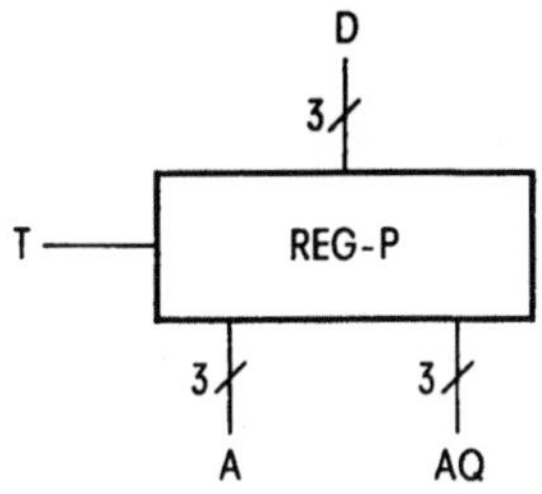

Fig. 1.22. Register symbol REG-P

level) which is used by the design engineer for generating his logic plan. The layout libraries contain layout segments realizing the NAND in CMOS as well as in ECL technologies.

The *verification libraries* contain simulation models which make it possible to automatically translate a design into the input language of a certain simulator. Secondly they contain collections of (horizontally applicable) design rules for the various design levels serving as parameters for the various "checkers". Thus, the same design rule check (DRC) program may be loaded with a run set first for 3-μm, then for a 2-μm technology, and thereafter may perform the corresponding checks. Figure 1.20 shows the assignment of the model libraries to the design steps.

In the following, we want to illustrate the various design and verification libraries, and their application by means of an example.

Our illustrative design procedure (Fig. 1.21) begins with the requirements of an architecture on a higher block level. Within this block representation a SUBSYSTEM is defined. The first step considered consists in constructing SUBSYSTEM by means of components on the register-transfer (RT) level. This step – last in the architecture design – is called *RT design*. The design engineer uses a library of design symbols (*RT block symbols*). In such libraries, for example, the following are available: symbols for arithmetic-logic units (ALU), memory blocks (such as RAM, ROM), buses, counters or registers. For our example, we assume that a 3-bit-wide register is used, with inverted and non-inverted outputs: it is given the individual name REG-P. The corresponding symbol is shown in Fig. 1.22.

Description of the Object to be Simulated

In order to verify the result of the RT design, an RT simulator is used. The simulation object is described in a PASCAL-like language as input for this simulator.

Just as a predefined symbol for the 3-bit-register is called-up for the design, an *RT simulation model* is available as component for the simulator input: the procedure REG-P. It forms the function table of the 3-bit-register:

```
Procedure REG-P (IN T:BIT;IN D:BIT(3);
      OUT A, AQ:BIT(3) );
 (***REGISTER IS EQUIVALENT TO D-FF, LEVEL-CONTROLLED (CLOCK ='1')***)
 VAR Q:BIT(3);  (* REGISTER CONTENT*)
 BEGIN
 (*IF CLOCK UNDEFINED: THEN Q UNDEFINED*)
```

```
IF UNDEFTEST(T) THEN Q: = ALL-X ELSE;
IF T THEN Q: = D (*CLOCK = "1"- > Q = D*)ELSE;
A: = Q;
AQ: = NOT Q;
END.(*E1*)
```

Inputs		Outputs	
0	X	A(i0)	NOT A(i0)
1	D(i)	D(i)	NOT D(i)
T	D(i)	A(i 1)	AQ(i 1)

Function table of a N-bit-register (for each register bit: $1 \leqq i \leqq N$).

During logic design, REG-P is resolved on the level of the logic symbols. For that purpose, the LATCH component given in the circuit symbol library can be interconnected three times. Since LATCH has no inverted output, the required AQ outputs have to be realized with inverters (see Fig. 1.21).

Description of the Logic Model

For verifying the logical correctness, a logic simulator is used. Again logic models are available as procedure building blocks. They contain, again in PASCAL-like language, the functional descriptions of the individual components (execution section). Aside from the purely logical verification, the examination of time conditions also plays a crucial role at this level. For this reason the logic model includes data on delay times as well as time conditions which, if violated, lead to error messages (assertions). The logic model of our LATCH cell is the CLA01A procedure:

```
(*FBDL-description of VENUS cell CLA01A *)
(*---------------------------------------------------------------------------------*)

PROCEDURE CLA01A         (IN  E1, E2 : BIT;            (* E1 : DATA INPUT *)
                          OUT A1    :BIT);             (* E2: LOAD INPUT*)
                                                         (*A1:OUTPUT*)
(*----------------------------------------*)
(*     DECLARATION SECTION  *)
(*----------------------------------------*)
CONST         TS     =  3000 PS;    (*   SETUPZEIT                              *)
              TH     =  1000 PS;    (*   HOLDZEIT                               *)
           TDLH1     =  2500 PS;    (*   L/H -                                  *)
           TDHL1     =  3000 PS;    (*   H/L - DELAY TIMES E1 -->A1 (E2 = H)    *)
            TDX1     =  2500 PS;    (*   L,H/X -                                *)

           TDLH2     =  4000 PS;    (*   L/H -                                  *)
           TDHL2     =  4000 PS;    (*   H/L - DELAY TIMES E2 --> A1            *)
            TDX2     =  4000 PS;    (*   L, H/X -                               *)

            TLDW     =  5500 PS;    (*   LOAD PULSE WIDTH                       *)
```

```
ASSERTIONS
FALL (E2) & ((TIME-CHANGETIME (E1))< TS ) = >                  (* MONITORING OF *)
      ERROR ('CLA01A : SETUP TIME VIOLATION');            (* SETUP TIME CONDITION *)

(ALTER(E1) & ((TIME-DOWNTIME(E2)) <TH)) = >                    (* MONITORING OF *)
      ERROR ('CLA01A: HOLD TIME VIOLATION');               (* HOLD TIME CONDITION *)

(FALL(E2) & ((TIME-UPTIME (E2)) <TLDW)) = >                    (* MONITORING OF *)
      ERROR ('CLA01A : LOAD-HIGH-PHASE TOO SHORT');           (* LOAD PULSE WIDTH *)

(*----------------------------------------*)
(*      EXECUTION PART                     *)
(*----------------------------------------*)
BEGIN
IF    TRANSITION (E2, "0X", "1")
THEN CASE E1 OF
      "0":   A1 : =       "0" AFTER (TDHL2);                 (* CONNECT  E1 TO A1 *)
      "1":   A1 : =       "1" AFTER(TDLH2);            (* IF E2 GOES from L OR X *)
      ELSE : A1 : =       "X" AFTER (TDX2)                               (* TO H *)
      END

ELSE IF ALTER (E1)
      THEN CASE E2 OF
            "0": ;                                      (* OUTPUT UNCHANGED *)
                                                               (* IF E2 = L *)

            "1" : CASE E1 OF
                  "0" : A1 : = "0" AFTER (TDHL1);            (* CONNECT  E1 TO A1 *)
                  "1" : A1 : = "1" AFTER (TDLH1);                   (* IF E2 = H *)
                  ELSE : A1 : = "X" AFTER (TDX1)
                  END;
            ELSE : A1 : = "X" AFTER (TDX1)                (* OUTPUT UNDETERMINED, *)
                                                        (* IF E2 UNDETERMINED *)
                                                       (* AND IF E1 IS CHANGED *)

            END
      ELSE
END.
```

Description of Stimuli and Result of Logic Simulation

In addition to receiving the model description of the object to be simulated, the simulator must also know which input signals are to be processed. A suitable programming language is available for this stimuli input. We will not, however, take a closer look at it now. The result provided by the logic simulator includes a list containing all violations of the monitoring conditions and a timing diagram of the input and output signals:

```
               CODE = BIN     SCAN        TIME = 0PS       STEP = 4200PS
-------------------+^--------------^----------------^----------------^--------------------------------
E1             !x...11xxx..xx111..1111111xxxxxxx........xxxxxxx1111111......
E2             !X11111111111111111...11.....11.....11.....11.....11.....111.
A1             !xx..111xx...xx11.....11111111xxxxxxx.......xxxxxxx1111111...
```

```
          CODE = BIN  SCAN   TIME = 7900PS STEP = 100PS
-------------------+Λ--------------Λ----------------Λ---------------Λ--------------------------------

E1           !....................11111111111111111111111111111111111111
E2           !11111111111111111111111111111111111111111111111111111
A1           !...........................................1111111111111
```

Description of the Fault Model

The logic plan is subject to another verification step, namely, the *testability analysis*. A fault simulator checks if certain fault types (such as short circuit to ground, short circuit to supply voltage, break) occur at the circuit outputs for given test patterns. The input for the fault simulator is the test model. It is identical to the logic model mentioned above, but is augmented by information on the kind of faults to be simulated (fault model). The fault model for CLA01A is described in the following procedure (see also Chap. 4):

```
CIRCUIT CLA01A;
ELEMENT CLA01AFDL;
    EXTERNAL
        INPUT = E1,E2;                     (*E1:DATA INPUT*)
                                           (*E2:LOAD INPUT*)
        OUTPUT = A1;                       (*A1:OUTPUT*)
        FAULT = (LOGA0,LOGA1,LOGZZ);       (*LOGA0 = SA0-FAULT*)
                                           (*LOGA1 = SA1-FAULT*)
                                           (*LOGZZ = OPEN-FAULT*)
        TECHNIK CMOS;                      (*TECHNOLOGY*)
    END;
END.
```

The result of the simulation is normally provided in form of a table like the following:

Fault simulation statistic

Simulation	next	last	all
Number of faults	0	18	18
Simulated	0	18	18
Not simulated	0	0	0
Solid detected	0	12	12
Potential detected	0	6	6
Undetected	0	0	0
Rejected	0	0	0

Simulation	next	last	all
Detecting rate in % (average/total)			
Solid	0	67	67/67
Solid + potential	0	100	100/100

Description of the Circuit Model

The goal of *electric circuit design* is the transformation of the logic plan to the transistor level. The basic components of the design are, in this case, the electric device symbols, such as n- and p-channel transistors of various dimensions, diodes, wires and resistors. If a verification is requested at this level, circuit simulators [1.4] are to be used. They solve the various differential equations which describe the components, and thus render very precise information on voltages, currents and their time behavior.

First, the circuit model must contain references to the transistor models used (in our example, this is: TR1, TR2) as well as their parameters. Then, the components used (in our example 8 transistors, 6 capacitors) are described regarding their dimensions and interconnections.

Finally, one must indicate the input stimuli, source voltages, initial conditions, and the nodes to be observed. The following SPICE input describes a simple CMOS inverter, consisting of four pairs of transistors connected in parallel.

```
$$$$ CMOS CELL CSD01A.TD.TYP
INPUT LISTING

****************************************************************

*    VDD =    0.500000E + 01V     VBG = 0.250000E + 01V
*    TOX =    0.400000E + 03A
*    CIRCUIT   =    CDS01A 1    ICMO
*
*    ***   NODE CROSS REFERENCE TABLE
*
****       A1        1
****       E1        2
****       VDD       3
****       BULK 4
****       GND       0
****       WELL 6
*
***  MOS  ELEMENT    DESCRIPTION
*
```

```
M000001  1  2  3  4  TR2     W =  3.600E-05 L = 3.000E-06
 + AD = 3.060E-10 AS = 1.710E-10 PD = 5.300E-05 PS = 9.500E-06
M000002  1  2  3  4  TR2     W =  3.600E-05 L = 3.000E-06
 + AD = 3.060E-10 AS = 1.710E-10 PD = 5.300E-05 PS = 9.500E-06
M000003  3  2  1  4  TR2     W =  3.600E-05 L = 3.000E-06
 + AD = 1.710E-10 AS = 3.060E-10 PD = 9.500E-06 PS = 5.300 E-0.5
M000004  3  2  1  4  TR2     W =  3.600E-05 L = 3.000E-06
 + AD = 1.710E-10 AS = 3.060E-10 PD = 9.500E-06 PS = 5.300 E-0.5
M000005  1  2  0  6  TR1     W =  2.000E-05 L = 3.000E-06
 + AD = 1.700E-10 AS = 9.500E-11 PD =  = 3.700E-05 PS = 9.500 E-0.6
M000006  1  2  0  6  TR1     W =  2.000E-05 L = 3.000E-06
 + AD = 1.700E-10 AS = 9.500E-11 PD = 3.700E-05 PS = 9.500 E-0.6
M000007  0  2  1  6  TR1     W =  2.000E-05 L = 3.000E-06
 + AD = 1.700E-10 AS = 9.500E-11 PD = 3.700E-05 PS = 9.500E-06
M000008  0  2  1  6  TR1     W =  2.000E-05 L = 3.000E-06
 + AD = 9.500E-11 AS = 1.700E-10 PD = 9.500E-06 PS = 3.700 E-0.5
*
***            PARASITIC CAPACITANCE DESCRIPTIONS
*
CW2            6      4      1.632E-13
CP3            1      4      3.909E-14
CP4            2      4      4.371E-14
CM5            3      4      3.930E-14
CI7            6      1      5.452E-14
CI8            2      6      4.940E-14
*
***            TRANSISTOR PARAMETER SETS:
*
**********************************************************************
*     NMOS4    PMOS4        PARAMETERSETS #8
**********************************************************************
*
.MODEL TR1 NMOS4 = a K0  = bE-6 PHI = c F = d L0 = e A = f
 + G = g H = h COX = iE-3 RSH = j CJ = kE-3 CJSW = lE-9 LUD = mU
MJSW = n IS = o IS = p OF = q
MODEL TR2 PMOS4 UTO = a' K0 = b'E-6 PHI = c' F = d' L0 = e' A = f'
 + G = g' H = h' COX = i' E-3RSH = j' CJ = k'E-3 CJSW = l' E-9 LUD = m'U
 + MJSW = n' JS = o' IS = p'E = q'

DESCRIPTION OF THE SIMULATION RESULT
.PLOT TRAN V(2/E1) V(1/A1) (-0.5,5.5)
*
***INPUT:
*
```

```
.TRAN 0.4233NS 50NS
VD 2 0          PWL(0 04N 08N 5 18N 5 22N 0 35N 0)
*
***LOAD:
*
CA1             1   0     225F
*
****            WELL VOLTAGE:
*
VWELL 6 0 DC 0V
*
****SUBSTRATE VOLTAGE:
*
VBULK 4 0 DC 5V
*
***SUPPLY VOLTAGE:
*
VDD 3 0 DC 5V
*
END
```

Figure 1.23 shows the result of a circuit simulation as provided by SPICE.

1.5.3 Technology Dependence

The transition from a higher to a lower design level leads to a representation which is more and more detailed and closer to production. The closer the design engineer comes to the layout level, the more production-related design rules he has to observe. On the other hand, even the highest design levels are – at the current stage of design technology – not independent from technological considerations. Figure 1.24 qualitatively shows the degree of technology dependence of the model libraries of the IC design process.

In order to explain the effects of technology dependence, we want to give an example of the interaction between block representation for the architecture and geometric representation for the layout design. For interactive layout, there are usually two steps:

- *floor plan:* In rough estimation, areas are allotted to blocks which have been defined in accordance with functional requirements; those areas are then connected via the wiring areas, keeping in mind data and control flows; thus, the overall chip design (*floor plan*) is created.
- *layout:* The transistors and their connections are placed into the thus created compartments in compliance with the layout design rules. Frequently, this in turn makes a change in the floorplan necessary. A demanding iteration process may develop until the optimal fit has been achieved.

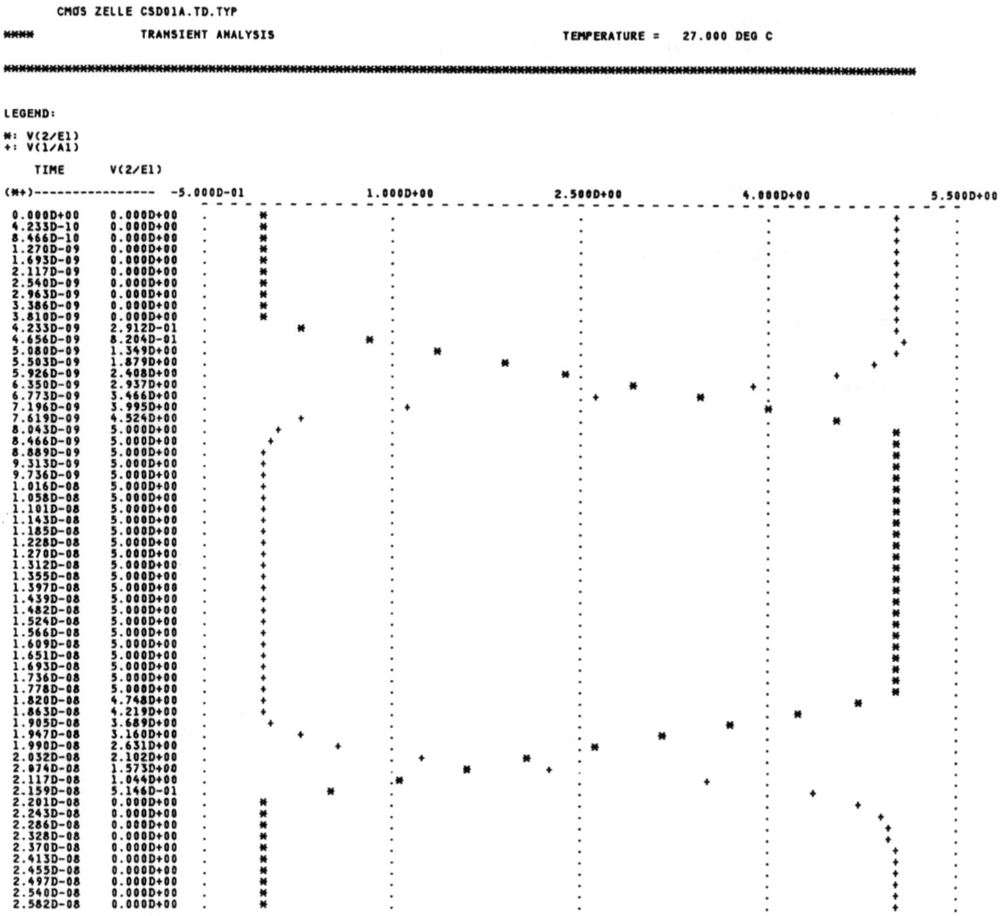

Fig. 1.23. Output voltage V (1/A1) (+ + +) as a function of the input voltage V (2/E1) (***) of an inverter

This kind of layout design requires, apart from specific technological knowledge, a lot of experience on the part of the design engineer. For this reason, CAD technology is directed at keeping the higher levels of design as independent from technology as possible, thus allowing the design engineer to completely focus his attention on the solution of his application problem. In order to reach this goal, the following two methods have been developed, and have been implemented in CAD systems:

- use of predefined and verified elementary functions,
- parameterization and automatic adaptation.

If the first method is used, the design engineer completes the top-down process before the geometric layout design begins: during this top-down design he finds predefined elementary functions on certain levels, for example the logic level, (for definition, see Chap. 1.6.1). If he uses them, and if he has the necessary placement and

Model	Degree of technology dependence
Block symbols	o
Register-transfer models	o
Logic symbols	o
Logic design rules	o
Logic models	o
Test design rules	⊗
Test models	⊗
Device symbols	⊗
Electric design rules	⊗
Circuit models	⊗
Stick elements	●
Layout contours	●
Layout design rules (geometrical and electric)	●
Layout segments	●
Oversize / undersize data	●
Test program models	●
Electric data	●

o independent
⊗ influenced
● dependent

Fig. 1.24. Degree of technology dependence of the model libraries

routing programs at his disposal, he comes to the end of his manual design process. The layout design is carried out automatically.

The elementary functions must first, of course, be prepared for various technologies and layout design methods. This task is fulfilled by the CAD engineer. He adheres to the bottom-up method, by interactively and manually developing cells which have the optimum size, delay times, power dissipation and scope of application (see Fig. 1.25).

Such a predefinition with extensive expenditure invested by the CAD engineer is justified, because the predefined models may be used by a large number of design engineers.

The second method, the parameterization, originated in so-called λ-design rules, suggested by Mead and Conway [1.5]: The basic idea consists of providing all layout dimensions with a scale factor λ. If the layout design rules are changed, then only λ has to be changed. A necessary prerequisite in this context is a layout description language. Examples of such languages are the *C*altech *I*ntermediate Format (CIF) and the Siemens HKP format (*H*och*k*omprimiertes *P*ositionstape = highly compressed positioning tape).

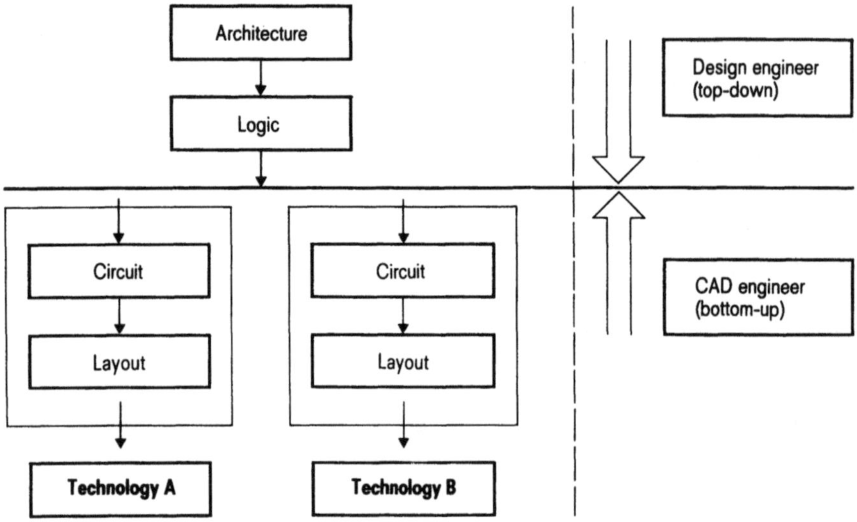

Fig. 1.25. Coordination of CAD engineer and design engineer

However, optimal use of the varying process design rules cannot usually be obtained with this simple type of λ-parameterization, as various experiments have shown. For this reason, procedural languages are being introduced which allow a formal definition and declaration that may also be applied on higher levels. The so-called chip compilers or chip generators are the relevant examples [1.6].

A solution to the problem of obtaining total technology independence will certainly lead to a combination of the two methods: the user will be able to use more and more powerful predefined and verified circuit components (macro cells, for definition see Chap. 1.6.1) which he can adjust to his needs by *function parameterization* or *declaration*. To the CAD engineer, the *technology-related parameterization* will be useful which facilitates a fast adjustment of cells to new process technologies.

1.5.4 Verification

In Sect. 1.5.1, we have roughly dealt with the methods of verification for ensuring consistency; first horizontally, i.e. on the same level of representation, and then vertically, i.e. across various levels of representation. Now, the verification process will be explained in more detail.

A design engineer with the task of designing a circuit on level n obtains, as target function, the result of the previous design step $n-1$. In a closed, computer-based process this target function is available in a data base in the form of a structured data set. Moreover, sets of design rules are given for each level. On the one hand they lead the design engineer through the design process, but, on the other hand, they also restrict his scope. At present, the design rules are only partially available in form of formalized data sets. The rest of it is still based upon the knowledge and experience of the design engineer. This part is not formalized and objective.

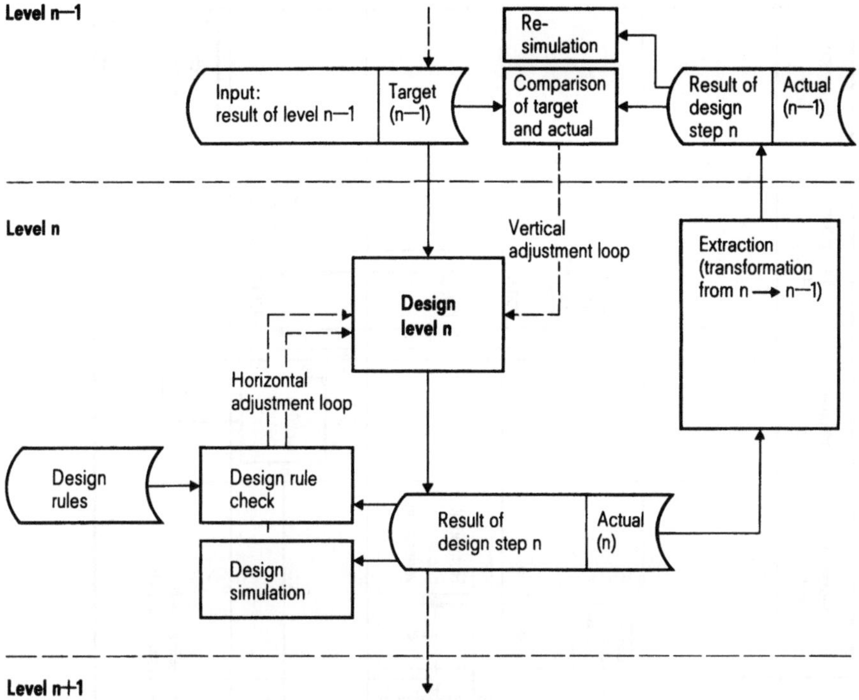

Fig. 1.26. Verification process

Let us consider, for example, the layout design: The target function (level $n-1$) should be a correct transistor circuit. It has been stored in the data base as a transistor net list. The layout engineer now applies his – subjective – knowledge about the basic structure of transistors for transformation into the layout level (level n), and thereby complies with the formalized, geometric layout design rules (such as "minimum distance between two aluminum wires") and the electric design rules (such as "connection of source areas with the supply voltage, if possible, via the aluminum level").

In order to check the design result, simulations are carried out on the higher levels. If formalized design rules exist, their compliance is verified by means of special check programs, or *checkers*. Design simulation and design rule check refer to the representation which is being designed at that time. They lead to a local or horizontal adjustment loop. This *horizontal verification* aims at *ensuring the exact reproducibility* (i.e. compliance with the abilities of the fabrication line) of a circuit by means of a closed chain of rule-checking steps. The principle of verification is shown in Fig. 1.26.

When the horizontal design consistency has been reached, one has to check if the design also complies with the target function of level $n-1$, i.e. if *vertical consistency* is also ensured. The *vertical verification* has to ensure *specification consistency*. To achieve this, there are programs which transform a design to a higher design level. They are called *extractors*. Their result, the actual design of level n transformed to level $n-1$, can then be compared with the target function of level $n-1$. Possible deviations lead to a vertical adjustment loop. Beside the comparison between target

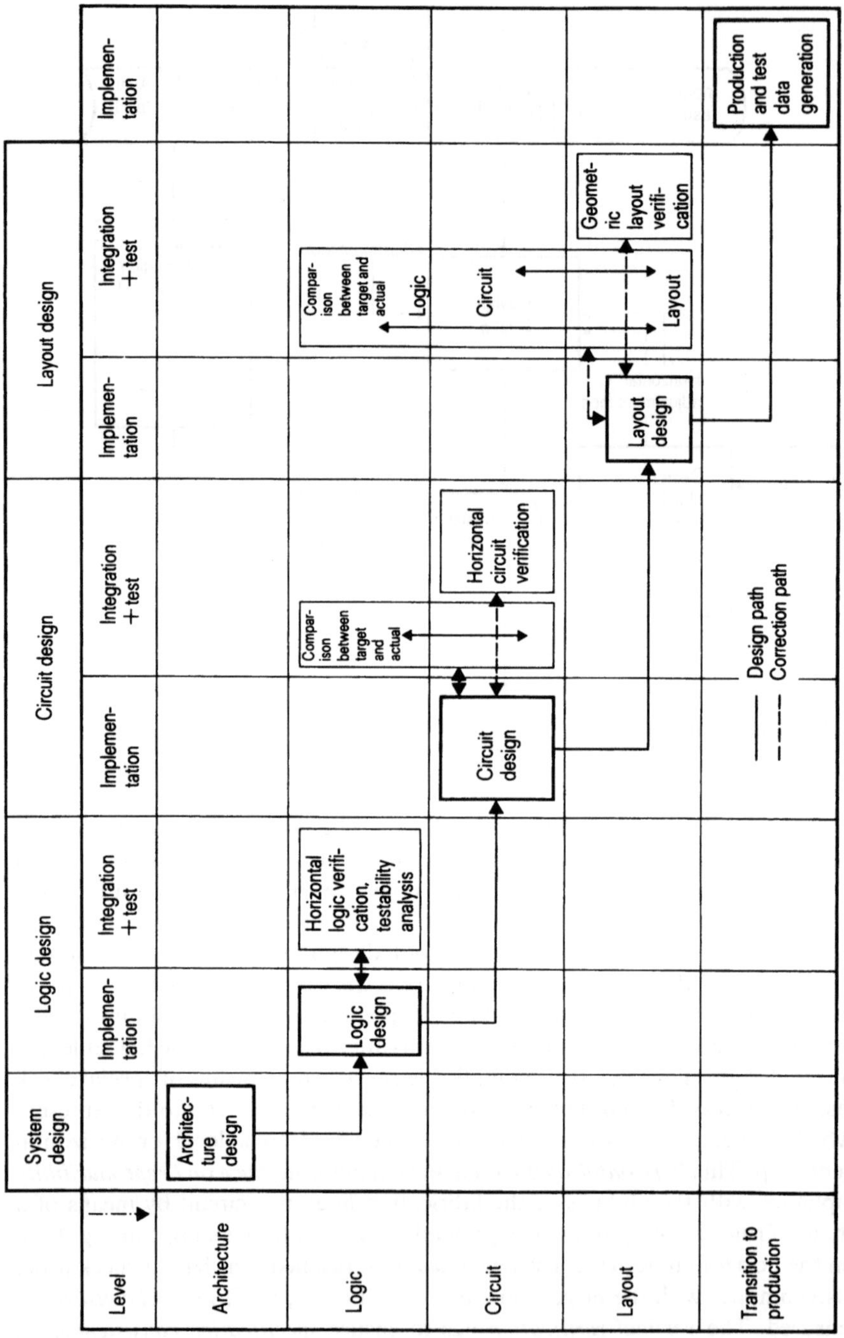

Fig. 1.27. Typical design and verification process

and actual results, a simulation may be carried out on level $n-1$ and the target function may be verified functionally, by means of simulation results.

For designs on a higher level, one at least implicitly assumes properties which have to be fulfilled by the designs on lower levels. Those characteristics can be checked only after the design on the lower level has actually been carried out. Programs which compute certain values, for example, by means of models, and supply these values to more accurate and realistic simulations (resimulation), are called *analyzers*. Deviations from the assumptions cause another entry into the vertical adjustment loop. This process of data feedback from lower levels to higher ones is called *back-annotation*.

An example for characteristics to be analyzed is the time delay of a circuit. During logic design, the geometrical wiring of the chip is not known. For this reason, the logic or timing simulation is carried out with estimated values for the wiring delays. If the actual time delay, based upon the completely layed-out circuit, has been ascertained by means of a time delay analysis program, a resimulation, and, if necessary, an adjustment may be carried out.

Of course, the ideal verification process shown in Fig. 1.26 is, in practice, not completely carried out on all levels. Rather, the automatic check of formalized design rules may be found on the lower levels, whereas the simulation occurs rather on the higher ones. Because structuring and automation of the design process progresses from lower to upper levels, there are still no sufficient formal description or effective simulation tools available for the architecture design phase. Another significant deviation from the idealized process is the extraction across several levels. A typical design and verification process is shown in Fig. 1.27.

1.5.5 Standard IC Design Procedure

Each standardization is meant to simplify. This goal is reached through the compulsory introduction of structures, procedures and interfaces. But this results in a reduction in degrees of freedom. If the various possibilities of feed-back loops are superimposed over the general design process, a confusing variety of conceivable procedures arises, as is shown in Fig. 1.28.

In order to standardize the IC design process three complementary measures are executed:

- Single procedure steps and partial procedures are automatized.
- Predefined and verified circuits are used.
- Certain combinations of procedure steps (permissible procedures) are enforced and supported automatically.

The standard procedure used in this book is based upon the assumptions that

- predefined cells are available below the logic level,
- the models for all design levels are predefined and verified[8],
- automatic design programs are available,
- certain design procedures are automatically supported and guaranteed.

[8] "Verified" means that the cells have been simulated, fabricated and measured, individually as well as in cell combinations.

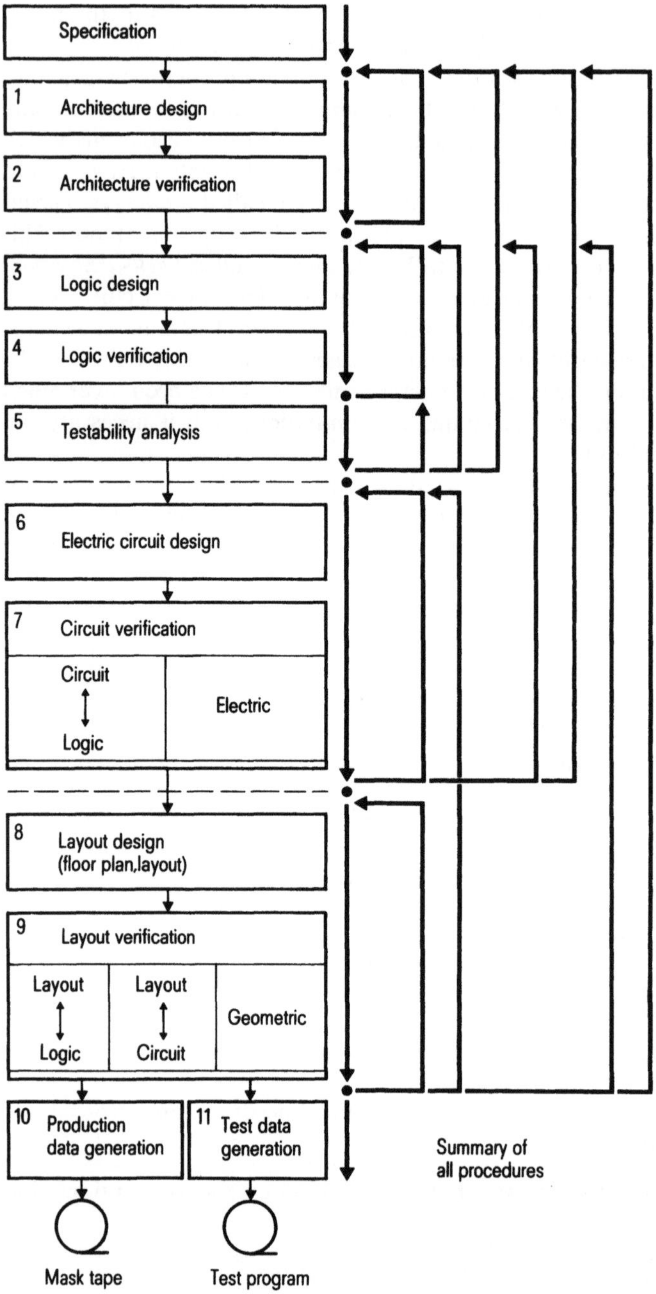

a

Fig. 1.28. Possible procedures as derived from the general design process

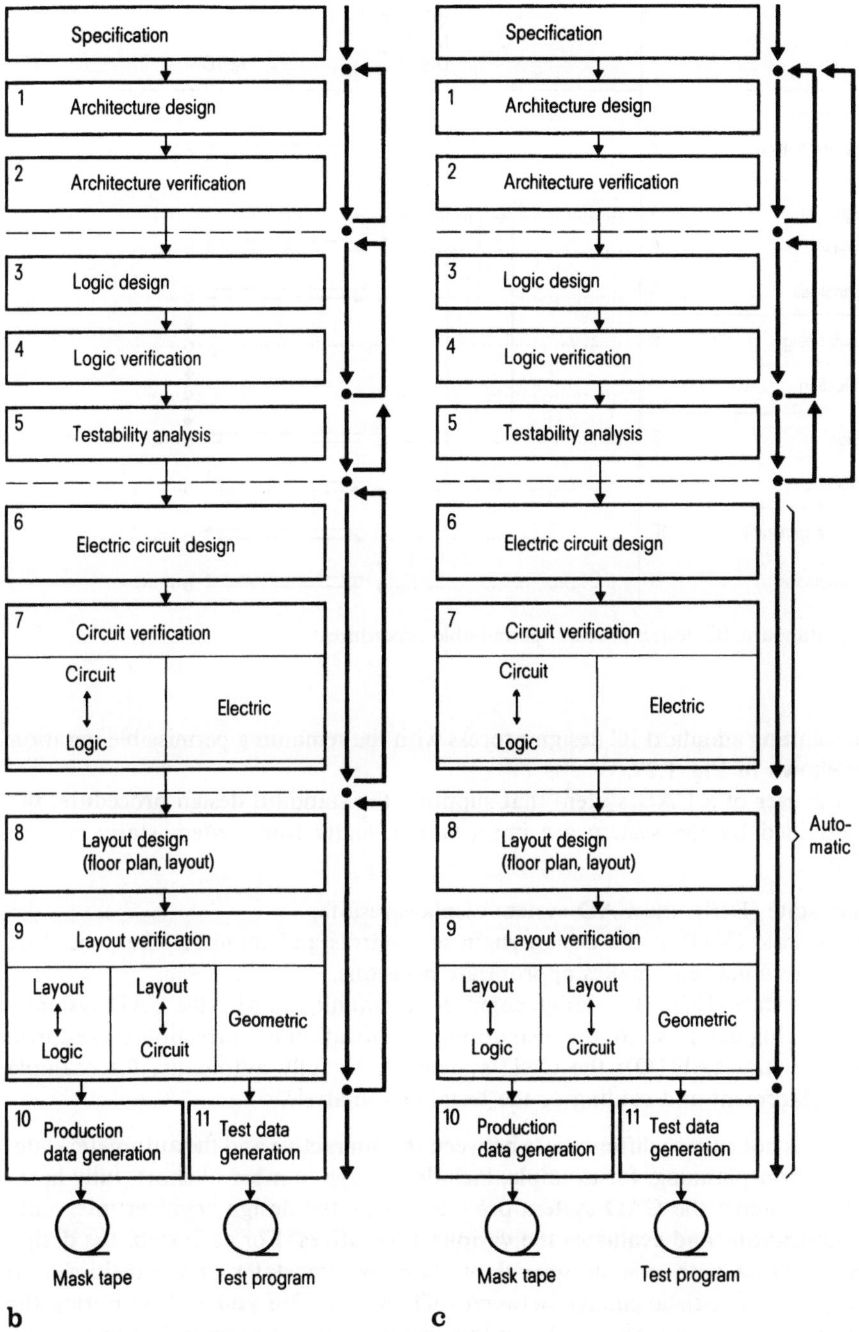
Specification
1 Architecture design
2 Architecture verification
3 Logic design
4 Logic verification
5 Testability analysis
6 Electric circuit design
7 Circuit verification
Circuit
Logic
Electric
8 Layout design
(floor plan, layout)
9 Layout verification
Layout
Logic
Layout
Circuit
Geometric
10 Production data generation
11 Test data generation
Mask tape
Test program
b
Specification
1 Architecture design
2 Architecture verification
3 Logic design
4 Logic verification
5 Testability analysis
6 Electric circuit design
7 Circuit verification
Circuit
Logic
Electric
8 Layout design
(floor plan, layout)
9 Layout verification
Layout
Logic
Layout
Circuit
Geometric
10 Production data generation
11 Test data generation
Auto-
matic
Mask tape
Test program
c

Fig. 1.28

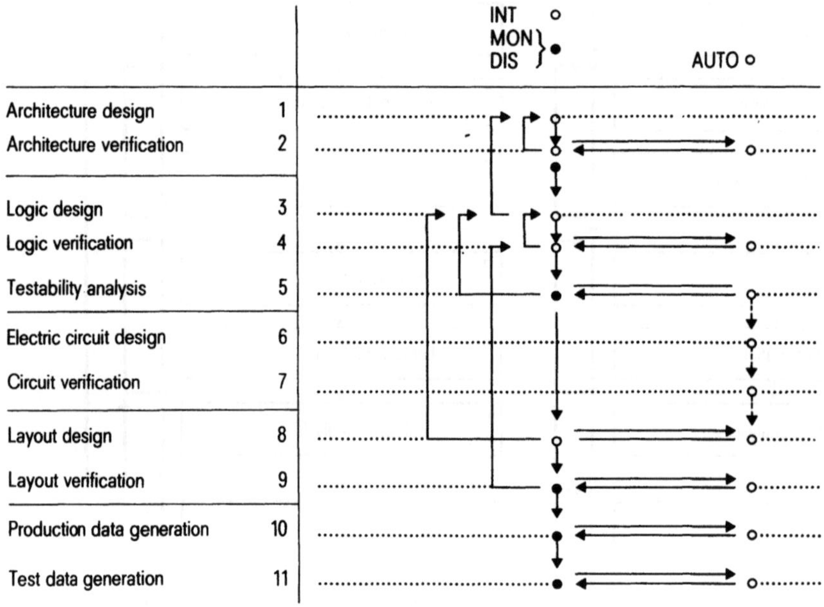

Fig. 1.29. Standard IC design process: permissible procedures

The resulting standard IC design process with the remaining permissible iteration loops is shown in Fig. 1.29.

For the user of a CAD system that supports the standard design procedure, the states provided by the system are important. Usually four system states can be identified:

- display state (DIS): the CAD system displays results,
- monitor state (MON): the design engineer controls and monitors the procedure from his terminal and makes appropriate decisions,
- interactive state (INT): the design engineer communicates with the CAD system – mainly during design steps, but also during modification or correction procedures,
- automatic state (AUTO): the CAD system works fully automatic, for example during placement and routing or the testability analysis.

Often it is not easy to differentiate between the interactive and the automatic state. Interactive floorplanning, for example, includes a large number of short, fully automatic design steps: the CAD system provides ideas, the design engineer intervenes with modifications and evaluates the various alternatives. For each step, the design engineer is faced with one or several of these system states and transitions. In Fig. 1.29, a characteristic change between INT, MON, DIS and AUTO during the standard design process is shown. From this representation, the simplified procedural structure becomes particularly clear.

It is the task of the CAD engineer to develop the CAD system, including the development of the SW system and the preparatory development of the cell components. He designs and verifies each cell in compliance with the steps of the general IC

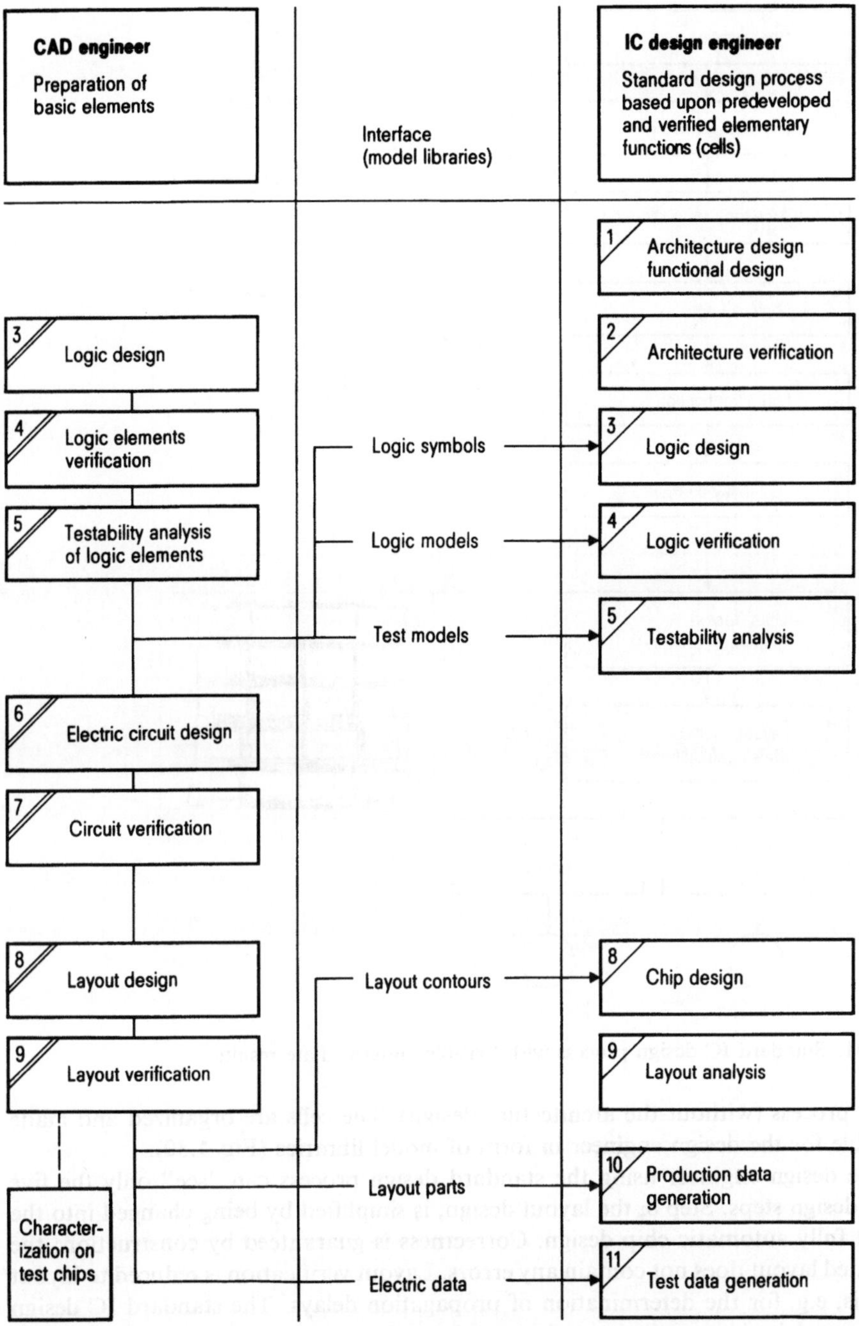

Fig. 1.30. Development of the standard IC design process from the general IC design process

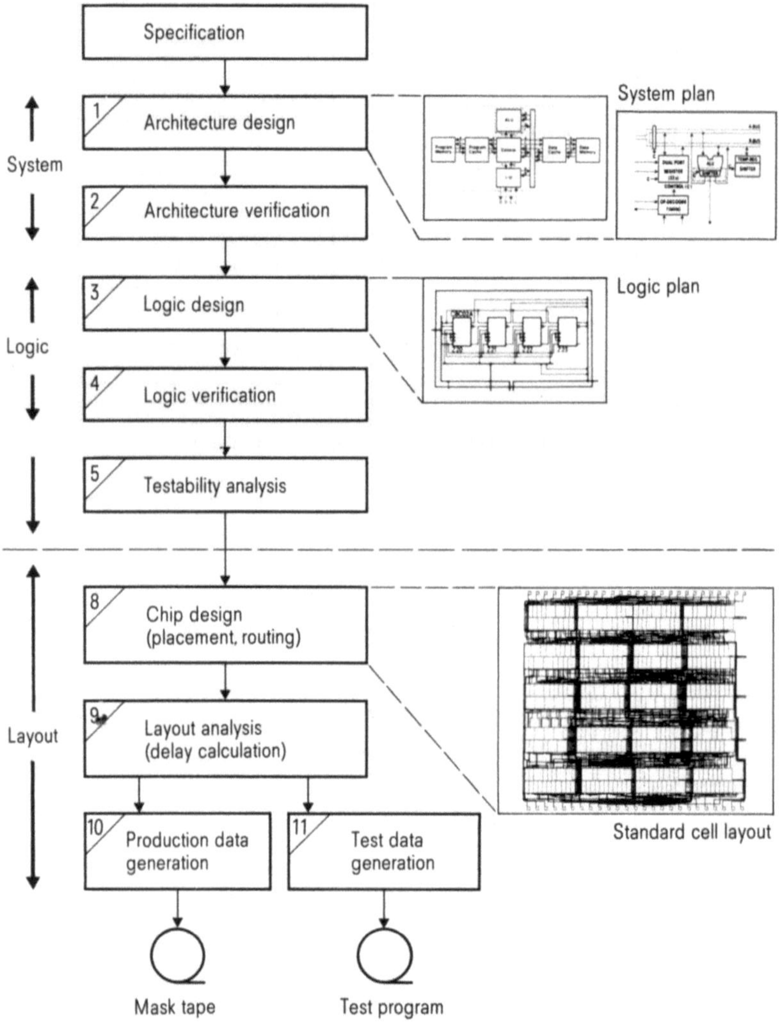

Fig. 1.31. Standard IC design process with "visible" intermediate results

design process (without the architecture design). The cells are organized and made available for the design engineer in form of model libraries (Fig. 1.30).

The design engineer using the standard design process can "see" only the five upper design steps. Step 8, the layout design, is simplified by being changed into the almost fully automatic chip design. Correctness is guaranteed by construction: the generated layout does not contain any errors. Layout verification is reduced to layout analysis, e.g. for the determination of propagation delays. The standard IC design process and its intermediate results are shown (using a standard cell library), in Fig. 1.31.

Summing up, one may say that the measures described above lead to the following effects which result in the simplification of the standard IC design process as compared to the general IC design process:

- Symbols and models are predefined and stored in libraries. Only a limited set is offered for use. Predefined parts are also increasingly applied for the general design process. However, there they are often modified or supplemented by individual additions.
- The predefined models are verified many times, first by the design engineer and later when in use. Their correctness is guaranteed. Errors due to individual modifications are impossible.
- Restriction to a limited set of components complying with uniform conventions concerning format, connections etc., makes it possible to automatize certain steps. The electric circuit design is omitted entirely. It is only implicitly included in the logic and the layout design.
- Correct programs for fully automatic construction exclude any error. Verification steps are left out completely or at least simplified (principle of "*correctness by construction*").
- Compulsory definitions of interfaces between the design steps enable the development of a "closed data structure space", the application of modern data description tools and software procedures which ensure vertical and horizontal consistency of data. This allows, for example, an automatic generation of input data for simulators. For the general design process, however, the input data is usually produced individually and interactively on the screen.

Figure 1.32 shows the possibility for the various design steps to be automatized, and for the models to be predefined.

To some extent, CAD systems for the standard IC design process already resemble the capabilities of compilers. They can be used to automatically generate the correct command sequences for controlling the production process and the automatic test equipment, starting from the formal, graphic-descriptive representations of the circuit. To achieve these advantages, the freedom of layout design has to be reduced; those restrictions will be introduced in the next paragraph, and they will be discussed in more detail in Chap. 3.

1.6 Introduction to the Standard Methods of the Layout Design

The purpose of the standard design procedure is to reduce development effort, development time and risk. This goal is reached, for example, by automating the layout design process. In addition, the degrees of freedom for the geometric configuration have to be limited.

Only a certain predetermined number of elementary circuit functions (cells) is available and released for application.

The conditions on which the standard design methods are based relate to:

- the cell scheme,
- the arrangement and connection scheme.

In the pertinent literature layout design methods are noted and provided with names which are derived partly from the cell scheme, partly from the arrangement and connection scheme. This introduction familiarizes the reader with a classification

Automation potential		Design step
General design process	Standard design process	
Interactive	Interactive	Architecture design
Interactive	Interactive	Architecture verification
Interactive		Logic design
Interactive/ automatic	Interactive/ automatic	Logic verification
Interactive/ automatic	Automatic	Testability analysis
Interactive	-	Electric circuit design
Interactive/ automatic	-	Circuit verification
Interactive	Automatic	Layout design
Interactive/ automatic	(Analysis: automatic)	Layout verification
Automatic	Automatic	Production data generation
Interactive	Automatic	Test data generation

Fig. 1.32. Possibility of design steps to be automatized and of models to be predefined

which helps him to find his way through the numerous methods. On the other hand, standard design systems provide performance features for the design engineer which enable him to change given cells in a certain way, or to design new, individual cell conglomerates, i.e. to use the cells in different ways. Each standard design system is thus characterized by specific

- application schemes.

The following paragraphs familiarize the reader with the current standard methods of the layout design.

Model library	Predefinition	
	General design process	Standard design process
Block symbols	Predefined	Predefined and verified
RT models	Individual	Predefined and verified
Logic symbols	Predefined or individual	Predefined and verified
Logic models	Predefined or individual	Predefined and verified
Test models	Predefined or individual	Predefined and verified
Electric device symbols	Predefined	-
Circuit models	Individual	-
Stick elements	Predefined	-
Layout contours	-	Predefined and verified
Layout design rules	Predefined	-
Layout parts	Individual	Predefined and verified
Model test programs	Individual	Predefined and verified

Fig. 1.32

1.6.1 Cell Scheme

The term "cell scheme" refers to the geometric cell form and the connective structure, i.e. the position of signal and supply terminals.

In the layouts of the seventies which had little structure, circuit sections with the same logic functions nevertheless used to have different geometric patterns, depending on their electric and geometric environment. The goal of using single design circuit sections several times implied a standardization of the external geometric form. Not only the outlines, but also the internal structure of those standard circuit sections plays a role in design: the uniform position of p- or n-channel transistors, for example makes possible a simple power supply structure.

Elementary functions which have to match must therefore be subject to a homogenous concept in terms of geometric form and connective structure. They are standardized. Aside from its numerous uses, this type of standardization allows automatic placement and routing: Such programs use abstractions of the circuit components consisting of outlines and connection terminals. The placement and routing programs can be kept simple only if one adheres to a regular and orthogonal grid structure. The higher the variability of cells, the more degrees of freedom have to be taken into consideration for placement and routing programs.

To give a precise definition: Components representing elementary functions and belonging to a homogeneous scheme are termed *cells*. Firmly predefined and verified cells are also called *cell primitives*. The entire complex of cells belonging to a certain cell-, arrangement- and connection scheme for a certain technology form a *cell library*.

For the design engineer, the cell primitive is the smallest visible unit. Often, however, cells which are generated or parameterized are designed by the CAD system of even smaller units, the *cell elements*. They are not explicitly available for the design engineer.

There are three different kinds of cell schemes: the wiring macros, the standard cells and the macro cells.

Wiring Macros

The most basic cell scheme leads to structures with *basic cells* located matrix-like at fixed predetermined places. They consist of few transistors (usually 2, 4 or 6) which are not, or only partially, connected with each other (Fig. 1.33). By supplementing the connections between the transistors in an appropriate way operative cells can be developed. The total of all wire sections and contacts which are used for personalizing are provided with a type name, and stored in a library. Because those abbreviations are quite similar to the command sequences abbreviated with one name in the SW technology ("macro"), the term "*wiring macro*" is used here. In this context it should be emphasized that the basic cells and their matrix-like arrangements do not only represent an abstract scheme, but are physically prefabricated and verified in the form of so-called "masters". A cell with a certain logic function is created by appropriately connecting transistors of one or several basic cells with each other. The corresponding intracell-connecting scheme has been prepared in the form of a wiring macro. However, the "pre-fabrication" of the wiring macros themselves is immaterial: wiring macros are called from a library by the design engineer and laid over the basic cell matrix. They lead to mask tapes for the last three or four production masks which result in individual logic functions during additional production steps (personalizing). The complexity of wiring macros typically is about 4 to 50 transistors. The wiring macros offered lead to functions such as AND, OR, multiplexer, counter etc.

Standard Cells

Standard cells are subject to another cell scheme. Omitting physical prefabrication, one can create a more variable cell scheme. It only needs to provide as simple a

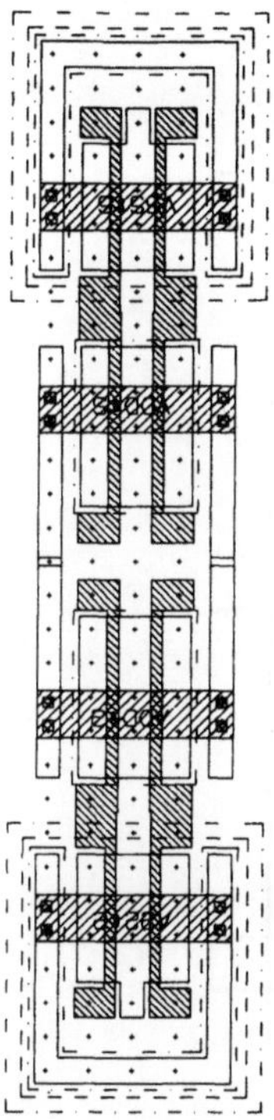

Fig. 1.33. Basic cell for wiring macros

juxtaposition of cells as possible, according to the principle of modularity. Standard cells are rectangular, all of the same height, and, depending on their complexity, of variable width. Current supplies are positioned horizontally at the same height in order to provide automatic connection when they are put together. Their inputs and outputs are located on only one side, or on two – the upper and the lower (Fig. 1.34). The complexity of standard cells can, to some extent, be compared with that of wiring macros. There is a tendency to offer more complex cells, such as n-bit-counters or n-bit registers.

To give a more precise definition of terms: Logic cells created in the form of standard cells or wiring macros are connected to each other via the *inter-cell wiring*. The term "intra-cell wiring" refers to the connections between the transistors within one cell. For prefabricated integrated circuits, one has to distinguish between that part of the intra-cell wiring that is prefabricated on the master, and that part which is used for personalizing, and stored in the library, i.e. the wiring macro. In practice, the terms "intra-cell wiring" and "wiring macro" are, however, often considered synonymous.

Macrocells

Macrocells have an additional degree of freedom. Their only restriction is rectangular outlines. Their dimensions in height and width are variable. Thus, they are not restricted by a standard height. Another restricting condition is the connective struc-

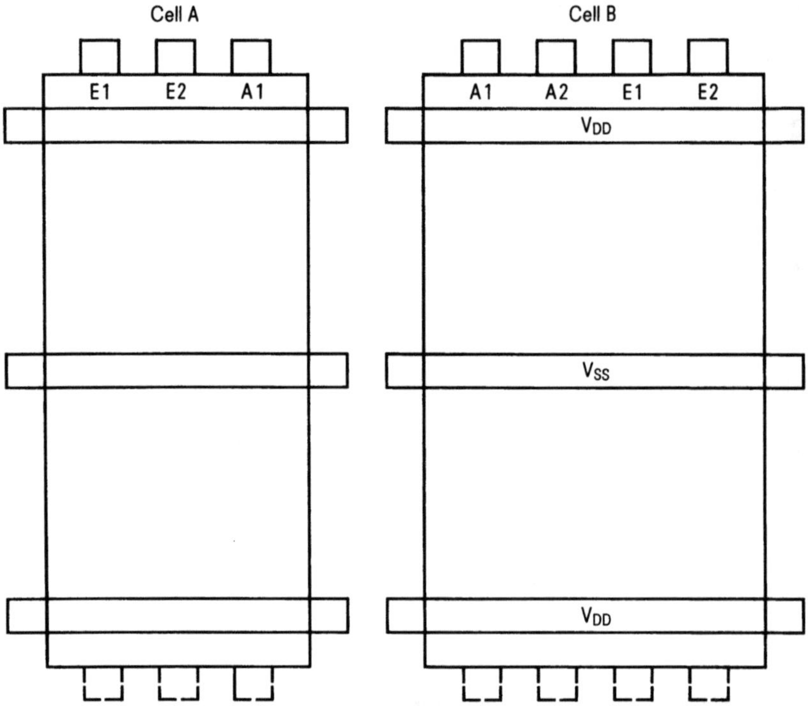

Fig. 1.34. Elementary forms of standard cells

ture of current supplies: they have to be positioned in such a way that V_{DD} and V_{SS} terminals can be separated from each other with one curve. The complexity of macrocells is theoretically unlimited: they may represent memories, programmable logic arrays (PLAs) or microprocessor cores.

1.6.2 Arrangement and Connection Scheme

In addition to the cell scheme, there are conditions established which refer to the possible arrangement and possible interconnections between cells. Three schemes are distinguished: matrix, linear and Manhattan scheme.

Matrix Scheme

Corresponding to the rigid basic structure of the wiring macros, horizontal and vertical channels with given widths are provided for their connections. This results in a matrix arrangement (Fig. 1.35a). The physically prefabricated master allows only a certain maximum number of circuits in each channel. For designs requiring a higher number, one has to use a more appropriate master. On the other hand, space is wasted for designs that would actually need less interconnection area.

Linear Scheme

Accomodating their homogenous height, standard cells are fit in an arrangement of single or double rows. The supply lines run through the cells horizontally. Between

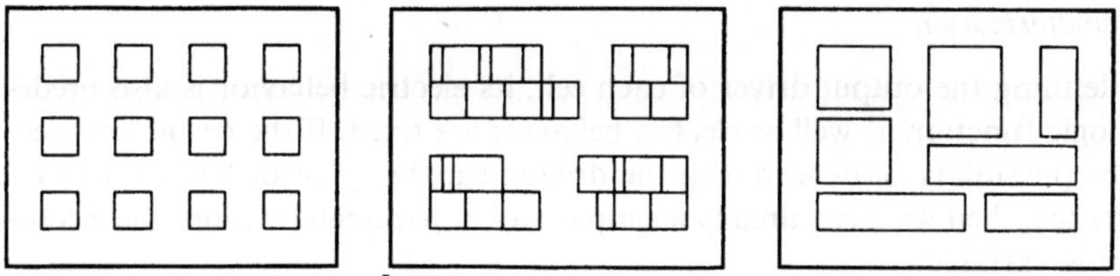

Fig. 1.35. Examples for matrix, linear and Manhattan arrangements

the cells space is left for horizontal and vertical wiring channels. Those channels are, however, variable, as no physical prefabrication has taken place, and their width is determined by the placement and routing program, according to design requirements. Figure 1.35 b shows such a linear arrangement.

Manhattan Scheme

If the cells are general rectangles, the arrangement ought to provide all adequate degrees of freedom. The only restriction is the horizontal or vertical orientation of the edges. The cells are positioned along the axes of a cartesian coordinate system (the grid may be arbitrarily fine). Corresponding to a layout style at transistor level, this scheme is called Manhattan scheme (Fig. 1.35c). The connective scheme also follows the cartesian raster. If the separation of the supply terminals is possible, a fully automatic construction of the interconnections is possible.

1.6.3 Application Schemes

Early cell libraries used to contain only unchangeable, i.e. rigid, components. For the design engineer it was only possible to use functions that were described in the catalogue. It was neither possible to modify individual cells nor to form application-specific groups by combining several given cells to one new, "individual" cell. But now, there are tools available for the design engineer to work more freely with the cells.

A rigid cell (cell primitive) is characterized by four features, namely:

- its logic function,
- its electric behaviour (as well as its production technology)
- its geometric form,
- its connective structure.

The design engineer currently has three possibilities available for more flexible cell use which would be more suited to his application:

- simple parameterization,
- recursive formation of cell blocks,
- functional parameterization.

Simple Parameterization

By strictly defining the output driver of each cell, its electric behavior is also predetermined. Logic function as well as electric behavior are fixed. If the design engineer is offered the possibility of determining the driving capability according to relevant load capacitance, then we have already a simple way of parameterization: the electrical behavior is variable.

Further examples for cells that can be parameterized are the n-bit counter, n-bit register cells etc., i.e. linear extensions described by one parameter. In this case the logic function is variable.

Both alternatives are referred to as "simple parameterization". For the design engineer, it is irrelevant if the CAD system calls up predefined cells, or generates cells on demand. From his point of view he calls only a generic cell which has been supplied with a global name and a parameter; in his circuit the cell is supplied with an individual name.

Recursive Cell Blocks

While parameterization modifies the characteristics of the individual cell, the design engineer can generate user-specific cells from simpler cells, by constructing cell conglomerates, so-called cell blocks.

For the most simple variant, a cell block appears only as an individual name of a cell group on a higher design level. For repeated use of the same group this name is used as abbreviation. The CAD system expands this name by calling the individual cells. The software equivalent is the inline-macro.

Of more interest, however, are those cell blocks which are not dispersed on a lower level, but are preserved as new user-specific units down to the layout level. Thus it is possible for the design engineer to keep control over certain global characteristics. CAD programs are available which make it possible to form cell blocks recursively (according to the cell scheme of macro cells) from standard cells or from macro cells. Thus, a hierarchy of cell blocks may be constructed.

Functional Parameterization

The more complex a cell becomes, the more specialized is its function. A cell that contains a microprocessor can only be used for more than one usage because its function can be modified according to the application, that is by means of a program. Programming a given hardware structure is one example of a complex functional parameterization. Such a functional parameterization, however, often requires – as opposed to the microprocessor example just mentioned – a far-reaching adjustment of the entire layout: For example, a RAM may have quite different aspect ratios, according to its size and organization. Recently synthesis programs have been developed which allow automatic generation of cells based upon a functional specification. Such synthesis programs are called *cell generators*. In practice, they are used today to generate for example PLAs, RAMs and ROMs.

Beside the three application schemes described above, methods have been developed allowing the design engineer to generate special cells by means of *symbolic layout methods* (stick diagrams) or even as manual layout according to his optimization goals, and to integrate those cells in a design process in such a way that the CAD system ensures consistency (*free macro cells*).

Higher degrees of freedom for the design engineer include, apart from technical difficulties (for synthesis), also organizational problems: Cell primitives are specified by the CAD system supplier. He assumes responsibility for their characteristics and proper functioning. The design engineer, on the other hand, is responsible for individual cells himself.

1.6.4 Compatibility

Naturally, not all cell schemes are compatible with all arrangement and application schemes.

The only arrangement that is suitable for wiring macros is the matrix. The wiring macros are connected to each other through an orthogonal grid with fixed channel width. The size of the chips is predetermined. Based on this idea, a design method was developed. It is called "*gate-array method*".

Standard cells can be arranged only in rows. They are also interconnected along an orthogonal grid which varies however, taking on the channel width actually

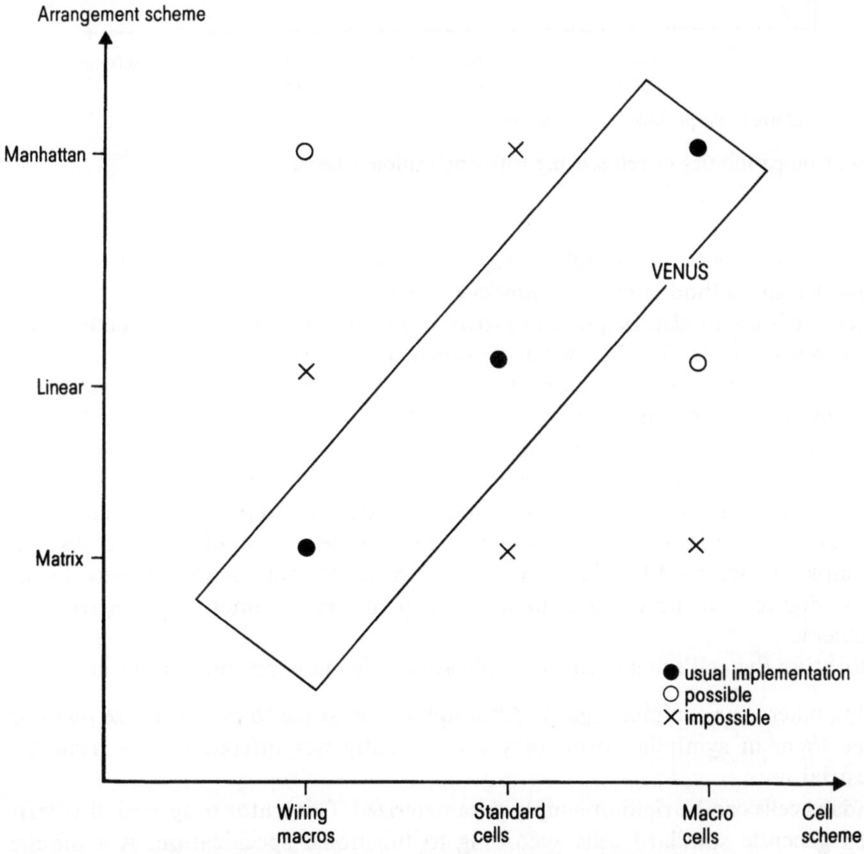

Fig. 1.36. Compatibility of cell scheme with arrangement scheme

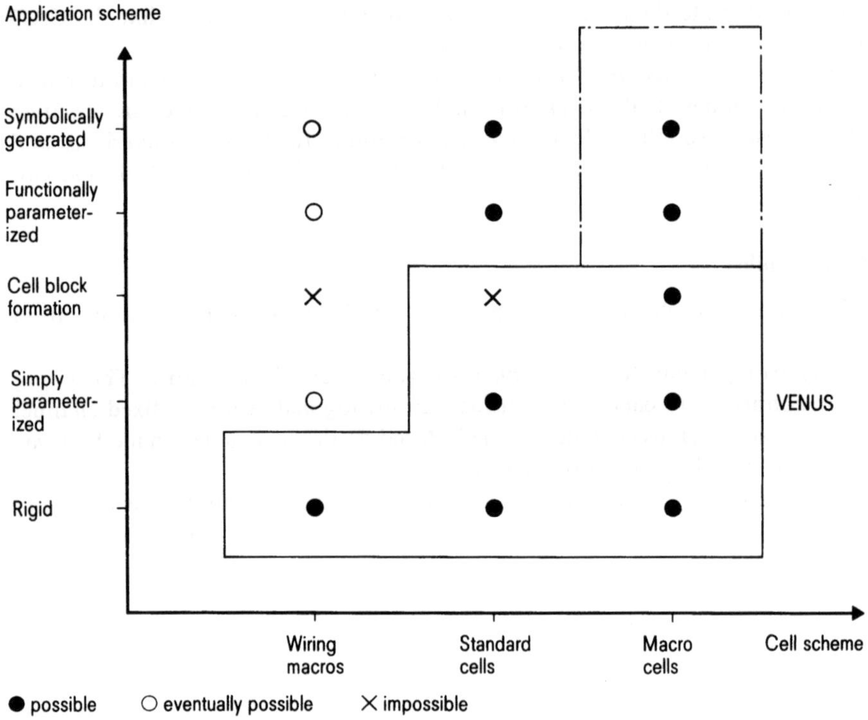

Fig. 1.37. Compatibilities of cell scheme with application scheme

needed. The chip size is variable. Only its aspect ratio may be predefined. The resulting design method is called "*standard cell method*".

Macro cells could also be placed in rows and properly connected. It makes more sense, however, to use the Manhattan arrangement, and connect the cells in compliance with the orthogonal grid. One must ensure that each macro cell receives sufficient current supply. If constant current density is required, the supply lines widen approaching the chip frame. The routing channels are also constructed according to the actual need, and so is the chip size. The resulting design method is called "*macro cell method*". It allows one to design cell blocks which are subject to the macro cell scheme, and to modify the design interactively under the control of the CAD system. An example of such modifications are pre-placement or construction of special connections. Figure 1.36 shows the compatibility of the cell schemes with the arrangement schemes.

Allocating the cell scheme to the application scheme is not quite as clear:

- Wiring macros are defined rigidly. Although it is possible to parameterize them or define them in symbolic form, they are currently not offered in this form for industrial use.
- Standard cells can be rigid or simply parameterized. Generator programs also exist which generate standard cells according to functional specification. A symbolic definition of standard cells is possible with certain restrictions.

- Finally, macro cells may either be called as cell primitves, or be constructed from standard cells and/or macro cells. Clearly they may be used for building hierarchies.

Figure 1.37 shows the compatibilities of cell and application schemes. Figures 1.36 and 1.37 also show those combinations supported by the VENUS CAD system.

1.6.5 User-Specific Definition of ICs and Cells: Review of Terminology

Figure 1.38 attempts to give an overview of the various cell-oriented design methods, and to summarize the terminology.

For cell-ICs, we distinguish between those with and those without prefabrication. Cell-ICs with prefabrication are called gate-array ICs. They consist of a master made of basic cells, and overlaid wiring macros. For CMOS gate-arrays wiring macros are identical with the logic GA cells. In the case of ECL gate-arrays, the logic GA cells are combined to groups which cannot be separated by the user (the physical GA cells). Finally, GA-ICs may consist of larger groups formed by wiring macros. Those GA macro cells may be specified by the user by simple or functional parameterization, or by constructing cell blocks consisting of wiring macros.

Cell-ICs without prefabrication may be built directly from standard cells, or from macro cells. If standard cells or macro cells are stored in a library in the form of rigid components, they are called cell primitives. Standard or macro cells with simple parameterization consist of cell elements. There is no clear distinction between simple and functional parameterized cells (generated macro cells). The term "cell generation" is mainly used for more complex cells which are specified functionally.

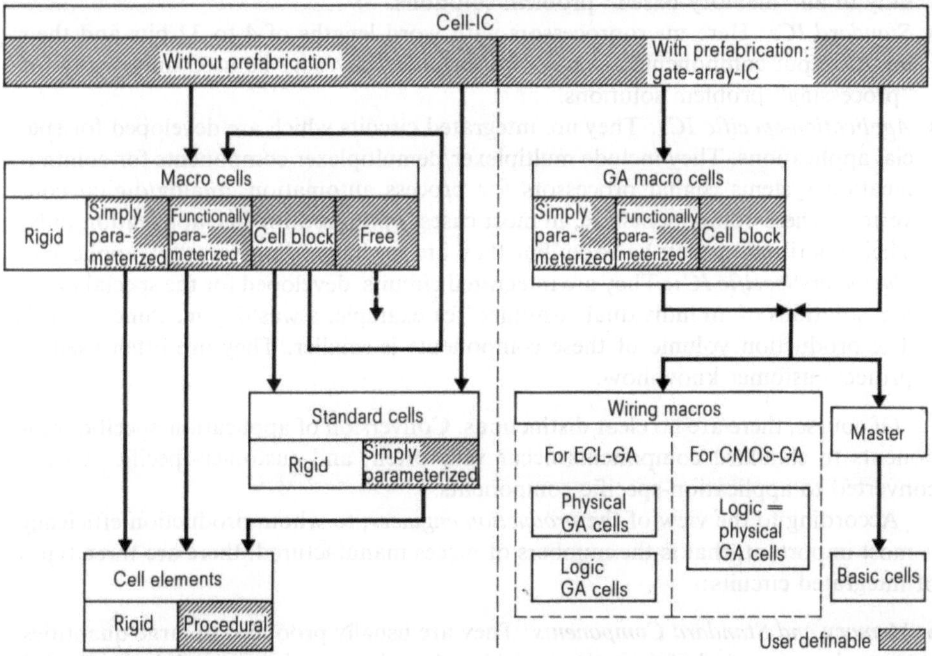

Fig. 1.38. Overview over the application-specific definition of ICs and cells

Cell elements used for simple parameterization usually are rigid, whereas those used for cell generators are often procedurally specified.

Macro cells built in the form of cell blocks are either composed of standard cells or hierarchically constructed again in form of macro cells.

At last, macro cells may be subject to free definition, that is, composed of stick elements, or in the form of a geometric layout.

1.7 Interpretations of the Terminology

In a technical field as new as design technology for highly integrated circuits, the terms used are, of course, not yet compulsory. The meaning of a lot of terms is unclear. Other terms have several connotations, and some may overlap. Although it is not our task to give precise and compulsory definitions of terms which is impossible at this early stage, the following explanations may help to provide an easier understanding. When an elucidation of terms is required, the classification of integrated circuits plays a major role; in this context, the viewpoint of the user is key.

According to the view of the *sales engineer*, for whom the range of application is most important, there are four different classes of integrated circuits:

- *Memory Components:* They include DRAMs (*D*ynamic *R*andom *A*ccess *M*emories), SRAMs (*S*tatic-*RAM*s), ROMs (*R*ead-*O*nly-*M*emories), EPROMs (*E*lectrically *P*rogrammable *ROMs*) etc. These components can apparently be used universally in all "memory-based" problem solutions.
- *Standard ICs:* Here microprocessors with word lengths of 4 to 32 bits and their input/output components are most important. They may be used universally for "processing" problem solutions.
- *Application-specific ICs:* They are integrated circuits which are developed for special applications. They include multiplexer/demultiplexer components for communication systems, signal processors for process automation, analog/digital converters. These components are, in most cases, optimized for one application only. Their distribution may be wide, but they are no longer "universally" applicable.
- *Customer-Specific ICs:* They are integrated circuits, developed for the special problem solutions of an individual customer, for example, a washing machine control. The production volume of these components is smaller. They are often used to protect customer know-how.

Of course, there are no clear distinctions. Conversion of application-specific components to standard components occur very often; and customer-specific ICs are converted to application-specific components.

According to the view of the *production engineer*, to whom production efficiency is most important, that is the numbers of pieces manufactured, there are three types of integrated circuits:

- *Memory and Standard Components:* They are usually produced in large quantities over a longer period of time, thus ensuring that piece cost is low, and the basic load of a production line is guaranteed over a longer period.

- *Semi-Custom Components:* They are components which are to some extent pre-manufactured and put on stock. The process of individualizing or personalizing them, i.e. adjusting them to the function needed, is carried out when they are actually required. The majority of production cost may be apportioned to a large number of pieces produced. The cost of personalization is attributed to each individual customer. Typical semi-custom components are the gate-array ICs.
- *Custom (Full-Custom) Components:* They are components which are produced to solve the special problem of an individual customer. They are usually produced in high numbers. Prefabrication is not possible. All mask levels are custom-specific.

The *design engineer*, who considers efficiency and high speed of development important criteria, divides the integrated circuits into two classes, namely those with an "optimized design" and those with a "semi-optimized" design.

The *optimized* components include the memory components and the standard components.

The optimization criteria for memory components are: highest density, i.e. minimum wasted space, and homogenous dynamics while keeping power dissipation at specified levels. Every memory cell has to be written and read as fast as any other memory cell; this characteristic must not be affected by the cell's position. The optimization criteria for standard components are: highest density for a given dynamic external behavior, and minimum power dissipation.

If application-specific or customer-specific circuits are expected to have a high production volume, they are also included in this class; the design engineer considers them "full-custom" components. In order to meet the optimization criteria mentioned above, greater design effort and a longer design period, including frequent redesigns, are accepted; of course, there are economic limits. The design is based upon basic circuits which are optimized particularly for the special product. Predefined basic circuits (cells) are, for the time being, hardly used.

The second class, the *semi-optimized* components, includes application-specific and customer-specific circuits having optimization criteria calling for little development time, much development safety, and simple redesign characteristics. In exchange, lower density of integration, i.e. a some waste of silicon area, lower speed and a higher degree of power dissipation is accepted. The design engineer calls these components also "semi-custom components".

The targets of optimization are reached by widely using predefined and verified circuit sections. Especially the gate array and the cell design methods have been developed for that purpose.

Criteria for the decisions that have to be taken by the design engineer are: integration density – dynamics – power dissipation on the one hand, and development expenditure – development time – redesign characteristics on the other.

The first three criteria have to be measured in *technological terms*. Designs optimizing those criteria are called optimized designs, this term comes from "historical habit", coined by physicists and engineers.

The last three criteria have to be measured in *economical terms*. Designs optimizing those are called semi-optimized design for the same "historical" reasons.

The process of removing these well-defined separations, however, has already been started.

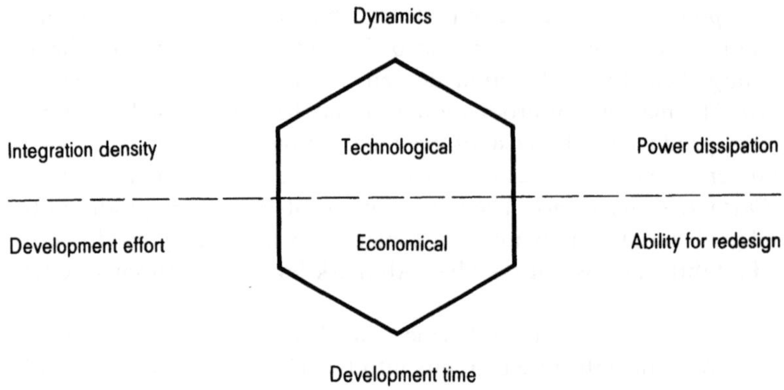

Fig. 1.39. Criteria for the decisions taken by the design engineer

This book is written from the position of the design engineer, as far as the application of terms is concerned: all methods based upon components which are materially or immaterially predefined are called semi-custom methods. They are equivalent to standard design methods. The general IC design process produces full-custom designs.

Literature – Chapter 1

1.1 Siemens Energie & Automation. Double Volume 2/1985.
1.2 Giloi, W. K.: Rechnerarchitektur. Berlin: Springer 1981.
1.3 Boehm, B. W.: Software Engineering Economics. New York: Prentice Hall 1981.
1.4 Nagel, L. W.; Pederson, D. O.: Simulation Program with Integrated Circuit Emphasis. Proc. 16th Midwest Symposium Circuit Theory, Waterloo, Canada 1973.
1.5 Mead, C.; Conway, L.: Introduction to VLSI-Systems. New York: Addison Wesley 1980.
1.6 Sandweg, G.; Schallenberger, B.: Das Chipgenerator-Konzept: ein Weg zum automatischen VLSI-Entwurf. Elektronik, Nr. 5, 1985, S. 169–174.

2 Introduction to Semiconductor Technology for Integrated Circuits

The semiconductor technology started its triumphant march in the beginning of the sixties; a few circuit functions were integrated on a monolithic silicon chip a few mm^2 in size. With the development of so-called MOS technology (MOS = *m*etal *o*xide *s*emiconductor) the key step towards extremely high integration densities was taken successfully in the middle of the sixties. In 1964, G. E. Moore, then director of the research department at Intel, predicted an annual doubling of the number of circuit functions per chip. The actual trend, a quadruplication every third year, is shown in Fig. 2.1.

The success is based mainly upon three factors: the capability of controlling an ever increasing chip area, the reduction in dimensions of the single components and the reduction of costs per circuit function or memory bit. On the average, costs declined approximately 25 % yearly. The increase in reliability was just as remarkable: Today's standard circuit works in a normal environment and does not require air-conditioning. However, the increase in development effort resulting from this higher complexity has a more and more restrictive effect. The development effort would increase overproportionately in comparison with the number of components per chip if cost – reducing measures were not applied. The VENUS design system is designed to alleviate some of the problems in this area. Primarily, the design engineer who has no knowledge of semiconductor technologies should be able to develop integrated

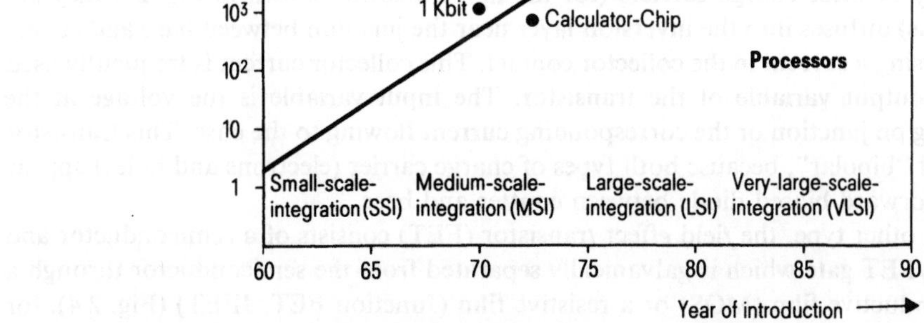

Fig. 2.1. The development of integration density per chip

circuits. The following paragraphs offer an introduction to semiconductor and integration technologies in two steps:

- a short "summary",
- and a more detailed description in the subsequent paragraphs.

2.1 Summary

The semiconductor technology has provided a high number of procedures for generating highly integrated circuits – so-called integration technologies. For the comparison of these integration technologies, we want to draw attention to parameters that are especially important for the design engineer. They are:

- integration densities,
- gate delay,
- power dissipation.

Because a reduction of delay times can only be attained for a certain technology in exchange for an increase of dissipated power, the *power delay product* (measured in pJ) is definitely a powerful parameter. If the relevant integration density is high the dissipated power may become a restrictive factor. In addition to the architecture-related measures for parallel or serial processing, the switching speed effects the processing power, i.e. the data throughput of a circuit. The integration is directly reflected in the cost as follows: the higher the integration degree, the lower the cost per gate function. The number of controllable interconnect levels also becomes more and more important to the integration degree. Chart 2.1 compares the most important parameters of the ECL, NMOS and CMOS technologies.

Figure 2.2 [2.1] summarizes the most important integration technologies. They can be divided in two classes, according to the utilization of physical effects. The base for one class is the bipolar transistor. The base for the other class is the field effect transistor. Both transistors are represented schematically in the Fig. 2.3 and 2.4 [2.2].

The bipolar transistor (Fig. 2.3) consists of a silicon-monocrystal, in which one generates negatively conductive (so-called n-conductive) regions by local doping with pentavalent atoms (phosphor atoms, for example) and positively conductive (so-called p-conductive) by doping with trivalent atoms (boron, for example). The transistor's functions are based upon the injection of charge carriers in the base. The majority of these charge carriers (for the npn-transistor shown in Fig. 2.3 they are electrons) diffuses into the inversion layer near the junction between base and collector, causing a current in the collector contact. This collector current is frequently used as the output variable of the transistor. The input variable is the voltage at the injecting pn junction or the corresponding current flowing to the base. This transistor is called "bipolar", because both types of charge carrier (electrons and holes) appear in the forward biased diode between emitter and base.

The other type, the *f*ield *e*ffect *t*ransistor (FET) consists of a semiconductor and a MOS/FET gate which is galvanically separated from the semiconductor through a non-conductive film (MOS) or a resistive film (Junction FET, JFET) (Fig. 2.4). for the insulating film structure, a *m*etal (or rather polysilicon), *o*xide, *s*emiconductor

Technology		Minimal dimension (µm)		Max. chip size (mm^2)		Max. number of trans. per chip (logic)		Max. number of memory bits per chip		Min. gate delay (ns)		Number of interconnect levels	
		84	90	84	90	84	90	84	90	84	90	84	90
bipolar	ECL	0.6 vertical	0.2 vertical	70	>100	–	20 k - 30 k	4 k (stat.)	16 k (stat.)	0.5	0.04	3	4
MOS	NMOS	2	< 1	60	>120	100 k	500 k	256 k (dyn.)	–	1	0.1 - 0.15	3	4
	CMOS	2	< 1	60	>120	80k	400 k	64 k (stat.)	4 M (dyn.), 1 M (stat.)	1	0.1 - 0.15	3	4

Chart 2.1. Comparison of important parameters of the ECL, NMOS and CMOS technologies as of 1984 (with the approximation for 1990)

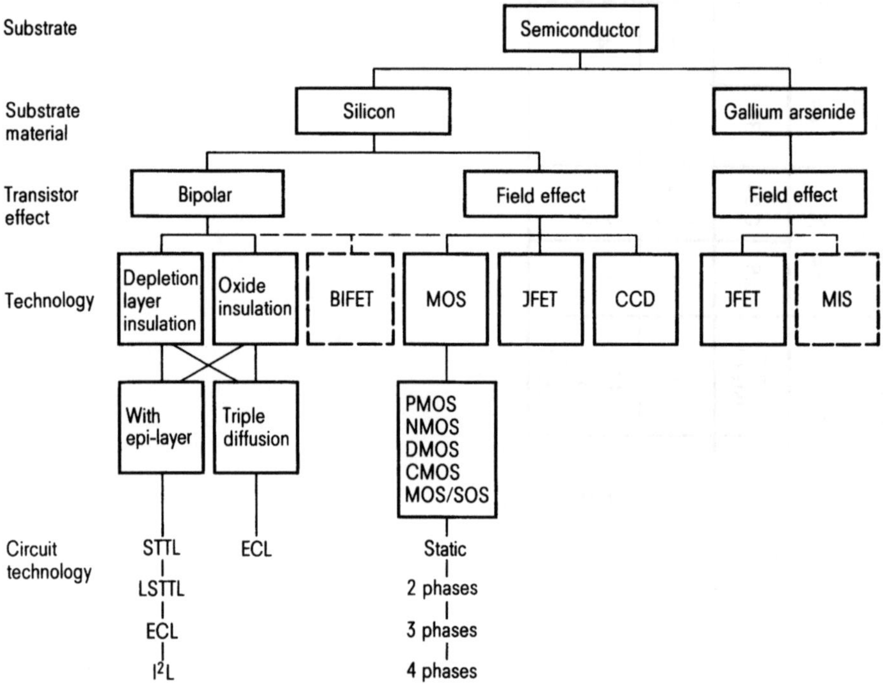

Fig. 2.2. Summary of the most important integration technologies [2.1]

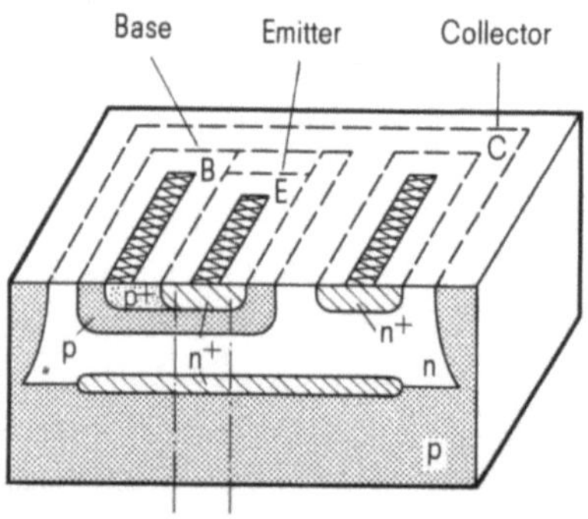

Fig. 2.3. Schematic representation of a bipolar npn-transistor

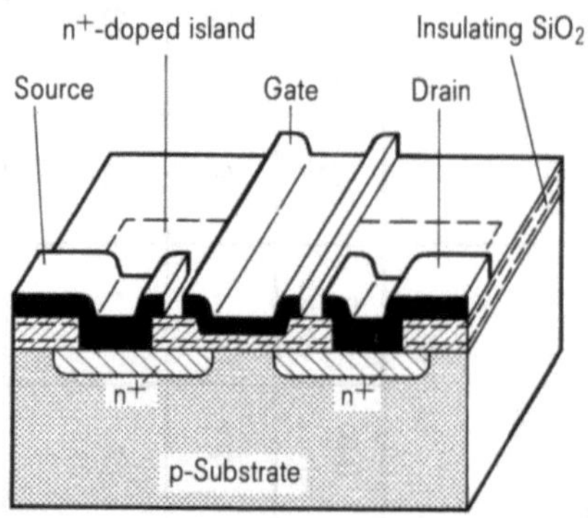

Fig. 2.4. Schematic representation of a MOS field effect transistor

arrangement is applied. This results in the acronym MOS transistor. It is also used as a collective term for field effect transistors. The semiconductor of the MOS transistor (Fig. 2.4) consists of two high-doped connective zones, which are called source and drain, and an intermediate low-doped section, with a doping opposite to that of source and drain.

Normally, no current can flow between the electrodes, because of the inserted p-region. If a MOS/FET gate which is set above the insulator is connected to a positive potential against source, negative charge is induced on the surface. This charge, which here consists of electrons, induces an n-conductive channel between source and drain. This transistor is termed n-channel-transistor; if the preceding signs of the charges, dopings and voltages are reversed, a p-channel transistor exists. For further information about semiconductor technology, the interested reader may turn to the following chapters of this book and the pertinent literature [2.5; 2.6].

For the IC designer the following connections are relevant:

High processing speed, i.e. low gate delays, can be reached with the bipolar technology. Contrary to MOS technology, however, one has to accept relatively low degrees of integration, i.e. comparatively high cost.

Because of its different structure, the MOS transistor requires ten to 100 times less space than a bipolar transistor. This is partially a result of the omission of the insulating pn-junction between collector and substrate of the bipolar transistor. The structure of the MOS transistor is self-insulating. Moreover the MOS technology provides a higher degree of flexibility for wiring on the chip, because the high-doped drain and source regions (and polysilicon for the silicon-gate process) can be used as a second or third wiring level without further steps of processing.

A much higher degree of integration can be achieved with MOS technology at a lower speed. Thus, rather complex circuits and memory components can be generated. Figure 2.5 illustrates this dependency.

This evaluation applies to current, practical application. Development results, however, indicate that eventually in MOS technology higher circuit speeds can be reached; thus the gain in speed of bipolar technology is partially lost.

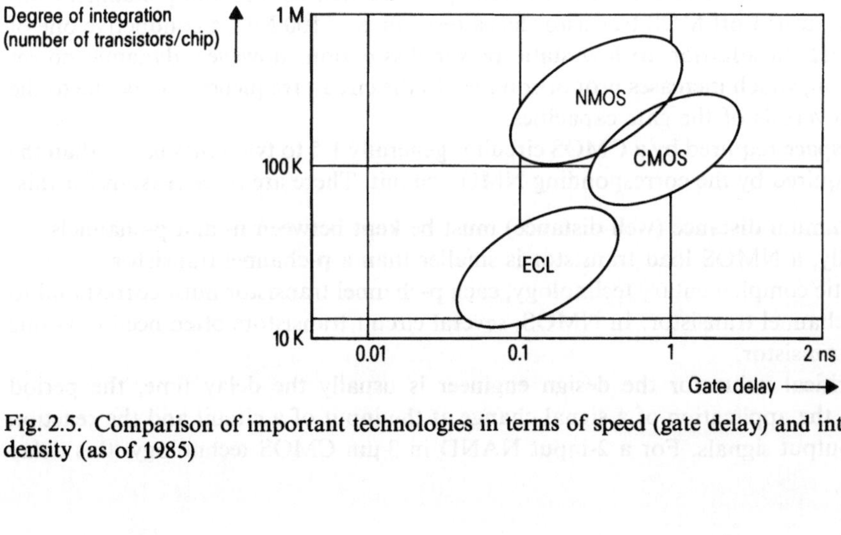

Fig. 2.5. Comparison of important technologies in terms of speed (gate delay) and integration density (as of 1985)

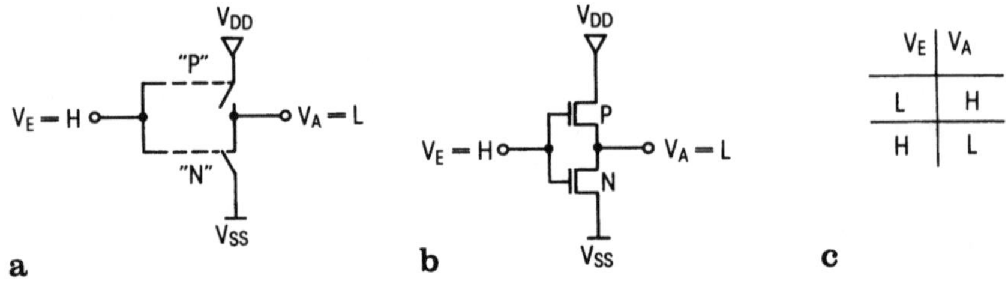

Fig. 2.6. Function of an inverter: a) switch model, b) CMOS transistor circuit, c) truth table

For the IC designer, it is important, that from the high variety of possible integration technologies (Fig. 2.2) two mainly have emerged for standard design procedures, namely CMOS and ECL.

In addition to the complementary variant of MOS circuit technology (CMOS), NMOS circuit technology has been used frequently for optimized designs. For semicustom elements however, CMOS is prevalent for the following reason: The NMOS technology has the general disadvantage of higher dissipation power. This is especially important in this context because the predeveloped cells have to be oversized in terms of their output power, in order to obtain satisfactory time delays even with an extreme output load. The static dissipation power, which exists for NMOS, but not for CMOS, grows proportionately with the output strength; when predefined cells are used, it leads to a higher power consumption than it would be the case for a corresponding manually designed and optimized NMOS circuit.

While only n-channel transistors are used for the NMOS technology, transistors of both polarities are available for CMOS (*complementary* MOS) transistors. CMOS circuits therefore come quite close to a non-dissipative switch, which is ideal for the digital circuit technology. Figure 2.6a shows an inverter realized with switches: If the input signal $V_I = H$ closes the "n-switsch", but keeps the "p-switch" open, the output signal is $V_O = L$. If one makes sure that the two switches are never closed (i.e. conductive), at the same time no current, apart from a short transition period during the switch over, will flow from V_{DD} to ground, no static dissipation power will arise. This ideal inverter (Fig. 2.6b) can be well approximated with n- and p-channel transistors: a turned-off MOS transistor has a resistance of 100 MΩ, a conductive one of several kΩ. In addition to low static power dissipation, however, dynamic power dissipation, which increases proportionately to the circuit frequency, exists due to the charge reversals of the gate capacities.

The space required by a CMOS circuit is generally 1.5 to two times larger than the space required by the corresponding NMOS circuit. There are three reasons for this:

- A minimum distance (well distance) must be kept between n- and p-channels.
- Usually, a NMOS load transistor is smaller than a p-channel transistor.
- In static complementary technology, each p-channel transistor must correspond to an n-channel transistor. In NMOS, several circuit transistors often need only one load transistor.

A critical value for the design engineer is usually the delay time, the period between the application of a signal change at the input of a circuit and the reaction of the output signals. For a 2-input NAND in 3 μm CMOS technology, this delay

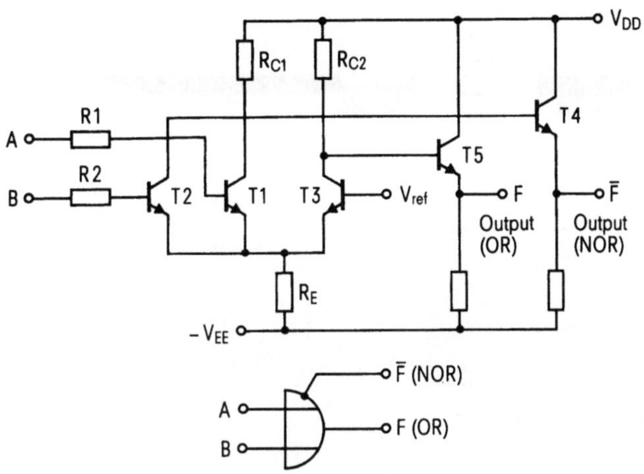

Fig. 2.7. The ECL-OR/NOR gate (basic circuit)

time is about 2 ns. For 0.7 μm technology, which is expected to become production technology by the end of the eighties, it will be only about 100 to 200 ps.

Finally the permissible noise levels have to be considered when a technology is chosen. In this context, CMOS is clearly superior to other technologies because of its high voltage swing and its symmetrical circuit characteristics.

In the context of computer development, mainly for large-scale computers, the demand for low power consumption is considered less important than the wish to reach high process speed, i.e. low delay times. This is the main application for bipolar circuits. Because ECL technology (*E*mitter *C*oupled *L*ogic) is the bipolar circuit technology with the highest processing speed, it is used here [2.4]. The OR/NOR circuit (Fig. 2.7) is used as basic gate.

2.2 Components and Technology

This paragraph will explain, for the designer who has not yet dealt with semiconductor technology, the basics of semiconductor technology *together* with possible circuit technologies. The explanation is kept concise. For a more detailed study of these principles, we refer the reader to the technical literature listed in the appendix.

2.2.1 The Bipolar Transistor

A. Function

The simplified cross-section of a bipolar transistor (npn-type) is shown in Fig. 2.8. The transistor consists of two n-doped regions, separated from each other by a p-doped region. Thus, two pn-junction are created. The left junction is connected to a voltage source in such a way that the pn-junction is forward biased; a high number of majority carriers (here electrons) therefore flows from the n^+-region (emitter of the

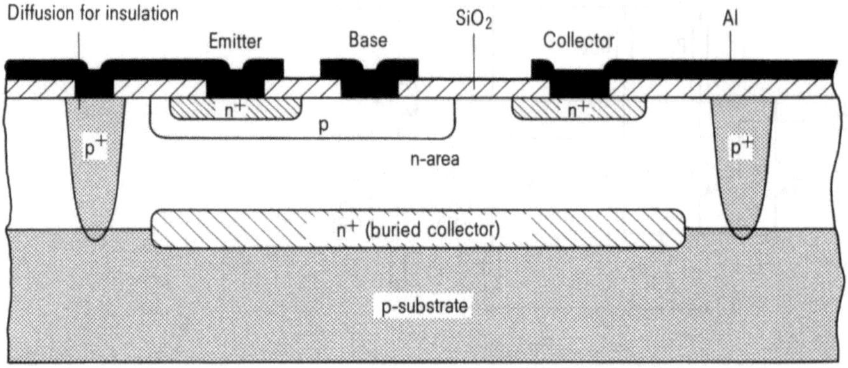

Fig. 2.8. Cross section of a bipolar npn transistor

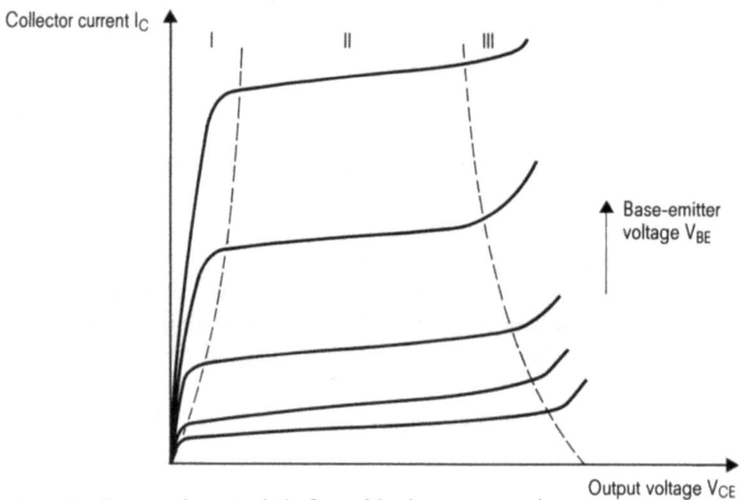

Fig. 2.9. Output characteristic for a bipolar npn transistor

transistor) to the p-region (base of the transistor). The right pn-junction is connected reverse bias with a battery: only a low leakage current flows. The electrons flowing from the left n-region to the base are now, however, attracted by the field next to the reverse biased pn-junction and reach the n-region of the collector. The number of electrons flowing to the collector is determined by the voltage and the length of the base regions. The relation between that part of the electron current coming from the emitter that reaches the collect and the total emitter current is slightly below 1, for presently used transistors usually about 0.98.

The output characteristic for an npn transistor is shown in Fig. 2.9. The collector current is plotted against the collector voltage (to be more precise: collector – base voltage), the base current is shown in as parameter. The characteristics can be divided into three sections:

- *The curve section for low voltage (I):* In this first section, a small increase in output voltage V_{CE} induces a large increase in the collector current I_C. In this section, the

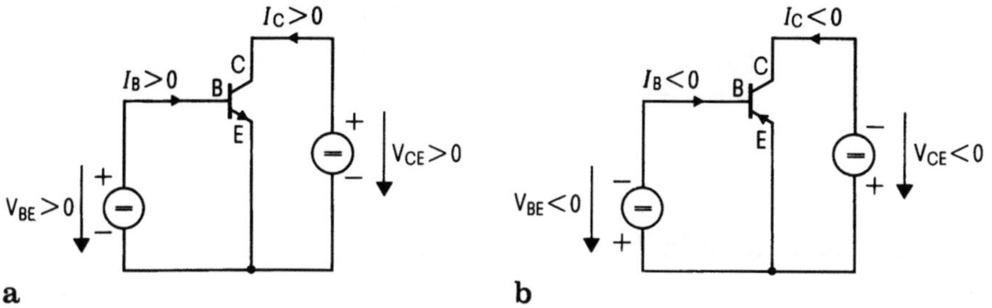

Fig. 2.10. Circuit diagram, voltages and currents of a bipolar transistor (npn-type (a) – pnp type (b))

voltage applied to the pn-junction between base and collector is so low, that only a small portion of the electrons in the base are attracted by the collector. If the voltage at the collector-base junction is increased, all electrons reaching the pn-junction at the collector are attracted, and the second section is reached. The voltage where the characteristic curve breaks is called saturation voltage $V_{CE\,sat}$.

- *The linear section (II):* Here an increase of V_{CE} induces only a small increase in current I_C. This increase is due to the fact that an increase in V_{CE} enlarges the depletion region, and thus the base is narrowed. More electrons are attracted by the collector.
- *The curve section for high output voltage U_{CE} (III):* At this point, the applied voltages are so high that a breakdown of the pn-junction occurs. This breakdown is caused by high electrical field strength in the collector's depletion region; it leads to a sudden increase in mobile charge carriers. During normal operation the transistor must not reach this section (specification of the maximum collector-base voltage V_{CEmax}).

In addition to the npn-transistor, whose functions are briefly described, there are transistors with inverse doping, i.e. pnp-transistors. Voltage is reversed and electrons are replaced by holes. The circuit symbols and the relevant voltage values are shown in Fig. 2.10a, b.

B. Fabrication

The fabrication process of integrated bipolar circuits and MOS circuits is based upon photolithographic processes. The structures are projected from a photomask to the semiconductor wafer. After the oxidation of the Si-surface, photo resist is distributed over the rotating wafer by means of a coating drum. The resulting thin film is dried and exposed through the mask containing the patterns that are to structure the wafer. The exposed film areas are removed during the subsequent development procedure. The non-exposed areas remain (positive film). This structured film serves as mask for etching the exposed SiO_2 with appropriate chemicals down to the Si monocrystal. Thus the structure on the mask is transferred into the SiO_2 layer through the photoresist. The remaining SiO_2 may now be used as a mask for the diffusion of doping atoms. For this reason, the described process occurs several times during the fabrication of integrated circuits (bipolar and MOS).

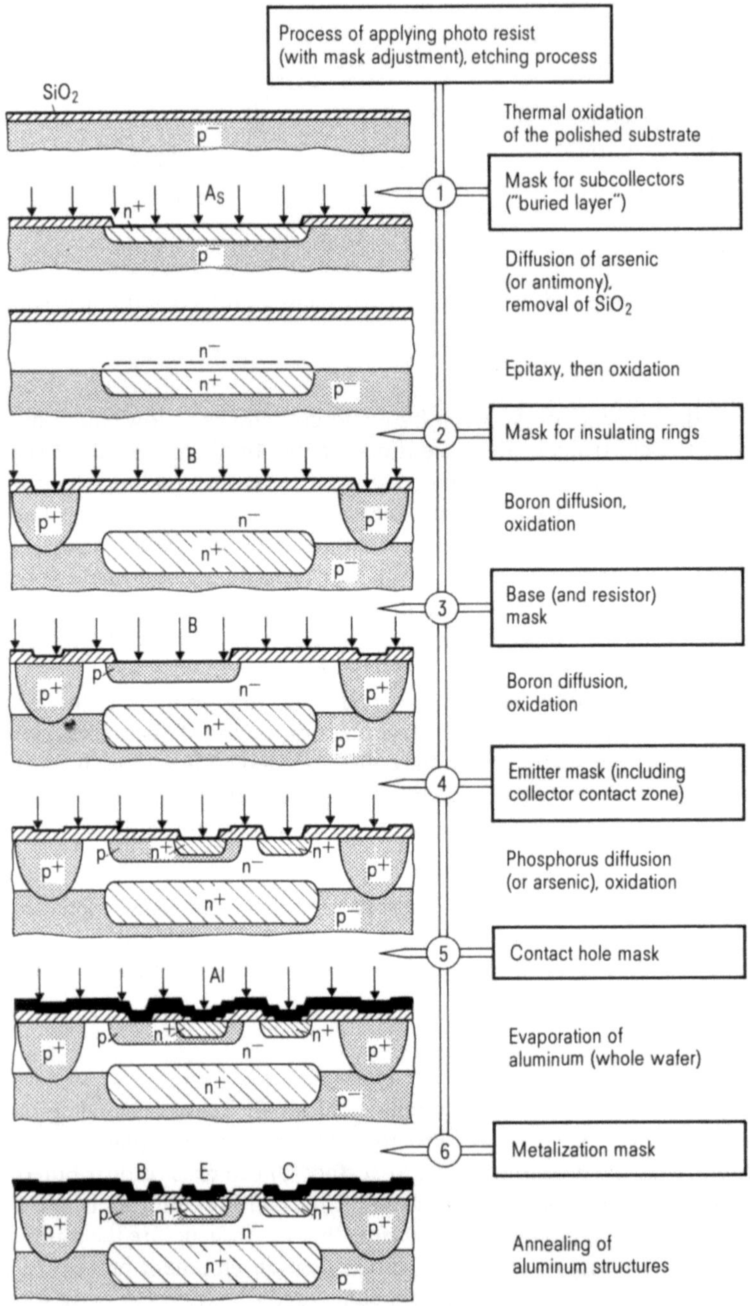

Fig. 2.11. Sequence of process steps for the generation of a bipolar transistor (npn-type) (according to [2.4])

Let us turn to the fabrication steps for an integrated (bipolar) npn-transistor (Fig. 2.11). First, a SiO_2-layer is grown on a p-doped substrate (wafer) by means of thermal oxidation. After the first step of the photolithographic process, an n^+-region is generated to serve later as subcollector. In the next step, the monocrystalline surface (entire wafer) is coated with a thin monocrystalline Si-layer. This step is called epitaxy. Subsequently, deep p^+-pockets are diffused in the n-epitaxial layer. They insulate the individual transistors on the common substrate. Now the actual transistor electrodes are generated. First, a p-section for the base is diffused. The n^+-region for the emitter as well as the n^+-contact region for the collector are generated during the same step. The base width is determined by the distance between n^+ and p (step 4 in Fig. 2.11). The dimensions depend on the diffusion mechanisms and are not only determined by mask structures (in contrast to the MOS transistor).

After these steps, the transistor's electrodes have to be connected with the other circuit components via aluminum wires (steps 5 and 6 in Fig. 2.11). With this sequence of process steps, resistors and capacitors can also be generated on the same wafer. The use of the latter will be explained in more detail in the paragraph about basic circuits.

2.2.2 The MOS Transistor

A. Function

MOS transistors belong to the group of voltage-controlled semi-conductor components, which are also called field effect transistors (FET). Two groups are distinguished: the junction-gate field-effect transistor and the insulated MOS/FET gate transistor. Because depletion layer FETs are not used for integration, they are not described in more detail.

Figure 2.12 shows the cross-section of a field-effect-transistor that can be integrated. On a p-conducting substrate there are reversely doped contact regions (n^+-regions). They are called source and drain regions. Between source and drain there is a thin SiO_2 layer on the substrate upon which aluminum or polysilicon is placed (the gate electrode). Outside the gate region the oxide film is considerably thicker. It is here that aluminum lines, which connect the single components, run. This thick insulating layer insures that no parasitic transistors arise between the components. It guarantees the insulation between the transistors.

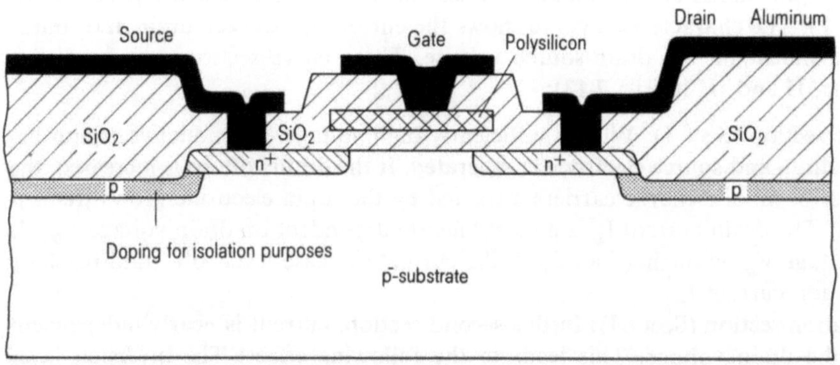

Fig. 2.12. Cross section of an MOS field effect transistor (n-channel-transistor)

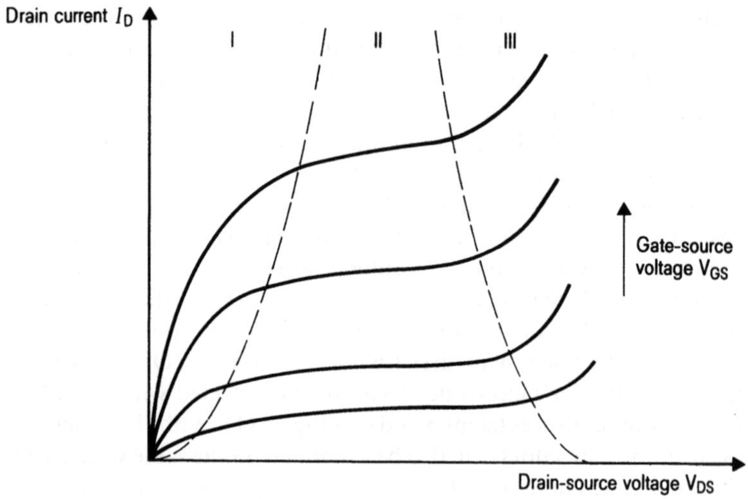

Fig. 2.13. Output characteristic of an n-channel MOS transistor

If a p-conducting region lies between the two n^+ diffusion regions then only a leakage current can flow between the n^+-regions when a drain-source voltage is applied. Both the drain and source regions are reverse biased against the substrate. When a positive gate voltage is applied (positive with regard to the substrate) the holes of the substrate are repelled from the surface by the electrical field. If the positive voltage is high enough, electrons are induced in a thin layer below the interface (inversion layer). At this point, an n-conducting region exists between the two n^+ diffusion areas; the transistor is conductive.

In addition to the n-channel transistor there is also a p-channel transistor, which is generated on an n-substrate with p^+-diffusions for drain- and source-electrodes. Furthermore, for both active elements, there are enhancement type transistors (if voltage is 0 V, no current flows) and depletion type transistor (by insertion of doping atoms below the gate electrode in the channel region, a conducting region between source and drain exists for 0 V gate voltage). SiO_2 is used almost exclusively as gate insulator. Other insulating materials are used only for special transistors.

The output characteristic of an n-channel transistor (enhancement type) is shown in Fig. 2.13. The characteristic curve shows the current I_D between drain and source to be dependent on the drain-source voltage. Three curve sections can be distinguished (I, II and III in Fig. 2.13):

- Triode section (Sect. I): With a sufficiently high voltage, a conductive region between drain and source electrode is generated. If the drain voltage is increased, the number of mobile charge carriers attracted by the drain electrode grows tremendously. The drain current I_D is almost linearly dependent on drain voltage V_{DS}. If the voltage V_{DS} is further increased, the current increase levels off, until reaching saturation current I_D.
- Saturation section (Sect. II): In this second section, current is nearly independent from the drain voltage. This leads to the following effect: The inversion layer between drain and source, which guarantees the conductivity of the transistor, is

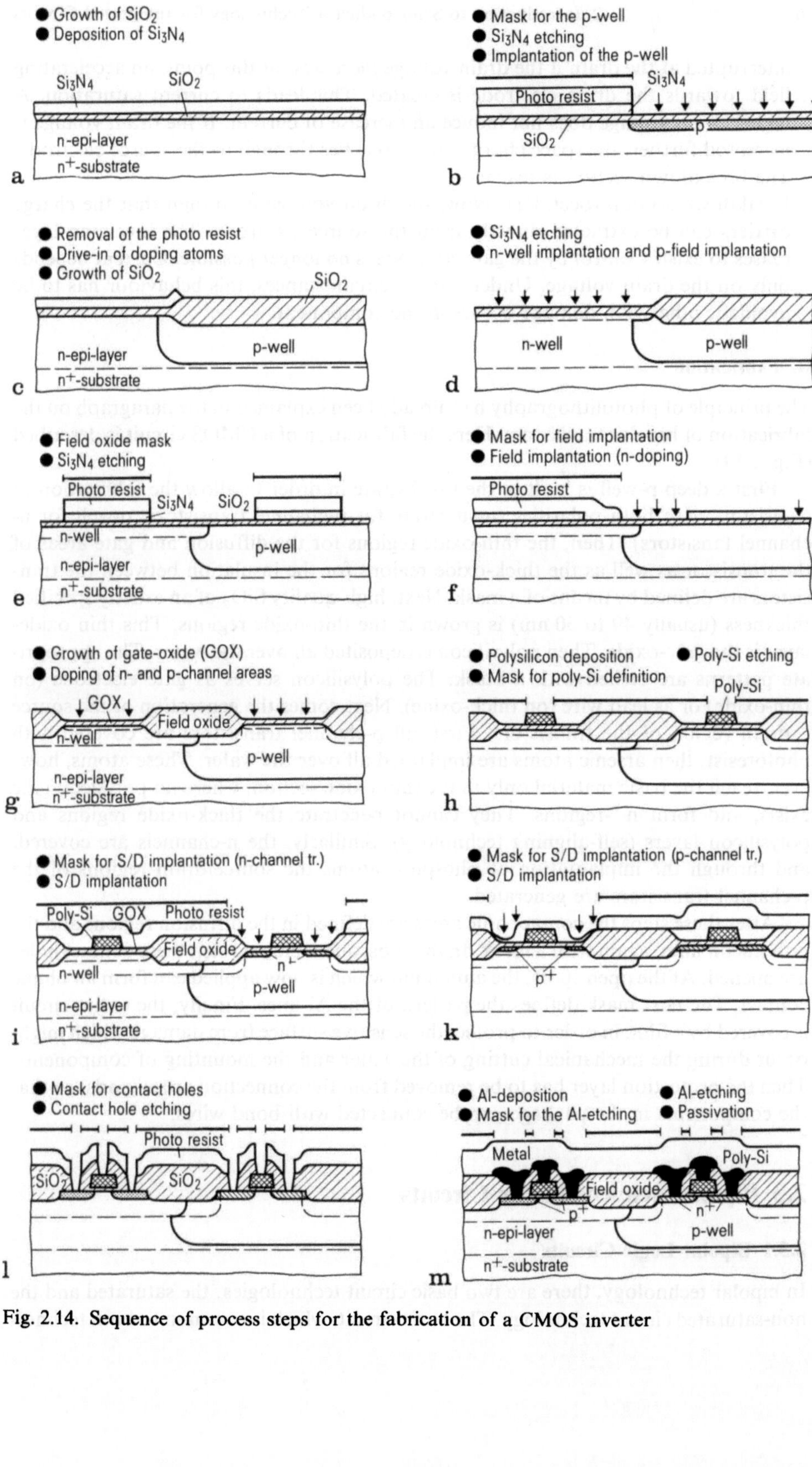

Fig. 2.14. Sequence of process steps for the fabrication of a CMOS inverter

interrupted at the drain if the drain voltage increases; at this point, an accelerating field towards the drain electrode is created. This leads to current saturation. A higher drain voltage does not induce an increase of current. If the drain voltage is increased further, the space charge region reaches through to the source electrode. The breakdown section is reached.

- Breakdown section (Sect. III): Now, the drain voltage is so high that the charge carriers can be extracted directly from the source electrode. The inversion layer ceases to exist. Control by the gate electrode is no longer possible. Current depends only on the drain voltage. Under normal circumstances, this behaviour has to be avoided; otherwise, damage occurs in the component.

B. Fabrication

The principle of photolithography has already been explained in the paragraph on the fabrication of bipolar transistors. Here the fabrication of a CMOS circuit is described (Fig. 2.14).

First a deep p-well is built in the n-substrate in order to allow the fabrication of transistors with both polarities (n-substrate for p-channel transistors, p-well for n-channel tansistors). Then, the thin-oxide regions for the diffusion and gate areas of the transistor as well as the thick-oxide regions for the insulation between the transistors are defined by means of a mask. Next, high-quality SiO_2 of an exactly specified thickness (usually 40 to 50 nm) is grown in the thin-oxide regions. This thin oxide-layer is the gate-oxide. Then polysilicon is deposited all over the wafer. The appropriate patterns are etched with a mask. The polysilicon serves as gate electrode (on thin-oxide) or as lead wire (on thick-oxide). Next comes the generation of the source – drain region of the transistors. First, all p-channel transistors are covered with photoresist, then arsenic atoms are implanted all over the wafer. These atoms, however, reach the basic material only in the thin-oxide section, where no polysicon gate exists, and form n^+-regions. They cannot penetrate the thick-oxide regions and polysilicon layers (self-aligning technology). Similarly, the n-channels are covered, and through the implantation of phosphor atoms the source/drain regions of the p-channel transistors are generated.

After these steps the contact hole areas are defined in the diffusion regions and the polysilicon lines, again with a mask. In the subsequent etching step these contact holes are opened. At the open spots, the aluminum which is now applied can form an ohmic contact. The next mask defines the pattern of the Al-lines. Finally, the entire circuit is covered by a film, in order to protect the sensitive surface from damage which might occur during the mechanical cutting of the wafer and the mounting of components. Then the protection layer has to be removed from the connection areas (pads) so that the components in the package can be connected with bond wires.

2.3 Bipolar Semiconductor Circuits

2.3.1 Bipolar Logic Circuits

In bipolar technology, there are two basic circuit technologies: the saturated and the non-saturated circuit technology. The basic circuit, which is common to both technol-

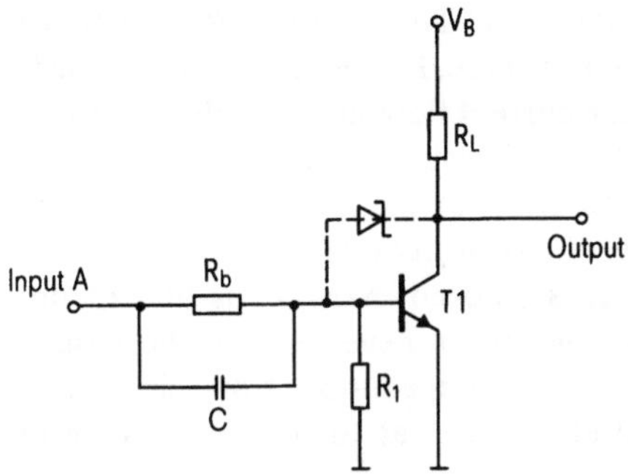

Fig. 2.15. Circuit diagram of an inverter with a bipolar transistor and an ohmic resistor as load

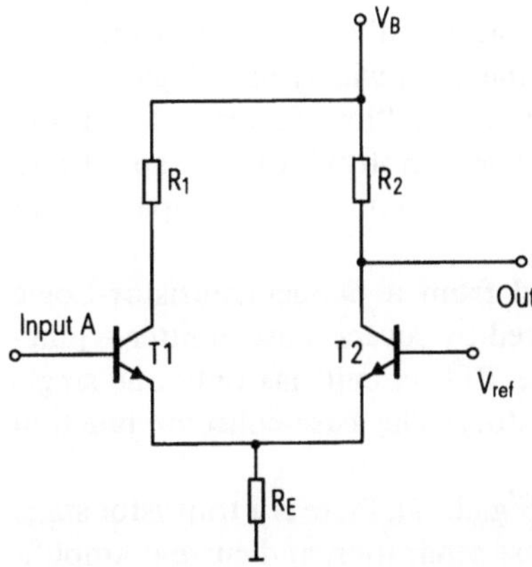

Fig. 2.16. Circuit principle of a differential stage with bipolar transistors (emitter-coupled amplifier)

ogies, is a simple inverter with a load resistor and a transistor (an npn-transistor, e.g.) as its active element. Its function is described (Fig. 2.15) in the following paragraphs.

If the base – emitter connection is short-circuited, the transistor is switched off: its output resistance is some MΩ. The output signal is the supply voltage level. If the transistor is switched conductive the output resistivity falls down to some Ω: the output potential is equal to the ground potential.

During the transistor operation, the transistor goes from "active" (non-saturated) condition to saturated condition. Because of the capacitator C, the transistor is driven with a higher voltage for a very short time. This considerably lessens the fall time of the collector potential. During the saturation state of the transistor, minority carriers are accumulated at the collector – base junction. This is why a certain period of time is necessary to bring the transistor back to off-state.

In order to attain very short delay times, the transistor has to work in the unsaturated region. The accumulation of minority carriers at the collector – base junction has to be prevented. This can be done by means of a Schottky-barrier diode: the normal pn-junction of a diode is replaced by a metal – semiconductor – junction. It has a minimum storage effect for minority carriers. The Schottky-barrier diode is placed between the base and collector of a transistor (marked with dotted lines in Fig. 2.15). Since $V_{BE} < V_{CE}$, the transistor is prevented from reaching saturation [2.7].

Figure 2.16 presents a second possibility for preventing an accumulation of minority carriers: Two transistors are connected as emitter-coupled amplifier (see also

Fig. 2.7). The base of T2 is set at reference voltage V_{ref}. The joint emitter resistance R_E, which may be considered as a constant current source, has relatively high impedance. R1 and R2 are chosen in such a way that the current flowing through R_E is twice as large as the current flowing through R1 and R2. If ground potential is applied to the input A, T1 is switched off, and T2 has to conduct the entire current of the constant current source. If the input is positive T1 is on (conducting). The emitter current of T1 increases and thus the T2 current is reduced. Now T2 is off, T1 on. Because of the limitation of the emitter currents the circuit never reaches the saturation level. The unfavorable, comparatively high output impedance of this circuit can be compensated by placing an additional emitter follower at the output (shown in Fig. 2.7).

For the circuit families TTL and ECL [2.4, 2.7], which will be explained in the following, we refer to the basic circuit versions, "saturated" and "non-saturated":

Transistor Transistor Logic (TTL) is presently the most widely used logic family. At this time, there are four types in the TTL line: standard, "low-power" (slower and lower power dissipation than standard), "high-power Schottky" (higher speed and higher power dissipation than standard), "low-power Schottky" (comparable in speed with standard, but lower power dissipation).

One can view a TTL gate as being generated from a Diode-Transistor-Logic (DTL) gate (see [2.7]) with the diodes being replaced by several base-emitter – junctions and one base-collector – junction. Thus, the TTL circuit has only one single transistor for several inputs (multi-emitter transistor). The base-collector junction serves as serial diode.

For the NAND gate of the standard TTL line (Fig. 2.17), there is a transistor stage added to the multi-emitter transistor stage, for phase separation and current amplification. This stage control a totem pole output. The short delay times of this logic family are due to the multi-emitter input, whereas the output stage improves the signal-to-noise ratio.

The circuit for the same function in Low-Power-Schottky TTL is shown in Fig. 2.18; for the transistors, saturation is prevented by means of Schottky diodes in the base-collector circuit.

The TTL family is continuously expanded, in particular through special MSI and LSI circuits. Complete functional blocks as well as arithmetic circuits, data paths and memories are available (see, for example, [2.8]).

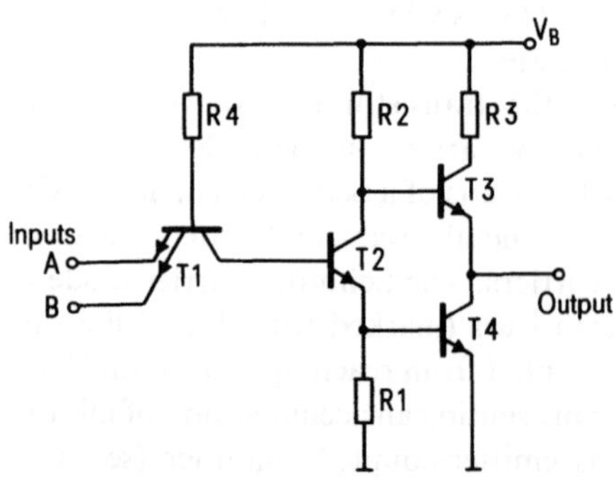

Fig. 2.17. Circuit diagram of a NAND gate in TTL circuit technology

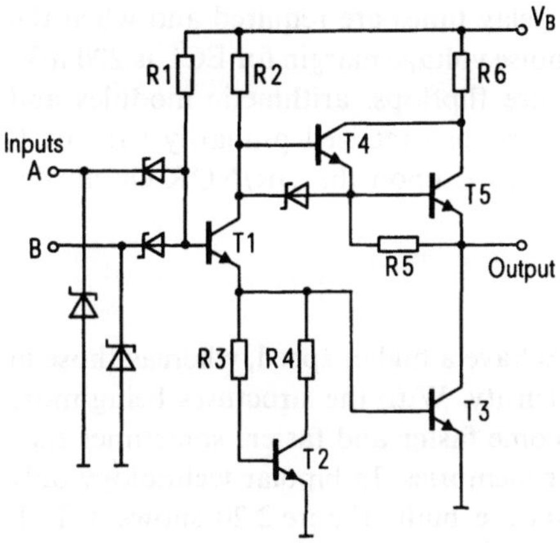

Fig. 2.18. Circuit diagram of a NAND gate in Low Power Schottky TTL circuit technology

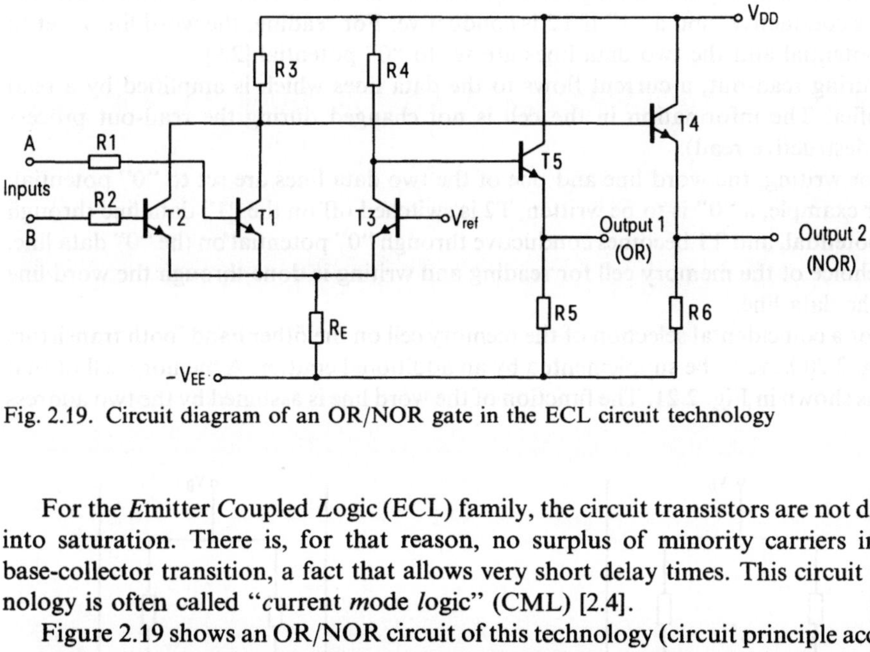

Fig. 2.19. Circuit diagram of an OR/NOR gate in the ECL circuit technology

For the *E*mitter *C*oupled *L*ogic (ECL) family, the circuit transistors are not driven into saturation. There is, for that reason, no surplus of minority carriers in the base-collector transition, a fact that allows very short delay times. This circuit technology is often called "*c*urrent *m*ode *l*ogic" (CML) [2.4].

Figure 2.19 shows an OR/NOR circuit of this technology (circuit principle according to the Fig. 2.7 and 2.16). The transistors T1 through T3 form the emitter-coupled amplifier. R_E works as constant current source. If both inputs are logic 0 (= − 400 mV) T1 and T2 remain switched off; their common collector potential is in this case V_{DD}, and T4 is conductive: thus the NOR output is at a high (H-)level. Because of V_{ref} (= ground potential) T3 is conductive, and the total current flows through T3 while T5 is off: the OR output is at a low (L-)level. If one of the inputs is switched to logic 1, the corresponding transistor is conductive; the collector potential decreases and T4 is switched off. The output goes to L-level. If the "logic 1" signal (= + 400 mV) is higher than V_{ref}, T3 is insulated, and the total R_E current flows through the conductive input transistor. Now T5 is conductive, its output is at H-level.

The ECL technology is used when short delay times are required and when the power dissipation is of little importance. The noise voltage margin for ECL is 200 mV. In addition to simple gate functions, there are flipflops, arithmetic modules and multi-functional circuits. Bipolar gate arrays are also realized primarily with ECL circuit technology. This realization is always based upon the OR/NOR circuit described in Fig. 2.19.

2.3.2 Bipolar Memories

Generally, one may say that bipolar memories have a higher speed, whereas those in MOS technology have a higher integration density. With the structures being more and more reduced in size, MOS has also become faster and faster; sometimes they reach speeds coming close to those of bipolar memories. In bipolar technology only memories with static memory cells or ROMs are built. Figure 2.20 shows a TTL memory cell. The cell consists of two multi-emitter transistors, one emitter each is required for data input and output. In quiescent state a "0" is stored on the word line if T1 is conductive, and a "1" if T2 is conductive. For reading, the word line is set to "1" potential and the two data lines are set to "0" potential [2.9].

During read-out, a current flows to the data lines which is amplified by a read amplifier. The information in the cell is not changed during the read-out process (non-destructive read).

For writing, the word line and one of the two data lines are set to "0" potential. If, for example, a "0" is to be written, T2 is switched off on the "1" data line through "1" potential, and T1 becomes conductive through "0" potential on the "0" data line. The choice of the memory cell for reading and writing is done through the word line and the data line.

For a coincidental selection of the memory cell on the other hand, both transistors of Fig. 2.20 have to be supplemented by an additional emitter. A memory cell of that kind is shown in Fig. 2.21. The function of the word line is assumed by the two address

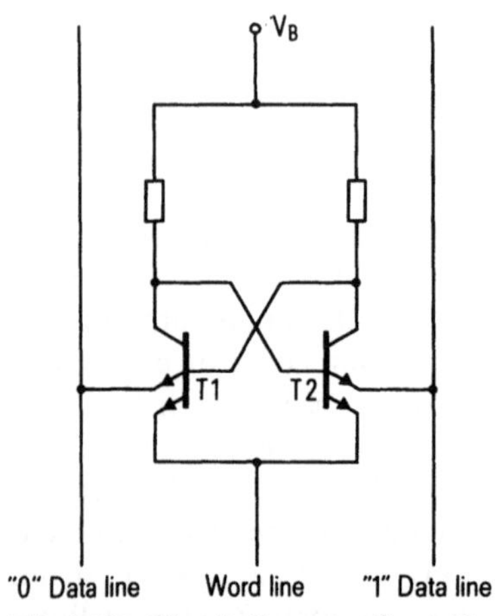

Fig. 2.20. Circuit diagram of a static memory cell in bipolar technology

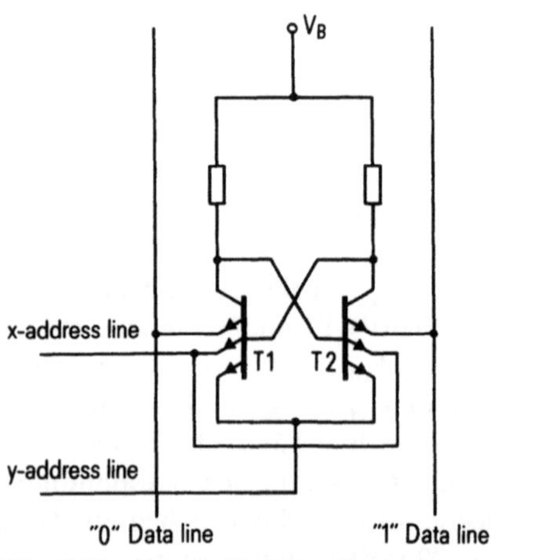

Fig. 2.21. Circuit diagram of a static memory cell in bipolar technology with coincidental selection

lines. If the cell is not in operation, the x- and y-address lines assume "0" potential. The cell is insulated through the AND connection of the multi-emitter transistors. Only when both address lines are set to "1" potential writing or reading of information, is possible [2.9].

In addition to the read/write memories described there are ROMs, which have a predetermined content that cannot be changed during use of the memory. A memory of that kind has only a reading function. ROMs can be subdivided in:

- reversible ROMs (currently only in MOS technology),
- irreversible ROMs (in bipolar and MOS technology).

Irreversible ROMs are programmed either directly by the manufacturer during production (ROM) or by means of a special programming process by the user (PROM). For this programming process, special programming devices are necessary.

If the programming of the memory component is carried out by the manufacturer the design engineer has to supply a list of the desired memory content. This list is implemented on the chip by means of a mask programming process. Because this programming procedure is economical only when high numbers of chips are produced, there are ROMs for laboratory purposes and low pieces counts which can be programmed by the desing engineer (PROMs). At the nodal points of a memory matrix, a serial connection of an active element (diode or bipolar transistor) is provided with a resistor (for example nickel-chrome). This resistor fulfills the role of a fuse and is destroyed during programming by an appropriate excess current. At that point, the connection is broken.

2.4 MOS Semiconductor Circuits

2.4.1 MOS Logic Circuits

The basic circuit for logic, shift registers, memories etc. is the inverter, i.e. a stage where output and input voltages have an inverse time dependency [2.5]. In addition to signal inversion, the signal is also amplified, therefore

$$\Delta V_a \geqq \Delta V_e .$$

The basic principle of an inverter can be recognized in Fig. 2.22. The drain of the MOS transistor T_S is connected with the supply voltage V_B through the load resistor R_L. Additionally, this connection is also the inverter output. The source is set to ground, the input voltage V_E is applied to the gate electrode. The substrate of the transistor is set to a potential V_{Sub} which provisionally is to be 0 V. If the input voltage V_E is lower than the threshold voltage of the transistor V_T, T_S is switched off. No current flows through the resistor R_L, and the output voltage V_A is at battery potential V_B. If, however, the voltage V_E is increased beyond the threshold voltage, the transistor T_S becomes conductive, and a current flows through the resistor R_L and the transistor; the output voltage decreases to a value which is determined by the voltage division between the conductive transistor and the load resistor. For MOS logic circuits, the output voltage V_A has to drop below the threshold voltage of the succeeding transistor so that the latter is turned off. This residual voltage depends on the

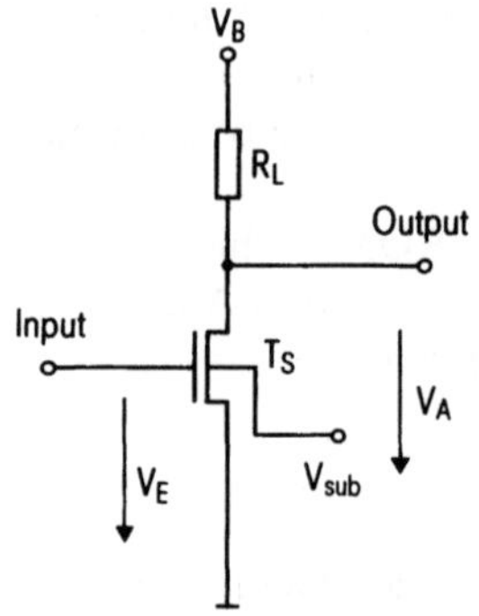

Fig. 2.22. Circuit diagram of an inverter with a MOS transistor and an ohmic resistor as load element

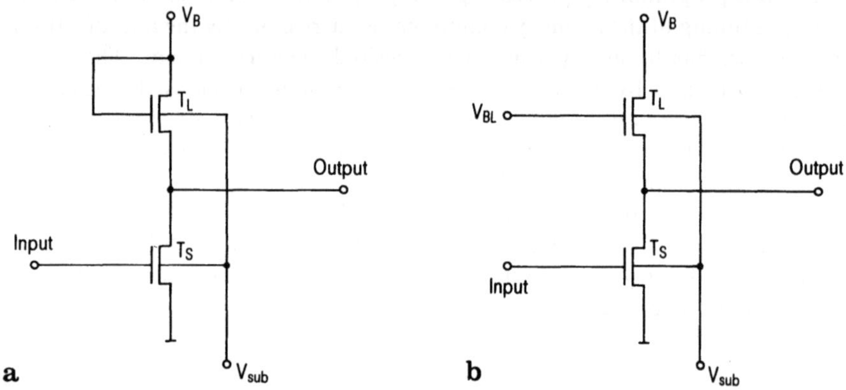

Fig. 2.23. Circuit diagram of an MOS inverter with a transistor of the enhancement type as load element

resistance ratio between transistor and resistor, as well as on the input voltage V_E applied to the transistor. In contrast to the bipolar technology, in integrated MOS circuits inverters with ohmic load resistors are hardly used, as the higher inner resistance of MOS transistors requires quite highly resistant load resistors. If those integrated load resistors are realized as diffused bulk resistors they require a lot of space. For this reason MOS transistors are used almost exclusively as resistors for integrated MOS inverters.

The possibility of combining MOS load and circuit transistors is shown in Fig. 2.23 to 2.25. The gate electrode can either be connected with the supply voltage (2.23a) or applied to a voltage source of its own (2.23b).

Figure 2.23a shows the enhancement type load transistor T_L. Its gate electrode is connected with the supply voltage V_B. Because the transistor is turned off as soon as its gate – source voltage is below the threshold voltage, the output voltage V_A can at best reach the value $V_B - V_T$. If one needs the entire supply voltage at the output, a voltage V_{BL} has to be applied to the gate electrode of the transistor, which is at least as high as the supply voltage plus V_T. The arrangement is shown in Fig. 2.23b. For depletion type transistors, for the most part, the gate of the load transistor is connected with its source. The transistor T_L therefore has a constant gate – source voltage

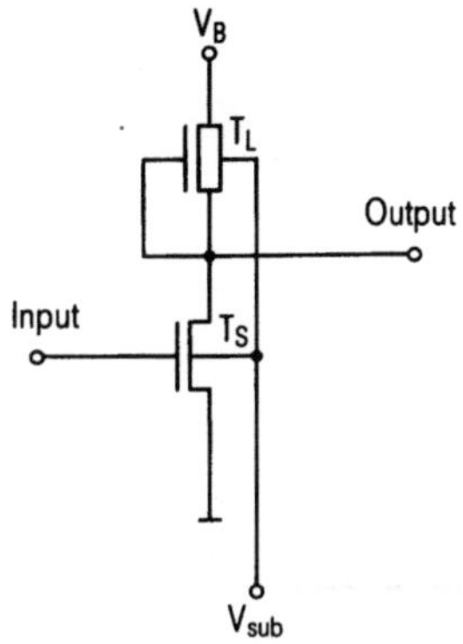

Fig. 2.24. Circuit diagram of an MOS inverter with a transistor of the depletion type as load element

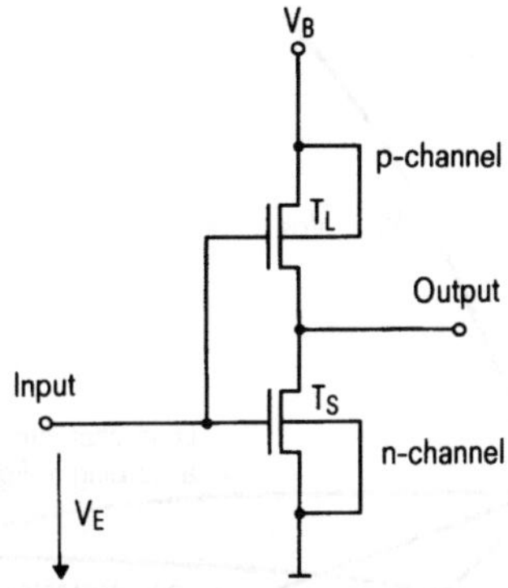

Fig. 2.25. Circuit diagram of an MOS inverter in complementary channel technology

of 0 V (Fig. 2.24). These three inverter circuits are similar in that by appropriately arranging the geometric ratios of circuit transistor and load transistor, one must be certain that the output voltage V_A for a conductive switching transistor is low enough to turn off the next stage. If, however, complementary transistors (n- and p-type) are on one chip, the inverter can be realized according to Fig. 2.25. The input signal controls both transistors, and for either one of the voltage values one transistor is non-conductive. Thus the full voltage wing is reached at output, and the geometric ratios do not effect the residual voltage.

In order to determine the transfer characteristic (input voltage to output voltage) of the various inverters, one has to draw the load element characteristic in the output characteristics of the circuit transistor. For the circuits in Fig. 2.22, 2.23a and 2.24, the load characteristics are drawn in the output characteristics of the switching transistor T_S (Fig. 2.26). All load characteristic curves intersect the switching transistor characteristic with $V_E = V_{GS} = V_B =$ const. at the same point S; the residual voltage V_R at output is equal for all of the three inverters. But the various load elements also have various load characteristics (see Fig. 2.26), which leads to varying slopes of the transfer characteristics between input and output voltage. These transfer characteristic curves are obtained by determining the intersection point of the switching transistor characteristic curve V_{GS} and the load characteristic. The pertinent section of the V_{DS} axis is the output voltage, the voltage V_{GS} of the switching transistor the input voltage.

For the complementary circuit according to Fig. 2.25, one has to draw the characteristic curve of *both* transistors, because the input signal controls both transistors. For each input voltage (V_{GS}) the intersections of the relevant characteristic curves are needed to determine the transfer characteristic curve of the inverter. These curves are shown in Fig. 2.27. It can be seen that the circuits, according to Figs. 2.24 and 2.25, have the highest slope (= highest degree of amplification) in the transition region between high potential (logic "1") and low potential (logic "0"). In addition to the high differential amplification, these two inverters also have the best load characteristic for a load capacity connected with output, and are therefore suitable for realizing

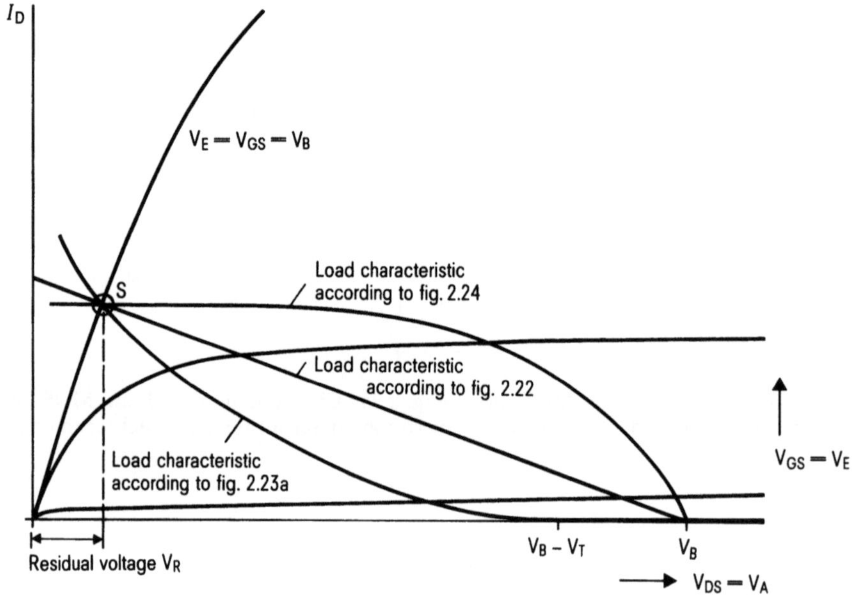

Fig. 2.26. Output characteristics of an MOS transistor with load characteristics drawn in

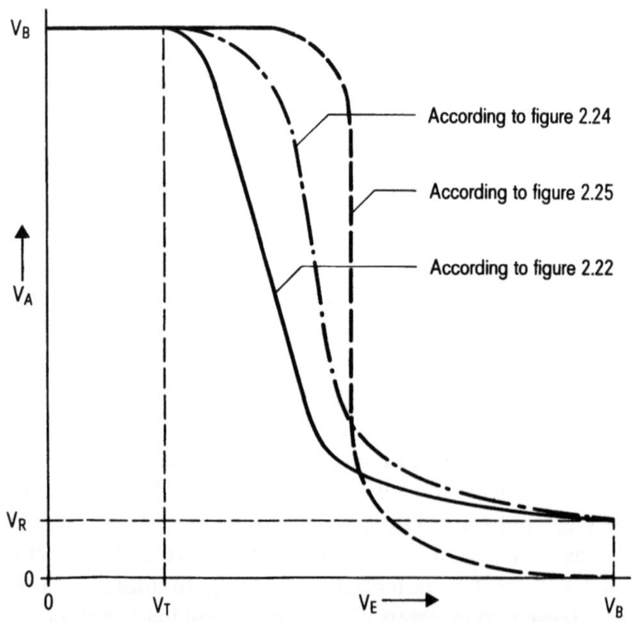

Fig. 2.27. Transfer characteristics of various MOS inverters

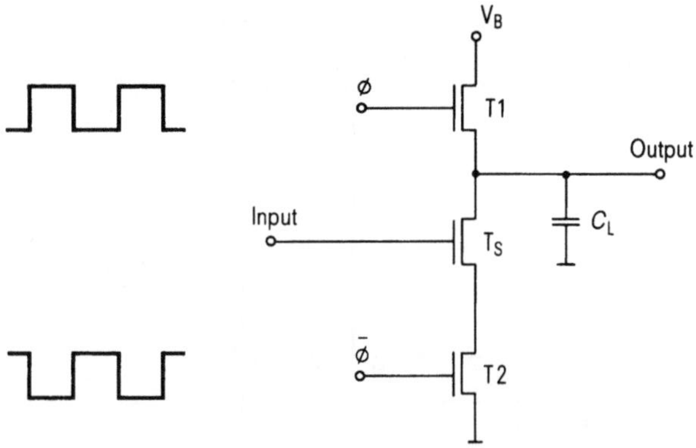

Fig. 2.28. Circuit diagram of a dynamic MOS inverter

fast logic circuits. For this reason, modern MOS logic components are exclusively realized in one of these circuit technologies [2.5, 2.11].

In MOS technology, due to the high input resistance of the MOS transistor (appr. 10^{12} to 10^{14} Ω) one can not only make use of static inverter circuits, but also realize an inverter by means of dynamic circuit technologies. Figure 2.28 shows such a dynamic inverter with the pertinent impulse forms. The principle consists in precharging the output nodes during the precharge phase. The state of the switching transistor T_S does not play a role as long as the transistor T_2 is turned off. When the precharge phase is finished, the transistor T_2 is conductive to ground, and, according to the input signal of the switching transistor T_S, the output node is either discharged, or it remains at precharge potential. The circuit technology is called dynamic, as the precharge voltage at the output node slowly decreases because of the leakage currents of the turned off transistors. For this reason, a dynamic inverter must not be operated below a set minimum frequency. With this dynamic technology, transistor geometries can be chosen with nearly random ratios (*"ratioless circuit"*), because there is always only one conductive path between supply voltage and output. With these inverters, logic circuits of various complexity can be designed [2.5].

Figures 2.29a and 2.29b show a 2-input NAND and 2-input NOR gate in complementary-channel technology. For the NAND gate the n-channel transistors are connected serially, the p-channel transistors in parallel. For the NOR gate the arrangement is reversed. Thus, gates of various complexity can be constructed. Even the combination of serial and parallel connections is possible and is applied quite often in the MOS technology. Figure 2.30 shows circuit diagram and logic symbol of a 4-input combination gate in CMOS technology [2.5, 2.10].

A basic circuit often used in MOS technology is the transfer gate. One takes advantage of the fact that drain and source are developed in the same way for a MOS transistor, and that it is therefore bidirectional. The transistor can be used as a switch. In conductive state, current can flow in both directions (according to the potential state at the electrodes), in off state, no current flows. A CMOS transfer-gate is shown in Fig. 2.31. The two transistors are controlled in anti-phase; either both are turned off or both are conductive [2.5, 2.11].

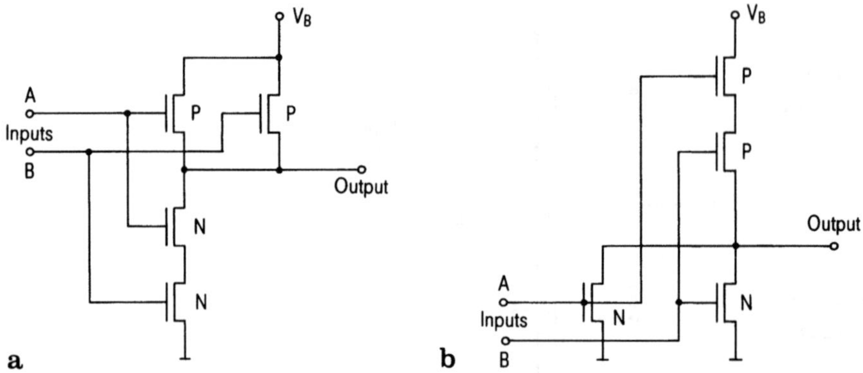

Fig. 2.29. Circuit diagram of an NAND gate (a) and a NOR gate (b) in complementary-channel technology (CMOS)

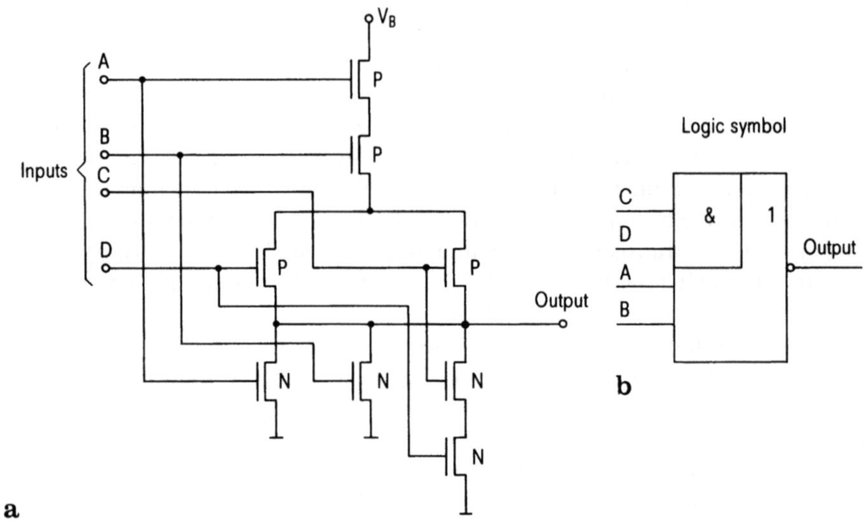

Fig. 2.30. Circuit diagram (a) and logic symbol (b) of an AND/NOR (2-2-4) combination gate in complementary-channel technology (CMOS)

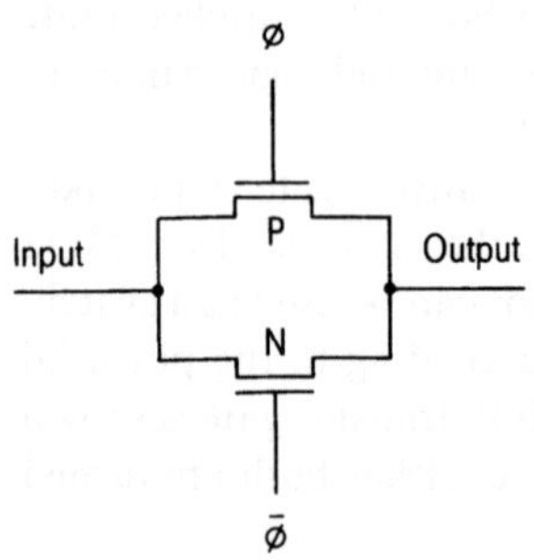

Fig. 2.31. Circuit diagram of a transfer gate in complementary channel technology (CMOS)

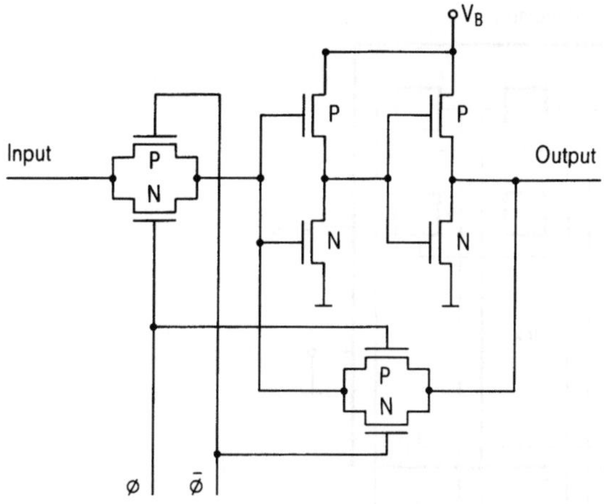

Fig. 2.32. Circuit diagram of a flipflop with breakable feedback (in CMOS technology)

The transfer gate (TG) is used, for example, for a flipflop whose feedback branch can be broken (Fig. 2.32). When the input TG is conductive, the other TG is open. The signal arrives at the output, twice inverted. When the first TG is open and the second one is conductive, no new information can be written in, but the former information remains at the flipflop output through the closed feedback branch. Such an element is generally called "latch".

2.4.2 MOS Semiconductor Memory

For digital semiconductor memories the information bit is stored in a semiconductor circuit, and when the bit is required it is read out. The information is written in electrically (write/read memory, electrically programmable ROMs) or during production (ROMs). Most of these memory circuits are integrated on one chip, and the large-scale integration is, to a very great extent, influenced and advanced by development in the memory field (see Fig. 2.1).

In organizational terms, memories can be divided into register memories and random memories. Register memories consist of shift registers. For reading out information, a certain number of shift impulses have to be applied until the required information bit arrives at the output. Thus, there is no random access to the memory locations.

The majority of the semiconductor memories are organized as memories with random access. The requested memory cell is selected by means of an address word. The address word determines, within a memory matrix, those memory locations that are to be activated for reading or writing information. A memory with random access includes an address decoder, the actual memory matrix and a write/read circuit (Fig. 2.33). For ROMs the write circuit is omitted. The address decoder is normally bipartite and has an x-decoder and a y-decoder. When the two adresses coincide, a certain memory cell is selected (bit-by-bit organization). When an x-address or a y-address selects more than one bit (e.g. 2, 4, 8 or more bits) we have a word-by-word organization.

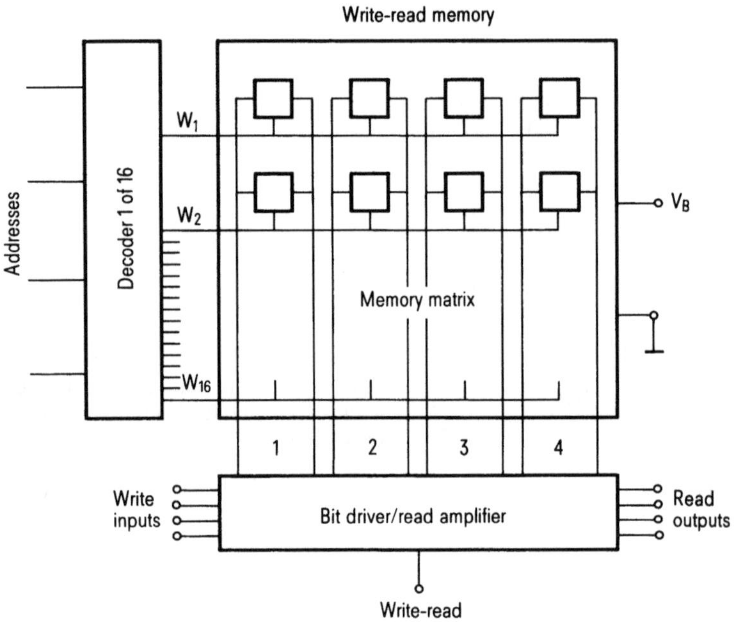

Fig. 2.33. Principle block circuit of a semiconductor memory

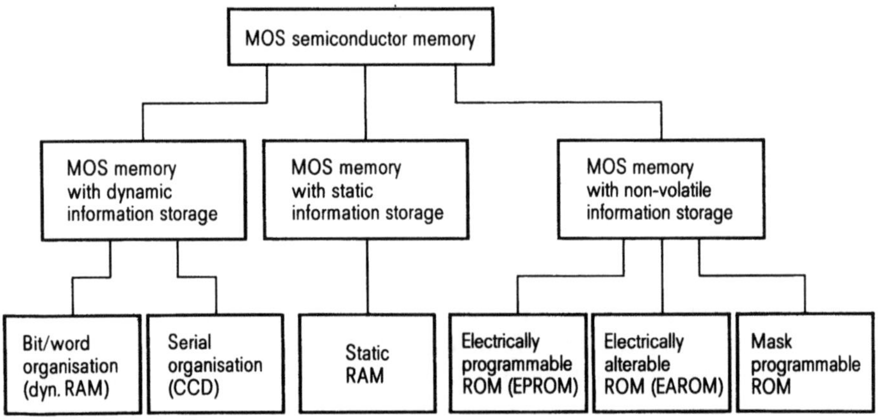

Fig. 2.34. Classification of semiconductor memories according to their way of information storage

Another design form of semiconductor memories is the associative or content-addressable memory which should also be mentioned here. The stored information is compared with the applied information, and when they match, the relevant address, or the whole information allotted to that address, is output [2.12].

Finally, semiconductor memories can also be divided in classes according to the way in which information is stored. There are *dynamic* memories, which need to have the information regenerated in predefined periods of time (mostly 2ms or 4ms) so it is not lost. Then there are *static* memories, which retain information as long as the supply voltage is applied. If it is switched off, the information is lost. Finally there are *non-volatile* memories, which retain information even during voltage failure. The individual MOS memories are now explained. The relevant memory cell is described and its application discussed (Fig. 2.34).

Memories with Dynamic Information Storage

In an effort to use as few circuit elements for information bits' storage as possible, engineers succeeded in progressing from the static six-transistor cell to the dynamic four-transistor cell (Fig. 2.35a), then to the three- (Fig. 2.35b) and finally to the one-transistor cell (Fig. 2.35c). In each of these cases, information is stored in an MOS capacitor. For dynamic memories the one-transistor cell is utilized; for this reason, only this type shall be described in more detail.

For storing information, only one memory capacitor C is necessary, and only one circuit transistor Tr1 is required for connecting the capacitor with the data line D (write/read line), or for disconnecting it. The number of elements necessary for storing one information bit has thus been reduced to a minimum.

If the information "0" is to be written in the cell of Fig. 2.35c, a voltage of 0 V is applied to the data line D, and at the same time the selected transistor Tr1 is activated through the word line W. The potential at the memory capacitor, which has been realized as a MOS capacitor, is set to 0 V. When the transistor has been deactivated the previous state remains, and the corresponding information is stored. Because this information is also equivalent to the equilibrium state of the MOS capacitor, this state need not be reactivated. If a "1" is to be written in, the write voltage V_S (mostly $V_S = V_B$) is to be applied to the data line, and the transistor Tr1 is switched on again. Now the memory capacitor is set to potential V_S. The mobile charge carriers at the Si/SiO_2 interface of the MOS capacitor are attracted through Tr1 by the voltage applied to the data line, and a deep space charge region is created below the memory electrode.

Because the MOS capacitor is no longer in thermal equilibrium, charge carriers are generated at the surface and in the area around the space charge region; they move to the interface and lower the potential in that area. If one waits long enough (some 100 ms to 1 s), the "0" state (equilibrium state) returns. Therefore the "1" state has to be reactivated periodically. A regneration is necessary also for the circuits in

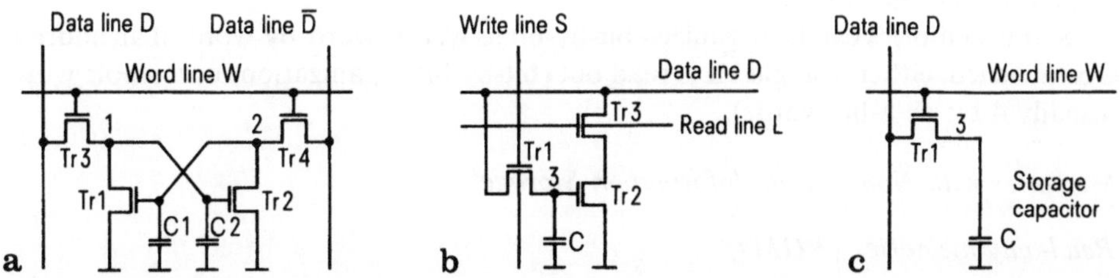

Fig. 2.35. Dynamic MOS memory cells a) Four-transistor cell, b) Three-transistor cell, c) One-transistor cell

Fig. 2.35a and b. It is realized through reading out, amplifying and subsequent writing in of information.

For reading out information, first a defined level of voltage is applied to data line D, which is then separated from the voltage source. Thereafter the selected transistor Tr1 is opened, and the charge in the memory capacitor and on the data line can compensate each other. This charge compensation leads to a voltage change on the data line, which is amplified when it reaches the output of the memory. Read out of information utilizing a one transistor cell is done destructively, i.e. after each reading procedure the information has to be written in again. This is achieved by means of a flipflop, which is used simultaneously as read amplifier [2.5].

Memory with Static Information Storage

The memory cell for static MOS memories is the crosscoupled flipflop (Fig. 2.36a to c).

The two selected transistors Tr1 and Tr2 provide the connection between the two bit lines and the flipflop nodes. Depending on the fact whether the left flipflop node is set to ground or to the potential of the supply voltage either a logic "0" or a logic "1" is stored (the allocation is optional).

There are various MOS technologies for building static memories. The technology with the least power dissipation is CMOS technology (Fig. 2.36c). In the two stable states no current flows in this cell between V_B and ground, because in both flipflop sections one of the transistors is turned off, the other is conductive. If, for example, the n-channel transistor is conductive in the left section, the corresponding p-channel transistor is switched off. In the right section it is reversed; the p-channel transistor is conductive, the n-channel transistor is off. If a single-channel technology (such as n-channel) is available there is always a current flowing in one of the two sections, which effects power dissipation; it must therefore be kept low. This can be achieved by means of load elements which are as highly resistant as possible. If MOS transistors of depletion type are used as load elements (Fig. 2.36a), these transistors need to have large channel lengths and small channel widths.

An elegant and space-saving solution is provided when polysilicon load resistors are used (Fig. 2.36b). Undoped polysilicon is used with a resistance of approximately 50 MΩ/□. Thus, not only is less power dissipation attained because the current is less than 1 μA, but also, less cell space is used because the flipflop can be constructed in a more space-saving manner [2.16]. As compared to dynamic memories, static memories generally have a bit-density which is four to eight times lower. Due to the omission of sensitive read amplifiers, the peripherals of static memories can, however, be built easier and faster.

Static memories can be organized bit-by-bit as well as word-by-word: if an address is established, either a single bit is read out (bit-by-bit organization) or a whole word (usually 4-bit or 8-bit words).

Memories with Non-volatile Information Storage

Read-only memories (ROMs)

The most basic form of semiconductor memories with non-volatile information storage are the so-called *r*ead-*o*nly *m*emories (ROMs). Here, the transistors are arranged

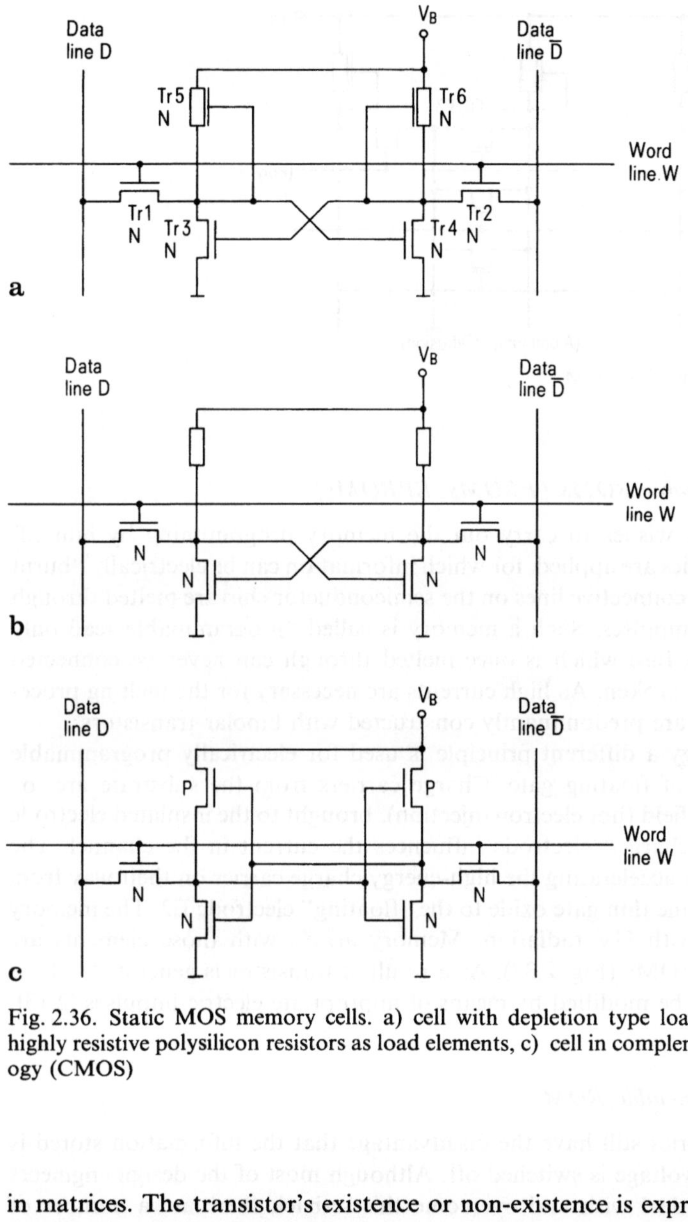

Fig. 2.36. Static MOS memory cells. a) cell with depletion type load elements, b) cell with highly resistive polysilicon resistors as load elements, c) cell in complementary channel technology (CMOS)

in matrices. The transistor's existence or non-existence is expressed by a stored "1" or "0" (Fig. 2.37). The position of the transistors and vacancies has to be known before the circuit is produced. The programming of the ROM can be carried out in various levels. The programming arrangement carried out in the diffusion level requires the least space. Programming may also be carried out in the metalization level; then more space has to be provided for each bit. Such memories are normally organized word-by-word, and are economical only when high numbers of pieces are produced. Classical fields of application for read-only memories are the storage of tables, code transformations or microprograms.

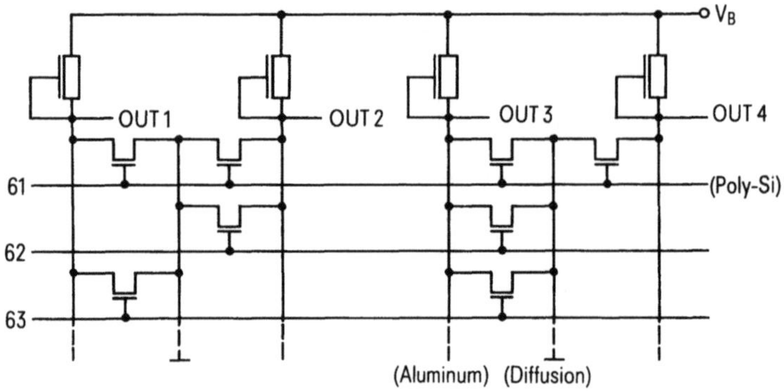

Fig. 2.37. Circuit diagram of a ROM

Electrically Programmable ROMs (PROMs, EPROMs)

If the design engineer wishes to carry out the memory programming by himself, semiconductor memories are applied, for which information can be electrically "burnt in". For that purpose, connective lines on the semiconductor chip are melted through by means of electric impulses. Such a memory is called "*p*rogrammable *r*ead *o*nly *m*emory" (PROM). A fuse which is once melted through can never be connected again, it is irreversibly broken. As high currents are necessary for the melting procedure, those memories are predominantly constructed with bipolar transistors.

In MOS technology a different principle is used for electrically programmable ROMs: the principle of floating gate. Charge carriers from the substrate are, by means of high electric field (hot electron injection), brought to the insulated electrode E2 (Fig. 2.38 a). The charged electrode influences the current in the channel. The electrode E1 serves for accelerating the high-energy charge carrier on their way from the substrate through the thin gate oxide to the "floating" electrode E2. The memory can also be deleted with UV radiation. Memory arrays with those elements are constructed just like ROMs (Fig. 2.37). As a result, a transistor is generated, whose threshold voltage can be modified by means of appropriate electric impulses [2.13].

Electrically Reprogrammable ROM

Semiconductor memories still have the disadvantage that the information stored is lost when the supply voltage is switched off. Although most of the design engineers have learned how to deal with and overcome this setback there are a number of applications for which easy programming with electric impulses together with non-destructible information storage is absolutely necessary. In the late 60ies, the principle of the MNOS transistor was discovered [2.12] (Fig. 2.38 b). For this transistor, the gate insulator is constructed out of two insulating films, a thin silicon dioxide film (appr. 2 nm) and a thicker silicon nitride film (appr. 40 nm). At the interface of the two insulating films, traps are created, which can be charged and discharged by means of electric impulses. The state of charge for the traps also affects the threshold voltage (and thus the current) of the transistor. Two logic conditions can be stored. Design

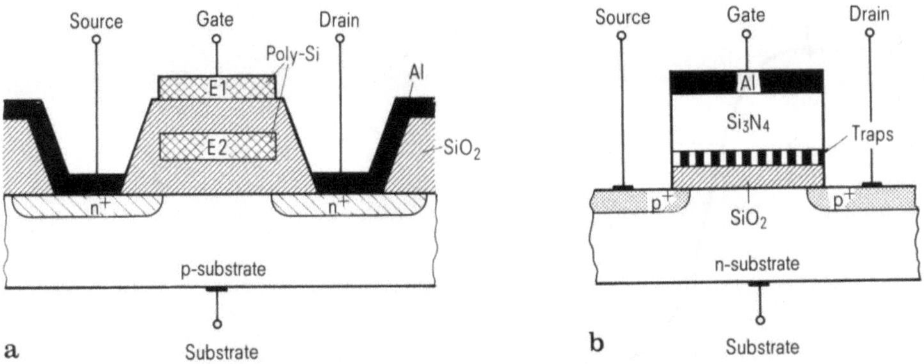

Fig. 2.38. Cross section of an electrically reprogrammable memory cell (a) and of an MNOS transistor (b) whose threshold voltage can be changed

and operation of a memory constructed with MNOS transistors correspond to that of a ROM or EPROM.

There are also several possibilities of erasing EPROM cells (Fig. 2.38 b) electrically. One of these possibilities is the SIMOS cell [2.14].

2.5 Analog MOS Circuits

2.5.1 Analog Cells

While it was possible to use MOS transistors as single components quite soon after their introduction as elements of analog circuits (RF-amplifier etc.), a few years ago almost all integrated MOS circuits were exclusively digital circuits. In view of the greater possibilities of integration, however, it has become more and more important that (single) analog circuits are integrated on one chip together with digital circuits.

The difference between digital and analog circuits will be clear as soon as the working points are compared by means of a transfer curve of an inverter (Fig. 2.39). If the input voltage is below the threshold voltage of the circuit transistor the latter is turned off, and the output level is equivalent to a logic "1". If the transistor is so controlled that the residual voltage at the output is below the threshold voltage of the succeeding transistor, this level is a logic "0". Working point A for analog operation is between the two levels. It has to be positioned in the linear section of the characteristic curve, if possible, and, in general, the differential slope (amplification) has to be as high as possible. While in digital operation the switching transistor is one time conductive, another time insulating, for analog operation load transistor as well as switching transistor are always conductive. This working point, where the amplification peak occurs, is set for a CMOS analog amplifier, too. A static leakage current flows.

First, simple amplifiers for analog signals in MOS can be built as already shown in Sect. 2.4.1 for the MOS inverter. Here, the inverter in CMOS technology reaches

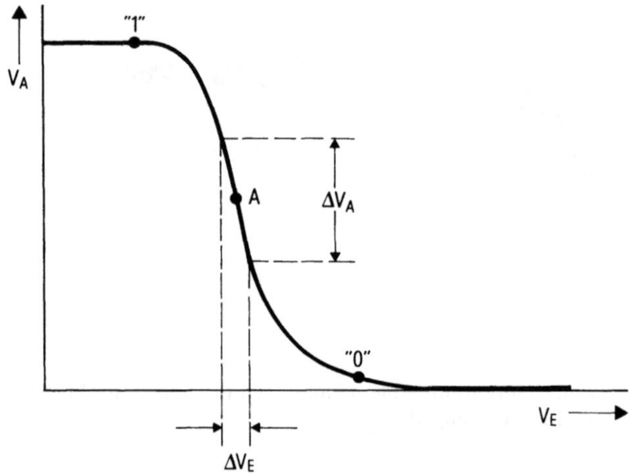

Fig. 2.39. Transfer characteristic of an MOS inverter with the working points for digital operation ("0" and "1") and for analog operation (A)

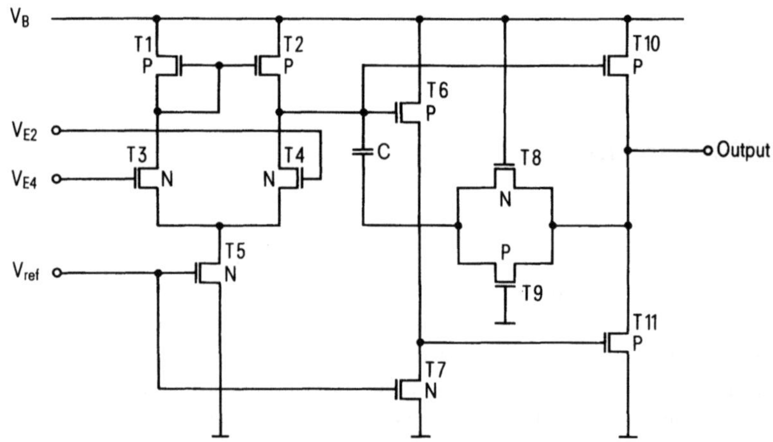

Fig. 2.40. Circuit diagram of a two-stage operational amplifier in complementary channel technology (CMOS)

its peak amplification. The small-signal amplification of analog amplifiers in single-channel technology is 10 to 100 times smaller.

The small signal equivalent circuit has to be used for the calculation of signal gain [2.3]. A high output conductance is essential for a high small-signal amplification: in the saturation section (Sect. II) of the characteristic curve shown in Fig. 2.13, the curve has to run as horizontal as is technologically possible. For this reason, the channel length of analog amplifiers is never made so small as would be technologically possible, but somewhat longer in order to keep the factor of channel length modulation as low as possible [2.15].

An operational amplifier, another analog circuit, is also described here as a basic component. In general, an operational amplifier has to have a high amplification. By external addition of passive elements the amplification can, thereafter, be adjusted to a given value, or mathematical operations such as addition or multiplication are performed.

Figure 2.40 shows an operational amplifier in CMOS technology. The core of the operational amplifier is formed by a differential amplifier. The source point of the two differential branches is controlled by the transistor T5. The gate voltage of this transistor is generated in a reference voltage generator. The transfer gate formed by the transistors T8 and T9 forms, together with capacitator C, and RC element for frequency compensation. In order to adjust the output level of the differential amplifier to a strong output driver stage, a level adjustment has to be performed first. This level adjustment is carried out by means of a source-follower amplifier. A powerful inverter can be used as output stage.

2.5.2 Analog – Digital – Mix

For a lot of applications, a combination of analog and digital cells is necessary. Even if cells appropriate for both alternatives are ready for use in the available technology, the design engineer still has to take into account other points of view. Here, these two rules are relevant:

- From the electric point of view it is extremely important that the supply voltage pads and lines used for analog cells are different from those applied for digital cells. If this step is not taken, perfect functioning can not be guaranteed.
- The interface of the two cell types has to be examined quite throughly. The electric behavior has to be defined precisely.

2.6 Mounting and Packaging

After the wafer has been structured in the process line it is put in an automatic tester. It has as many measuring needles as the circuit has connections. The tester has two functions, one mechanical and one electric. It sets the needles at the connection areas of the circuit and holds them in this position until the test has been accomplished (for example, 0.3 s or 1 s), lifts them and moves the silicon wafer until the measuring needles are above the next die. Then it lowers the needles, and the test procedure is repeated. The electric function consists in testing the circuit by means of a test program all within a fraction of a second, and in providing the chip with a colored spot in case of failure.

The term "yield" leads us to the key economic factor in the field of integrated circuit production. The production process described necessitates a lot of individual steps. During each process step errors may occur. Furthermore, the base material silicon is not ideal; there may be faults in the crystal lattice of the silicon material, or the doping may not be absolutely homogenous. This is why not all circuits on the silicon wafer, sometimes even only a few percent, pass the test. Then it is quite difficult, often even impossible, to find the source of the failure. In such a case one tries to find the source of the failure by means of the so-called process control.

Fig. 2.41. Integrated circuit in open package

One has, for example, a test wafer running parallel during diffusion and measures the sheet resistance, before subjecting it to further process steps. Or test structures (resistors, transistor, diodes) are added on the wafer, which are then measured during production or wafer test. Additionally, so-called visual inspections are inserted in order to discover visual errors. If one of the processes described, which are carried through as random test or as 100% test, leads to the discovery of a faulty wafer, one disposes of this wafer in order to save the succeeding processes, or one corrects the faults, if possible. It is possible, for example, to strip the aluminum film and repeat the evaporation process.

After the functional test has been caried out on the wafers, each wafer is notched with a diamond pin, then it is broken and thus separated into single dies, or it is separated into single chips by means of a saw. The good circuits (not marked with a colored spot) are inserted in a package or a spider by means of alloying or gluing. Then gold wires (thickness approximately 25 μm) are connected with the pads on the chip and with the package by means of thermocompression or ultrasonic welding (Fig. 2.41). Then the package is closed or the spider is coated with a synthetic material (Fig. 2.42).

Fig. 2.42. Chip in a closed package (finished component)

The isolation ability of silicon dioxide is very good. The resistance between gate electrode and silicon may be as high as $10^{15}\ \Omega$. For this reason, a charge-up of the gate electrode can easily occur due to the charges existing in the air. The breakthrough field strength for thermic SiO_2 is $(4 \ldots 6) \cdot 10^8\ Vm^{-1}$. The electric break through may occur for gate voltages above 60 V, which leads to the destruction of the gate oxide. Protection measures therefore have to be provided, which keep the input gate electrodes from being charged with too high a voltage.

Within the circuit there is no danger because the gate electrodes are always connected with the drain or source contact of a previous or a succeeding MOS stage, and are thus connected with the substrate through a pn-transition.

In IC technology, there are a large variety of packages which go beyond the aims of this book. Basically, there are packages made of plastics and those made of ceramics with numbers of connections ranging from 14 to more than 300. The connections are called pins. The majority of the packages, however, has between 14 and 68 pins. Another criterion for choosing the appropriate package is its resistance to thermal conductivity. The higher the power dissipation of the chip, the better the package has to conduct the generated heat to its surroundings. This is particularly important for bipolar circuits, which quite often use packages with applied heat sinks.

Literature – Chapter 2

2.1 Goser, K.: Vom Transistor zum System: Der Mikrocomputer. In: Mikroprozessoren und ihre Anwendungen 2.

2.2 Müller, R.: Grundlagen der Halbleiter-Elektronik, 4. Auflage, Springer Verlag, Berlin, 1984.

2.3 Müller, R.: Bauelemente der Halbleiter-Elektronik, 2. Auflage, Springer Verlag, Berlin, 1979.

2.4 Rein, H.-M.; Ranfft, R.: Integrierte Bipolarschaltungen, Springer-Verlag, Berlin, 1980.

2.5 Weiß, H.; Horninger, K.: Integrierte MOS-Schaltungen, Springer-Verlag, Berlin, 1982.

2.6 Ruge, I.: Halbleiter-Technologie, 2. Auflage, Springer-Verlag, Berlin, 1984.

2.7 Zuiderveen, E. A.: Handbuch der Digitalen Schaltungen, Franzis-Verlag, München, 1981.

2.8 The TTL Data Book, Texas Instruments, Fifth European Edition, 1982.

2.9 Waldschmidt, K.: Schaltungen der Datenverarbeitung. Stuttgart: Teubner 1980.
2.10 Siemens SEMICUSTOM: Zellorientierter Baustein-Entwurf. Handbuch 1, Schaltungsentwicklung, Ausg. April 1985. Siemens, München.
2.11 Hodges, D. A.; Jackson, H. G.: Analysis and Design of Digital Integrated Circuits. New York: McGraw-Hill 1983.
2.12 Kohonen, T.: Content-Addressable Memories. Berlin: Springer 1980.
2.13 Sze, S. M.: Physics of Semiconductor Devices, 2nd ed. New York: Wiley-Interscience 1981.
2.14 Rößler, B.; Müller, R. G.: Erasable and Electrically Reprogrammable Read-Only Memory Using the n-Channel SIMOS One-Transistor Cell. Siemens Research and Development Reports (1975), pp. 345–351.
2.15 Gray, P. R.; Hodges, D. A.; Brodersen, R. W.: Analog MOS Integrated Circuits. New York: IEEE Press 1980.
2.16 Capece, R. P.: Elektronik 20 (1979), p. 39.

3 Layout Design Methods

A basic idea common to all standard design procedures is, as previously explained, the preparation of reusable, prefabricated and verified circuit components [3.1]. The design expenditure for the layout is saved. And the design engineer can work at a higher abstraction level. Transistors and the technological limitations – described in terms of process technology related design rules – no longer form the design basis; rather it is logic elements such as gates, flipflops and memories. These elements are called cells. That is why standard-design procedures are also called *cell-oriented* design procedures.

There are also *transistor-oriented design procedures*. The most striking difference between transistor-oriented design and cell-oriented design is found in the design of the layout. Transistor-oriented means that the layout of the individual transistors is generated and optimized manually-interactively, with the aid of appropriate CAD tools.

Cell-oriented design methodologies offer considerable advantages in coping with the increasing complexity of VLSI circuits. They are, however, subject to restrictions. In Chap. 1.6, the characteristic restrictions in terms of cell geometries and wiring schemes are shown in a fundamental presentation. These basics are expanded here in order to supply an appropriate information base for the design engineer's choice of design procedure by means of the layout design methods used. For some design methods, the presentation includes only the detailed description of principle. For others, it provides additional working instructions, which allow for the utilization of the VENUS design system. The layout design methods presented also include the transistor-oriented IC design method, because on the one hand it is the basis for generating VENUS cells, on the other hand it is the basis for embedding free cells into cell-oriented standard procedures.

3.1 Cell-Oriented Design Methods

3.1.1 Standard Design Methods Using Libraries

The cells and cell libraries of standard design systems are designed in compliance with prescribed rules in such a way that the layout generation can be automated by means of CAD programs. The cells are characterized by their logic functions, their electric behavior, their geometric shape and their contact structures. The basic circuits for a cell library are selected on the one hand according to the universal application possibilities, on the other hand according to the library's manageability. An excessive variety of highly specialized cells reduces the ease of handling. The VENUS cell libraries, for example, contain about 100–150 different elements per design method

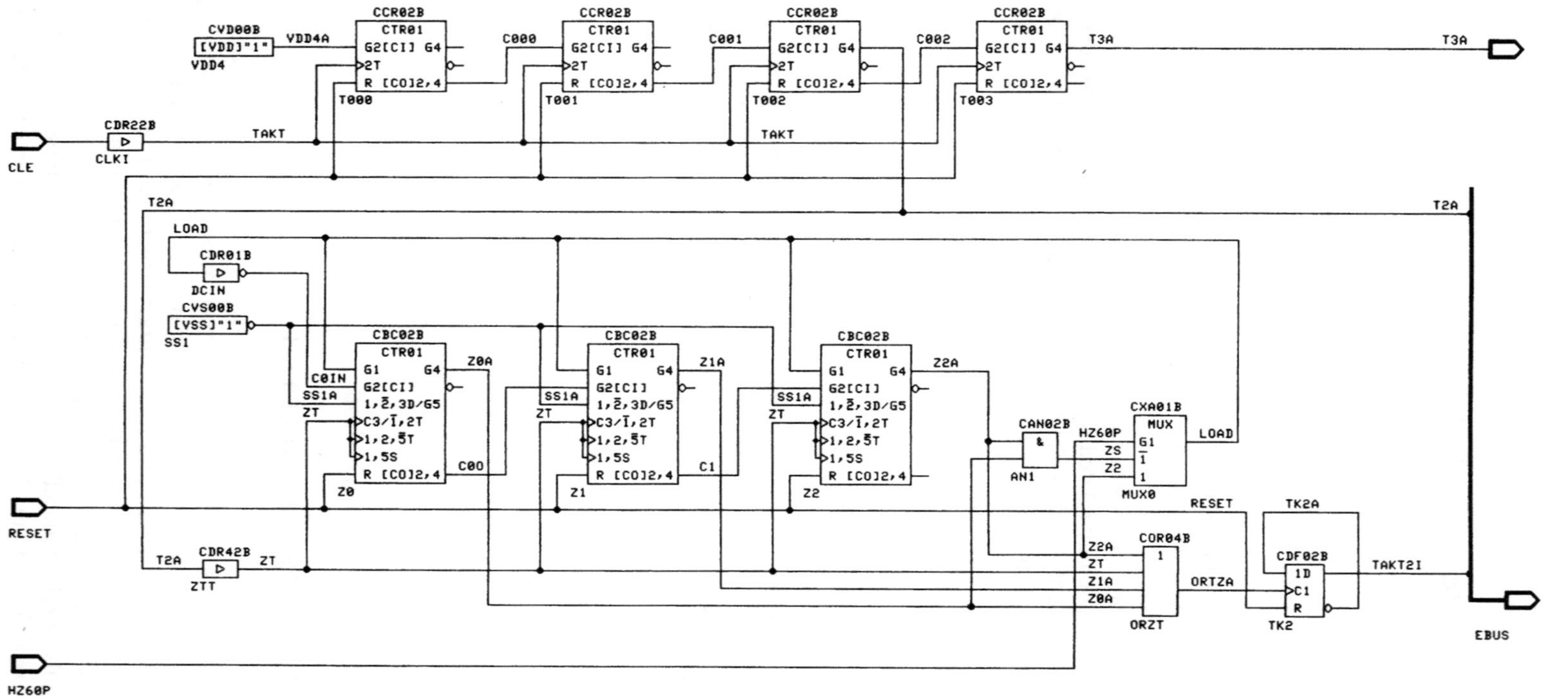

Fig. 3.1. Schematic generated with library elements at the CAD workstation

and technology variant. One cell may include up to about 50 transistors. The function and all design-related characteristics and parameters are described in a data sheet.

For the design engineer, the standard design methods have the advantage of simplifying the design process by reducing it to architecture- and logic design; often, this is also called functional design. It is carried out on the level of logical SSI (small scale integration) and MSI (medium scale integration) functions, with which the majority of design engineers are familiar because of their experience in PC-board design. For this reason, the cell catalogues in VENUS are also organized like SSI/MSI data books.

The functional design in VENUS is supported by a graphic-interactive logic design input (schematic capture), carried out at a graphics workstation; the user calls and connects the appropriate cell symbols using a tablet. The tablet contains a menu with all cells of the library chosen. Figure 3.1 shows a section of a schematic developed in this way: From the schematic, the CAD system generates a net list which is the basis of the circuit's logic verification through simulation. Changes which are required due to the simulation results are carried out interactively-graphically in the schematic. This iterative construction and verification continues until a functionally correct net list has been attained. It is the basis for generating the layout. The detailed design process in VENUS, including testability analysis and all other design steps required for the design of circuits, is shown in detail in Chap. 6.

The process of layout generation is almost entirely automated for cell-oriented design. The overall layout of a circuit is generated by placing and wiring the cells in accordance with the circuit logic described in the net list. The design engineer can intervene in the design process by setting up placement requirements or by interactive correction.

Earlier disadvantages of the automated design as compared to the manually optimized design of the layout, namely ineffective space utilization, low processing speed and high power dissipation have by now been removed through improvement and further development of the cells and the CAD programs. This is why the undisputable advantages of the standard procedures, namely high design safety, short design time and relatively low cost of design, are acknowledge more and more.

With the introduction of standard procedures, the main task for a design engineer has shifted from the layout design to the functional design. He no longer has the tedious work of generating correct geometric structures and can concentrate on the design of a well structured, regular and easily testable circuit. Today, the state of the art is a design based on rigid cell libraries and a generally applicable CAD system. Procedures which are continuously refined by completion with new cells that can be generated and with cells for analog functions, as well as by inclusion of functional programs which support the flexible, hierarchical block formation, are being introduced. These features are the subject of the following chapters.

3.1.2 Placement and Routing Procedures

Common to all systems supporting automatic layout design is the application of automatic placement and routing programs. For the gate array method as well as for the cell-oriented design without prefabrication, predesigned circuit components must first be placed and then wired. Although the various methods have different characteristics, the placement and wiring process is, in any case, based upon the same basic principle. It is roughly outlined here. Process-specific differences are explained in the relevant paragraphs.

Physical design – on cell level – is broken down into placement, global routing and local (or detailed) routing.

From the algorithmic point of view, *placement* is considered as an optimization task. Various methods differ in the choice of a suitable target function that can be minimized. The majority of these functions are wire-length oriented. An exact solution of the placement problem, which is often modeled as a quadratic assignment problem, is not possible for reasons of excessive run time, which makes the application of heuristic algorithms necessary. In the VENUS design sytem, the so-called Min-Cut procedure is mainly applied [3.2]. Thus, all cells can be processed simultaneously. First, an initial placement is generated; then iterative correction is performed. Initial placement is attained by division of cells into subsets in such a way that only a small number (as low as possible) of the wires connect the two subsets. The wire density in the layout is thus minimized.

During *global routing*, a rough assignment of connections to wiring channels is set up, which typically requires the placement to be adjusted.

The subsequent *detailed routing* determines the exact position of all channel sections and contact holes of the given connections, according to the geometric design rules. For that purpose, fast algorithms called "channel routers", are applied [3.3, 3.4].

The automatic placement and wiring programs allow the design engineer to further modify the result. On the one hand this is possible by a set of boundary conditions for the program run, on the other hand by interactive correction. The following meausres can be taken:

- By assignment of weights to nets and by specification of affinities between cells during schematic entry a cluster formation of cells can be obtained; individual cells can also be preplaced; likewise, after placement critical nets may be pre-wired manually.
- An interactive correction of the wiring results is possible at a graphic terminal; adherence to design rule and logic plan are enforced automatically and on line.

An interactive correction may become necessary if the wiring cannot be performed completely by the program. "Open connections" are some times left over. With the routing algorithms applied in VENUS, however, they occur only in a "local environment" – i.e. without global effects –, which makes graphic-interactive completion a simple task.

3.2 The Gate-Array Design Method

The gate-array design method is based upon a predetermined, regular structure of so-called basic cells and intermediate wiring channels. Basic cells consist of 2 to 8 transistors. The matrix-like basic-cell arrangements (masters) are currently offered with complexities between 1,000 (1K) and 20,000 (20K) gate equivalents or equivalent gate functions (*ä*quivalente *G*atter*f*unktionen = äGF). One gate equivalent corresponds to four transistors. By means of the wiring patterns stored in libraries, logic cells (gates, flipflops, multiplexers, counters) are generated out of basic cells. These are on different routing levels automatically connected according to the requirements specified in the schematic; thus, the final circuit is formed. In order to actualize the customer's circuit ("*customization*"), production steps for two aluminum layers and one contact layer only have to be performed if typical available technoloy is used.

The first step in the development of a gate-array design procedure is the definition of the basic cell, according to the number and arrangement of the transistors. Figure 3.2 shows the layout of a gate-array double basic cell, which is used within the VENUS design system for the 1K and the 2K master (master with 1,000 or 2,000 äGF).

For the construction of the masters, the size gradation for the different masters (in terms of equivalent gate functions) has to be determined with respect to the whole range of masters to be provided. A fine size gradation requires more effort for storing the various masters, waste of silicon, however, is minimized. The width of wiring channels is chosen in such a way that, on the one hand, wiring-intensive circuits can still be realized, on the other hand chip area is minimized. Currently, the VENUS system offers masters for several variants of the CMOS technology:

- for 3 μm structure width, there are 1K and 2K masters with various numbers of pads,
- for 2 μm structure width, there are a total of six masters with 2K to 10K äGF.

The realization of a certain circuit on the master chip is carried out in two steps:

1. Transistors of several neighboring basic cells are, by means of individual wiring macros, interconnected to cells with specific logic functions, such as AND gates, NOR gates, D-flipflops etc. The variation of this so-called intracell wiring brings up various logic functions. This intracell wiring is generated for a fixed set of about 100–150 logic cells per technology variant at layout level, and is stored in a library. The cells are verified to ensure correct logical and electrical behavior. Through simulation, the switching times for the data sheet are determined, which are verified by measurements. Currently the VENUS system offers two families of gate arrays for CMOS technology with a structure width of 3 μm and 2 μm, which are applied in combination with a range of suitable masters.

The result of personalizing several basic cells to one cell with a specific logical function is also termed *wiring macro*. This term mainly reflects the technological background of personalizing. Because the wiring macros are not different from the standard cells from the *logic* point of view, the term "*gate-array cell*" is also used. Figure 3.3 shows the example of a D-flipflop and its localization on a schematically presented master with a double-strip arrangement of the basic cells (in this presentation, the size ratios of wiring macro and master are not drawn to scale). The development of special functions for pad cells is achieved in the same way.

2. The second step of the circuit realization on a gate-array is the *placement* of the wiring macros on the grid of basic cells and *intercell wiring*. Through intercell wiring the logic cells as well as the pad cells are connected, generating the complete circuit. While the wiring macros are stored rigidly in a CAD system library, the intercell wiring is deduced from the schematic of the circuit that is to be developed, and is generated automatically. This procedure guarantees that, from the design engineer's

point of view, there is no difference between a gate-array design and a standard cell design.

The wiring patterns for both steps are realized with one or several metalization masks as well as the corresponding contact masks on the master. Today, there are masters available for practically all digital high performance technologies; these masters may also contain additional analog or memory components. The application of CMOS and ECL technologies, however, is favored.

3.2.1 Basic Cells and their Arrangement on the Master

There are various possibilities for the shape of basic cells as well as for their arrangement on the master. For CMOS, basic cell structures with two or three transistor pairs are predominant. If complex wiring macros (for example, counter cells) primarily occur in a circuit, it is more appropriate to choose a larger basic cell, because in this case, less effort is necessary, in principle, for the intracell wiring of a given function. The degree of space utilization, however, is low when small wiring macros and simple gates are mainly used, because the degree of effective utilization of subcells is lower. In the latter case, it is better to choose a smaller basic cell. Because, however, in most cases, wiring macros of various complexities exist in one circuit simultaneously, the arrangement of the basic cells on the master is more important than the size of the basic cells. The size is normally determined according to an average value, which is tailored to the requirements of the CAD system used for circuit development.

Due to the supply voltage connections and the ground connections, the basic cells are arranged in stripes on the masterchip. Single and double stripes, as well as double stripes with back-to-back placing of the basic cells are distinguished, as is shown in Fig. 3.4.

The space between the basic cell stripes contains a standard grid for intercell wiring, which is determined by the wiring program. The space to be provided is 30% to 50% of the entire chip space. Complete wiring is possible in most cases. For the wiring process, the circuits with the highest possible number of wires have to be considered; that means that for circuits with lower numbers of wires more or less chip space remains unutilized ("waste").

3.2.2 Placement and Routing of a Gate-Array Design

In order to develop circuits with gate-arrays of a complexity of 10,000 gate equivalents or more fast and safely, efficient CAD tools have to be applied. Prefabricated wiring macros stored in a cell library, as well as standardized intercell wiring grid facilitate the design of a gate-array circuit.

The development of a function at a certain place of the gate array is equal to the placement of a wiring macro from the library on this place.

According to the circuits net list stored in the data base of the CAD system, all required wiring macros have to be placed on the gate array and subsequently wired to each other; the space available for the entire process of wiring is fixed. For the placement of wiring macros, up to 90% (depending on the master used) of the predefined components can be utilized. In the VENUS design system the gate-array routing is performed with the program AULIS (*Au*tomatisches *L*ayout für *I*ntegrierte

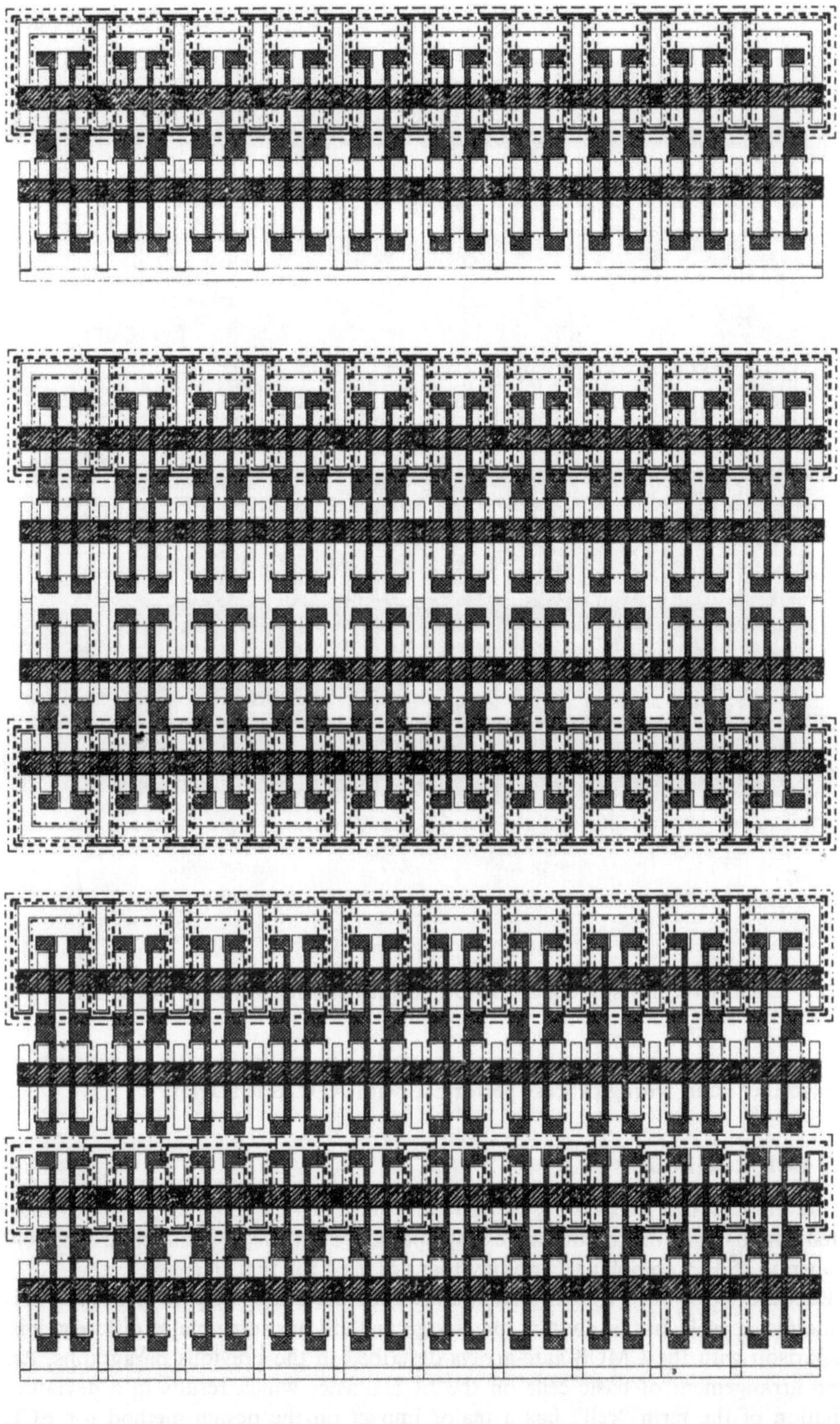

Fig. 3.4. Arrangement of basic cells in single and double stripes, as well as double stripes with back-to-back placing of basic cells

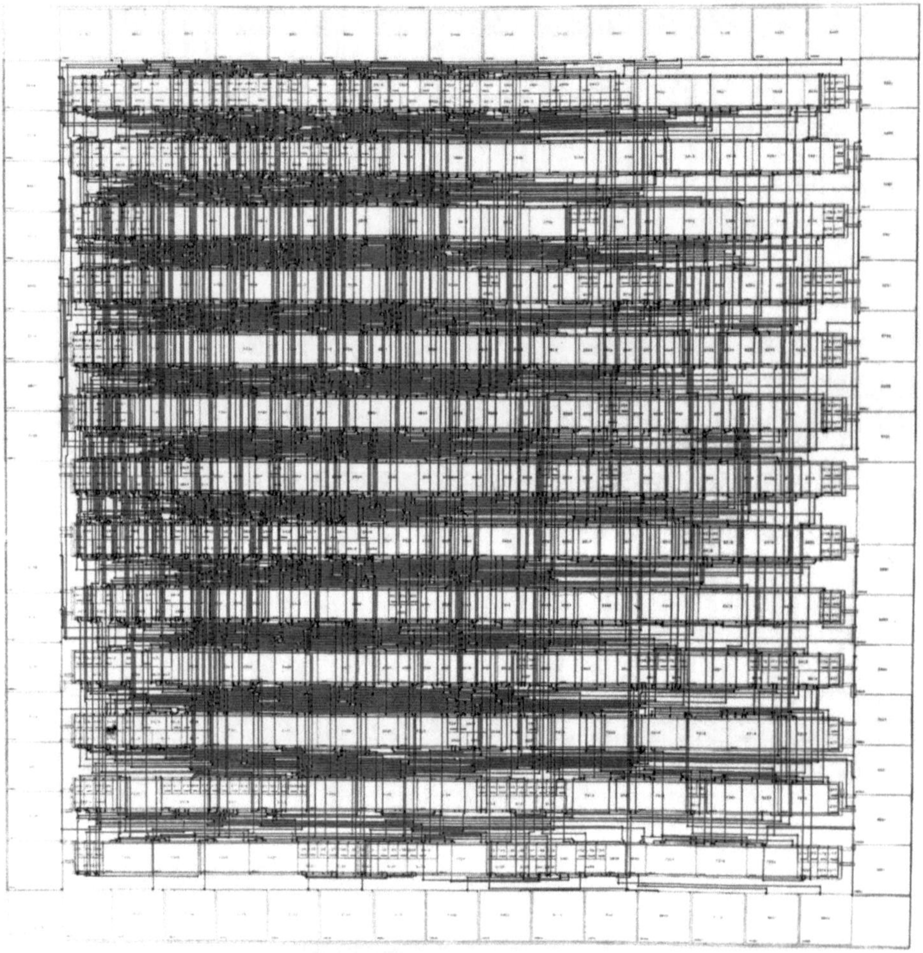

Fig. 3.5. Example for the layout of a CMOS gate-array circuit carried out by AULIS

Schaltungen = automatic layout for intregrated circuits). In Fig. 3.5 an example of the placement and wiring process performed with AULIS is shown.

3.2.3 ECL Gate-Arrays

The development of processors, in particular large-scale computers, requires integrated circuits with maximum processing speed. Nowadays, it is almost exclusively gate-arrays in ECL bipolar technology that are used for that purpose.

Beside CMOS, the VENUS system allows for the design of gate-arrays in ECL technology as well. In this context, the design method has some major differences in comparison with the CMOS gate-arrays described in the previous paragraphs. An island arrangement of basic cells on the ECL master which results in a deviating definition of the term "cell" has a major impact on the design method for ECL gate-arrays.

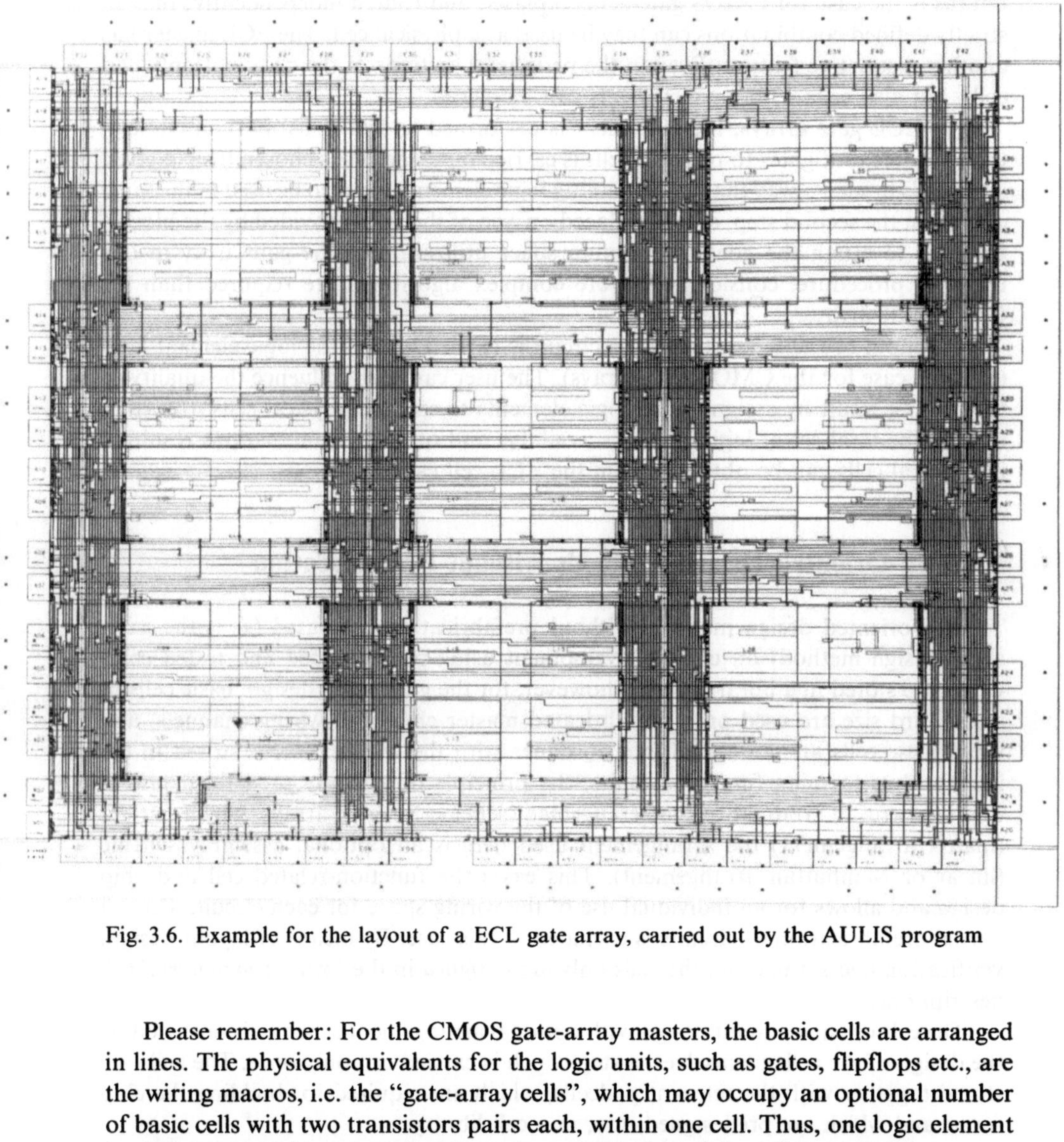

Fig. 3.6. Example for the layout of a ECL gate array, carried out by the AULIS program

Please remember: For the CMOS gate-array masters, the basic cells are arranged in lines. The physical equivalents for the logic units, such as gates, flipflops etc., are the wiring macros, i.e. the "gate-array cells", which may occupy an optional number of basic cells with two transistors pairs each, within one cell. Thus, one logic element is always mapped on one physical unit, and the size of the cell is optional.

The wiring example of a 36-cell ECL gate-array in Fig. 3.6, however, shows the block structure of the master with its 9 basic-cell island and the wiring channels surrounding them. Each of these islands segmented in four quadrants. The physical units, i.e. the ECL cells, may occupy one, two (horizontally or vertically adjacent) or all four of the quadrants of one block: only four different cell sizes are possible. For the master shown in Fig. 3.6, the maximum number of cells is therefore 36, if all quadrants are occupied by one cell each. Each quadrant has a complexity of 17 gate functions like NAND or NOR. For that reason, a physical cell generally has to contain several logic elements for optimum space utilization. The logic elements are not (as is the case for CMOS gate-arrays) placed and routed independently; only in strictly defined combinations can they be used as a physical cell. The ECL master has another distinctive feature: Even in the peripheral sections of the chip, not only pad cells, but also cells with logic elements can be placed.

For ECL gate arrays, the placement is performed in two steps: in the first step, the assignment of gates to physical cells is carried out. The cell is chosen from a given set of cells in such a way that the remaining logic elements can also be used in the same circuit. In the second step, cells are placed on one of the slots provided by the blocks. In order to attain the goal of optimum space utilization of the gate array in this two-step procedure, considerably more complex algorithms are required than for CMOS gate-arrays.

The input of the schematic of the circuit is carried out on the logic element level (as is the case for the CMOS gate-arrays). The user can then influence the quality of the routing result by assigning the logic elements to specific physical cells. Information on the total set of logic elements available and on their combinations resulting in physical cells can be obtained from the ECL cell catalogue.

3.3 Cell-Oriented Design Methods without Prefabrication

The cell-oriented design methods without prefabrication are based (as is the gate-array design method) on circuit development using predeveloped and tested cells, which are stored in a library. While, however, for the gate-array design, basic cells of a standard size are used on a prefabricated master chip with wiring channels, and while these cells are connected in accordance with the wiring macros stored in the library, thus forming functional units, the principle of a firmly predefined grid is omitted here. In contrast to the arrangement scheme for gate-arrays, which is rigid (matrix arrangement) the arrangement used here is an optional assembly scheme (linear or Manhattan arrangement). This eases the function-related cell and chip design and allows for an individual use of the wiring space for each circuit. Thus, a partial prefabrication of circuits is excluded in this case. By means of layout design, verification and simulation, the cells only are *designed* in the form of *immaterial* cell descriptions.

The cell-oriented circuit design with optional arrangement schemes is preceded by the design of the cells themselves, a step which is carried out only once. The cells are stored together with the relevant padcells, which are required for building the chip frame and which are predesigned, too, in a cell library.

Standard cells represent basic functions such as NOR- and NAND-gates, drivers, inverters, flipflops etc.; they are therefore equivalent to the low-integrated standard components in the current series of logic components. They are offered to the design engineer in sufficient number, primarily as rigid functions in a system library.

The expansion of these simple functions leads to more complex macrocells such as RAMs, ROMs or PLAs, and is equivalent to the transition from MSI to LSI components. Partly, macrocells are specified by the design engineer himself during the circuit design process and stored in a project-specific cell library. For that purpose, he is offered so-called generator programs in VENUS. But there are also rigid macrocells, such as microprocessor cores.

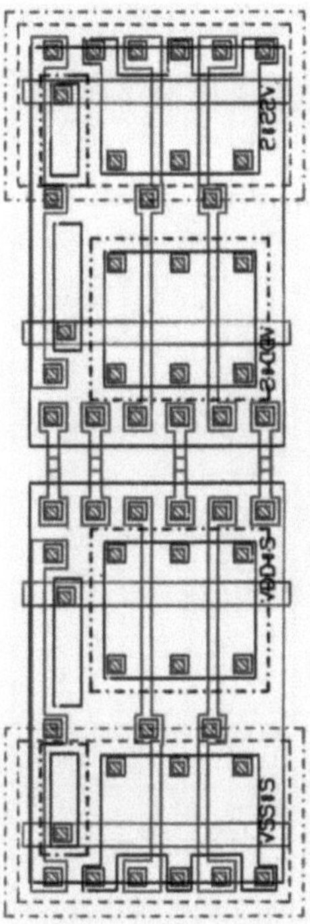

Fig. 3.2. Circuit and layout of a CMOS gate-array double basic cell formed by four transistor pairs, including polysilicon feedthroughs

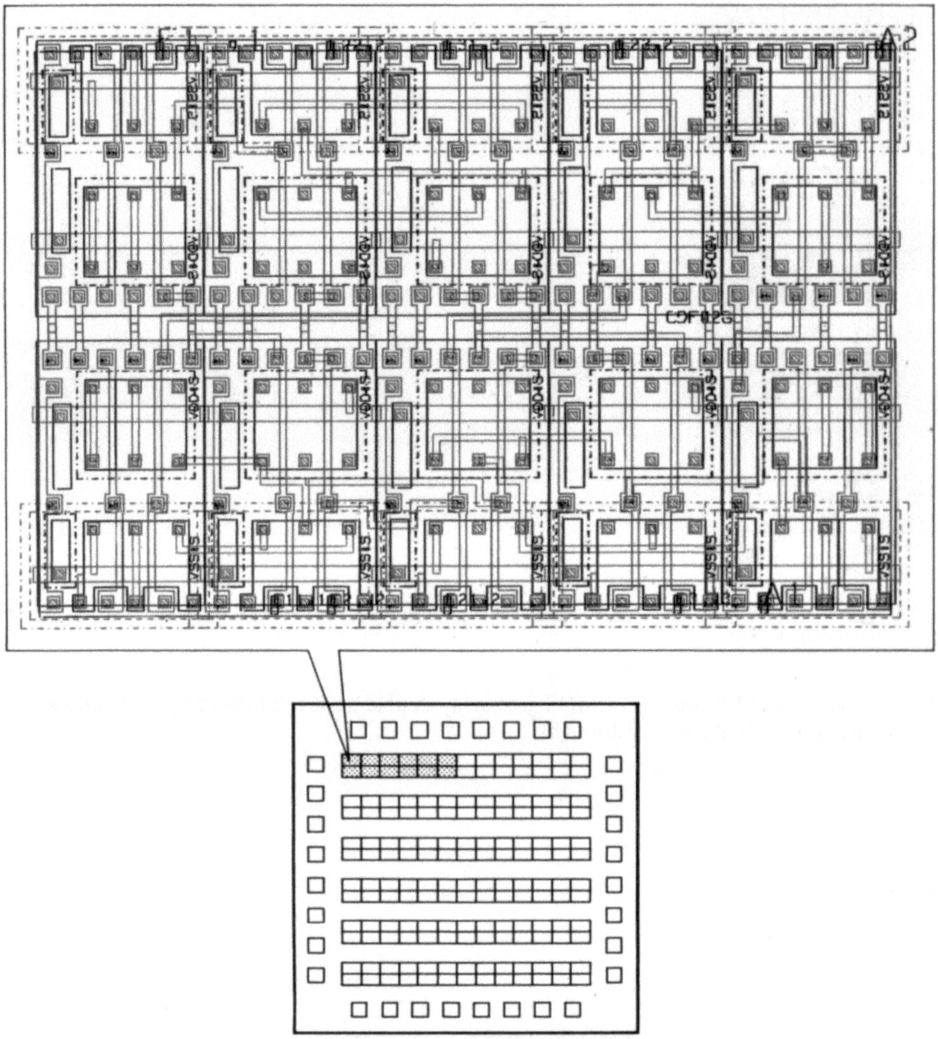

Fig. 3.3. Arrangement of a D-flipflop and its localization on the schematically presented master (principle)

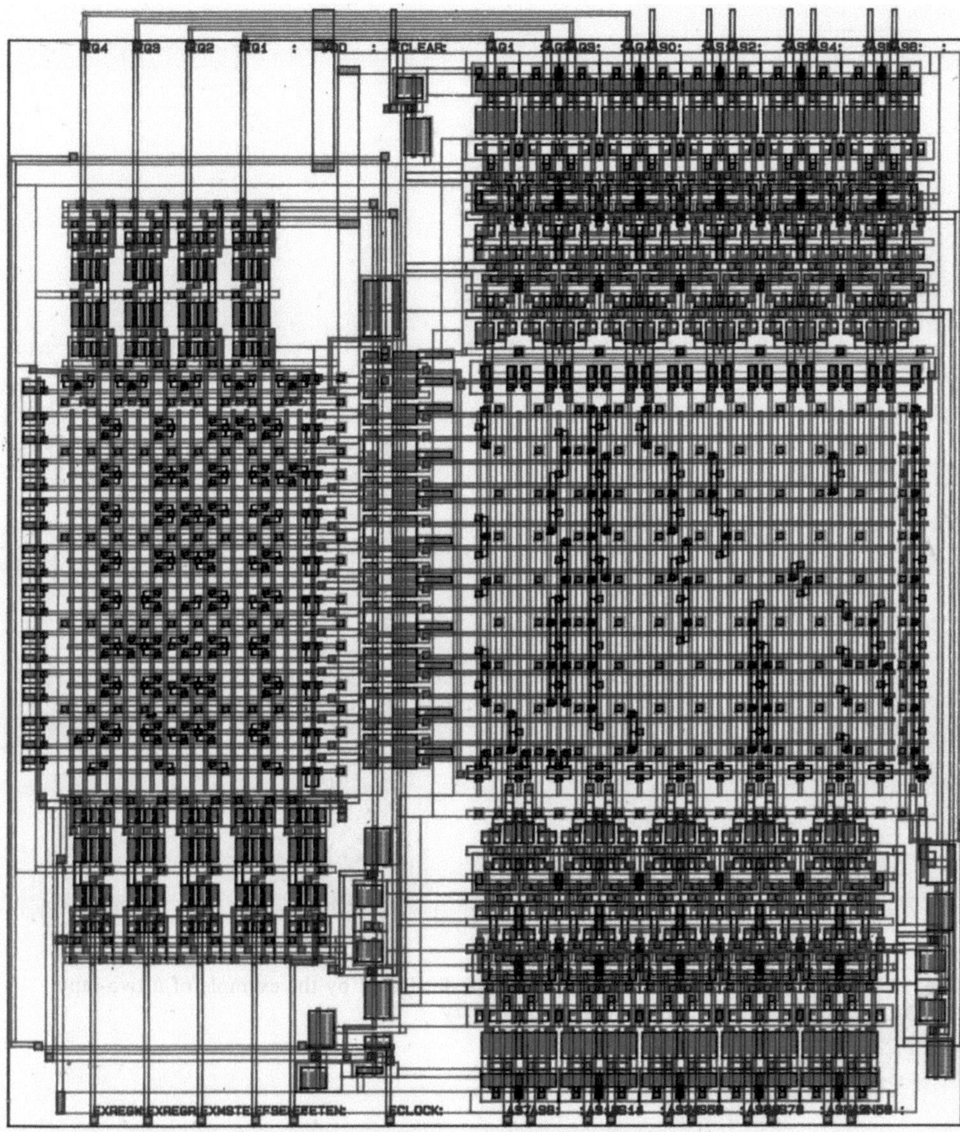

Fig. 3.16. Example for an automatically generated PLA structure

Function table

E1	E2	A1
0	0	0
0	1	0
1	0	0
1	1	1

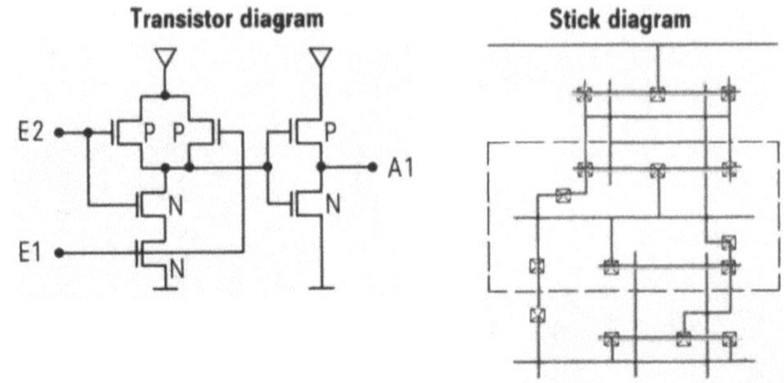

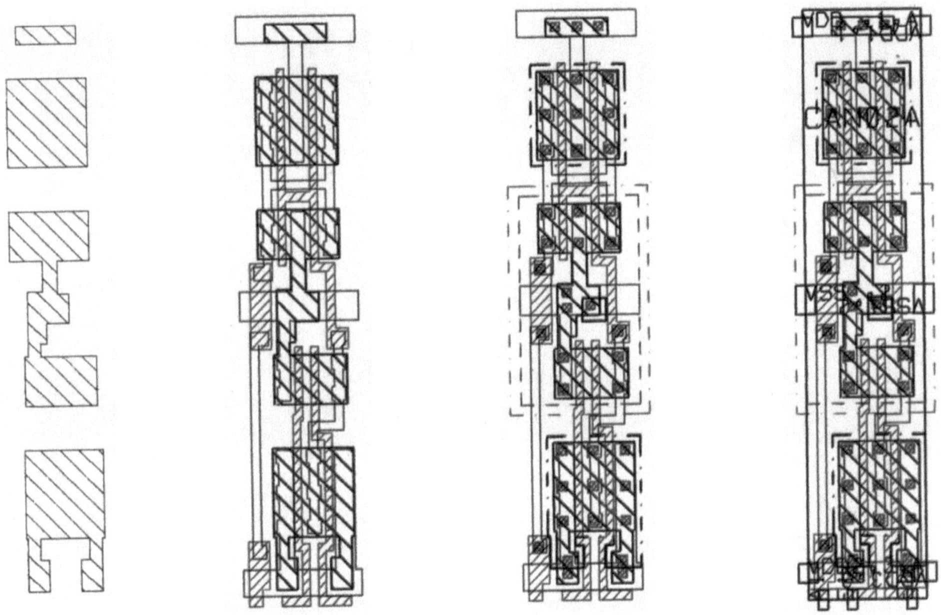

Fig. 3.18. Four steps of the manual layout design illustrated by the example of a two-input-AND-CMOS standard cell from the VENUS library

Mask	Structure light/dark	Colors plot	Minimum distance	Minimum width	Overlap
1 — A Locos	Dark	Black	a	b	c
3 — C Polysi 1	Dark	Red	d	e Of thickoxide	f
5 — E Contact	Light	Green (hatch)	g h	i x j	
6 — F Metal	Dark	Blue	k	l	m
10 — K p-well	Light	Blue short dashes	n	o	N+ P+ p q

Fig. 3.19. Examples for geometric design rules

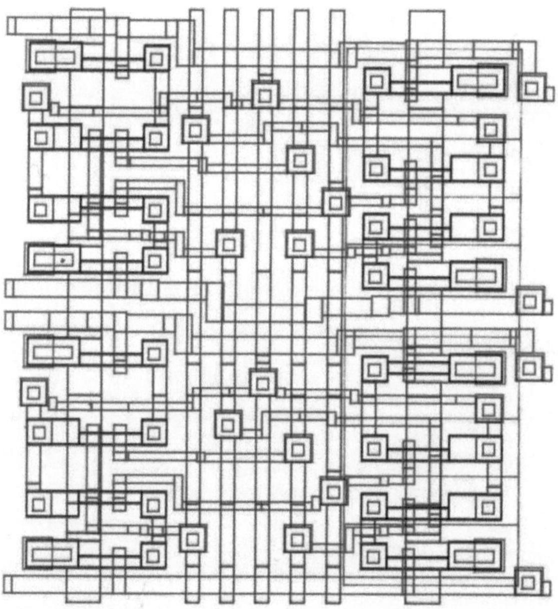

Fig. 3.20. Layout made of procedurally described cell components

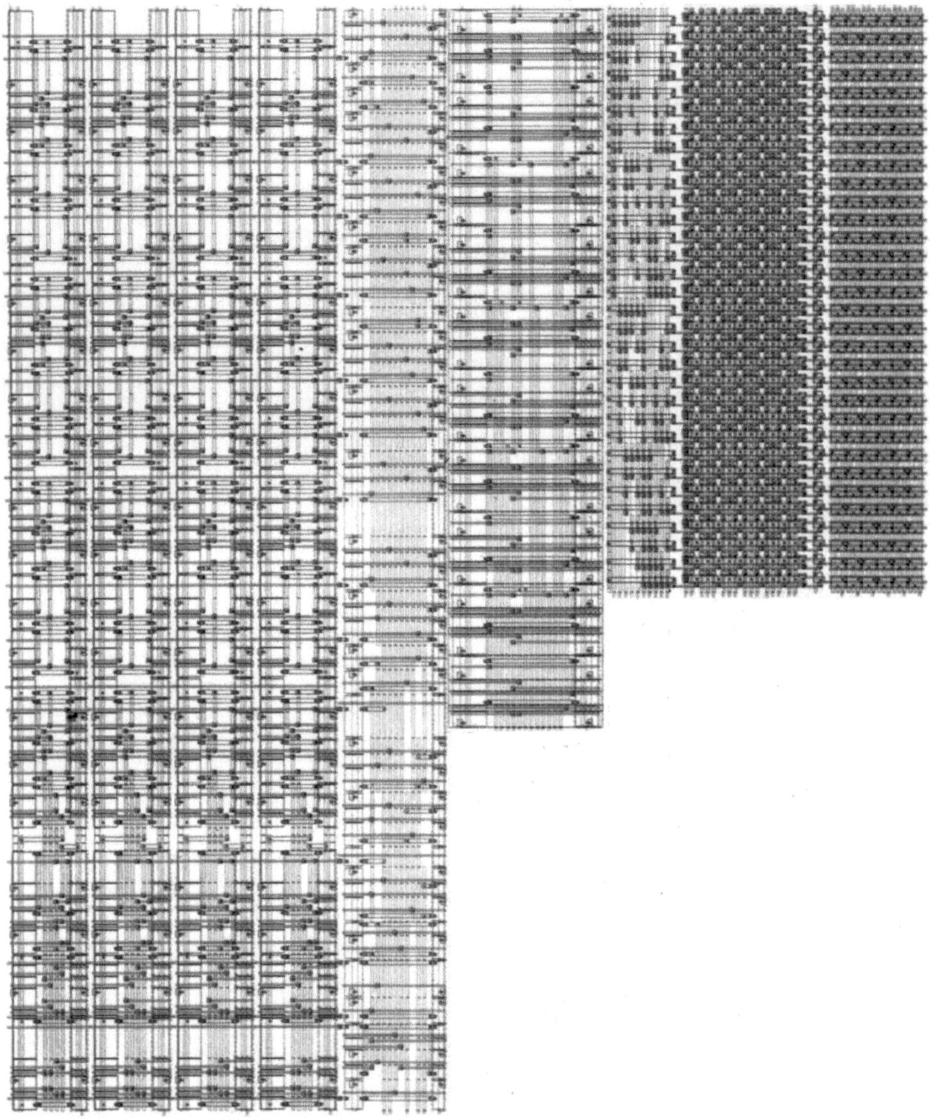

Fig. 3.21. Generated layout of a model processor

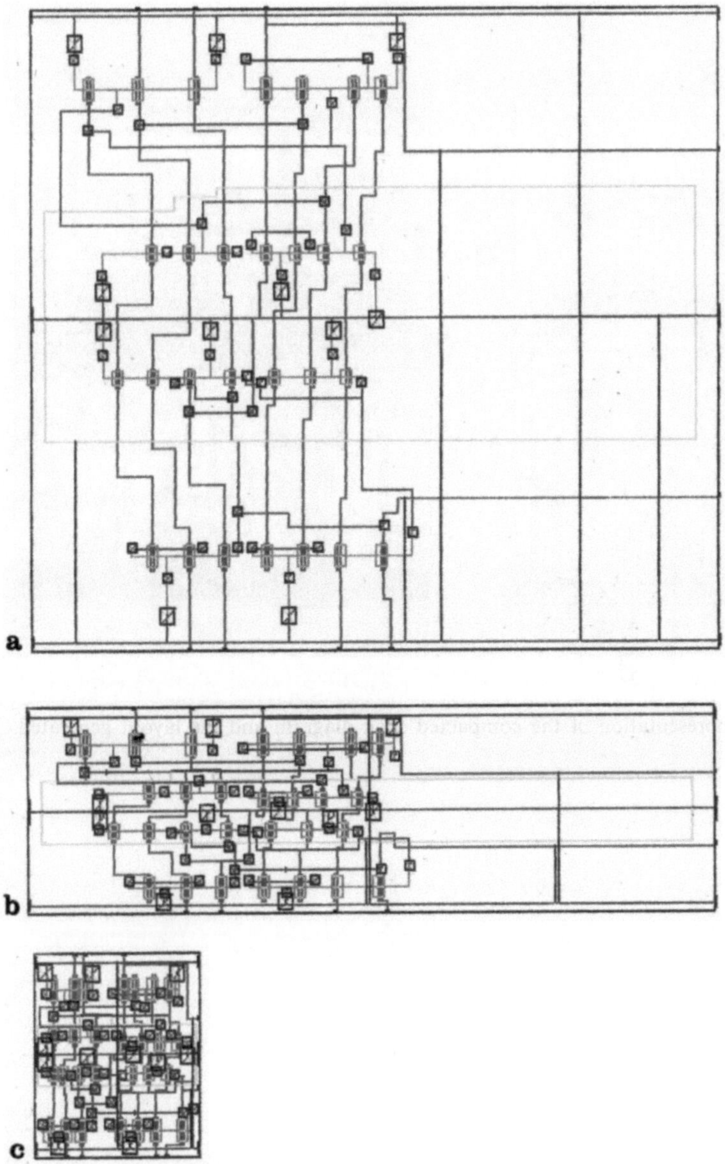

Fig. 3.23. Stick diagram (a) with vertical (b) and additional horizontal (c) compaction

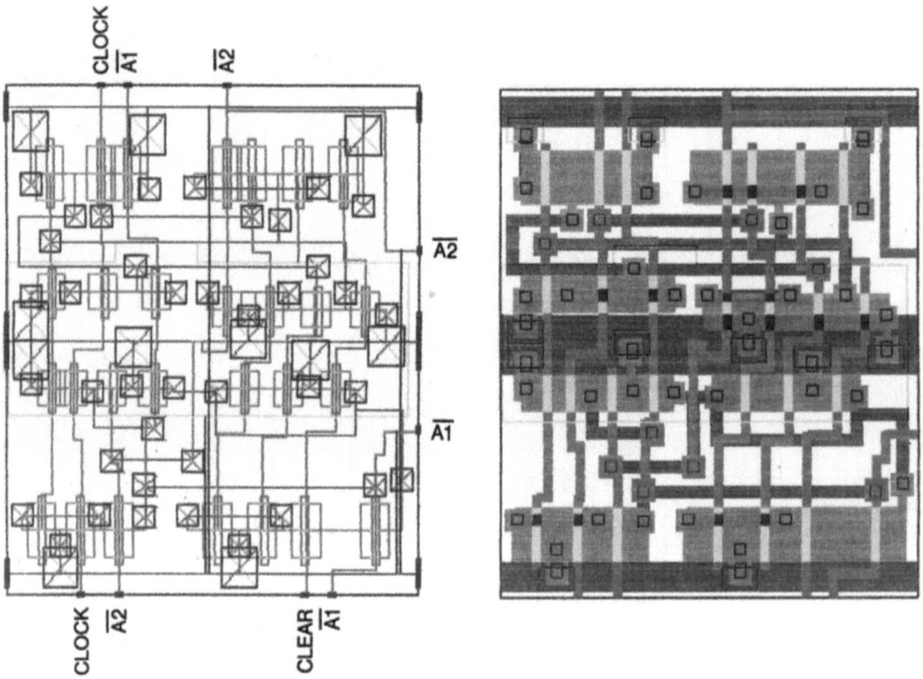

Fig. 3.24. Enlarged representation of the compacted stick diagram and the layout generated

The development of a chip is carried out by constructing the requested circuit from cells; the circuit developer normally needs only to know the data sheets of the cell catalog. The cell outlines are standardized; this allows the application of CAD wiring programs for automatic placement and wiring of the cells and connections.

3.4 Standard Cells

The distinguishing feature of standard cells is uniform cell height with variable cell width. They are designed according to a standard cell scheme: the supply lines are always at the same place, and all terminals are at defined grid points on the cell edges. Thus, standard cells can easily be connected with each other, and their wiring can be automated. Figure 3.7 shows the cell scheme of a standard cell used in the VENUS design system; Fig. 3.8 shows the layout of a 2-input-NAND standard cell; Fig. 3.9 shows part of a routed standard cell circuit.

There are standard cells with connections only on one side and standard cells with connections on two sides. In both cases one obtains, as is shown in Fig. 3.10, a simple arrangement scheme; one avoids having unused chip space. The standard cells are arranged in rows; supply and, for some systems, clock lines, too, are included within these cells, and are automatically wired. Two rows of cells include one horizontal wiring channel for the intercell connections. This scheme is suitable for technologies which, in addition to a metallic wiring level, have only one more wiring level with a high bulk resistance (for example, polysilicon). The intercell connections are provided by longer metal wires parallel to the rows and through numerous high impedance cross over wires.

3.4.1 Defining Standard Cells

For standard cells with terminals only on one edge, two rows each are connected to one block; they include a horizontal wiring channel, the width of which depends on

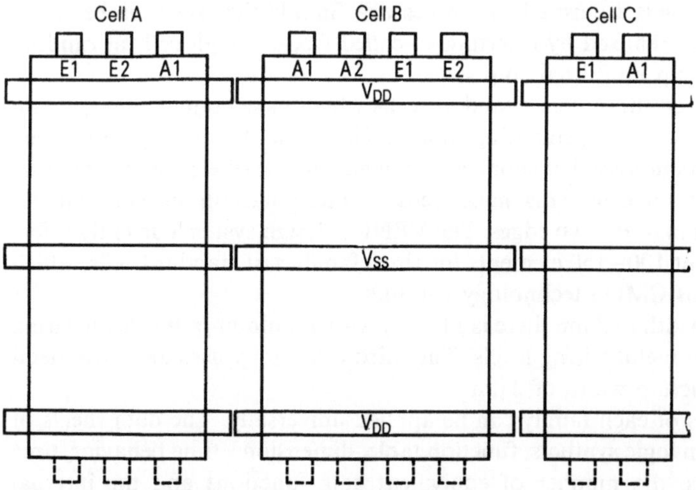

Fig. 3.7. Cell scheme of VENUS standard cells

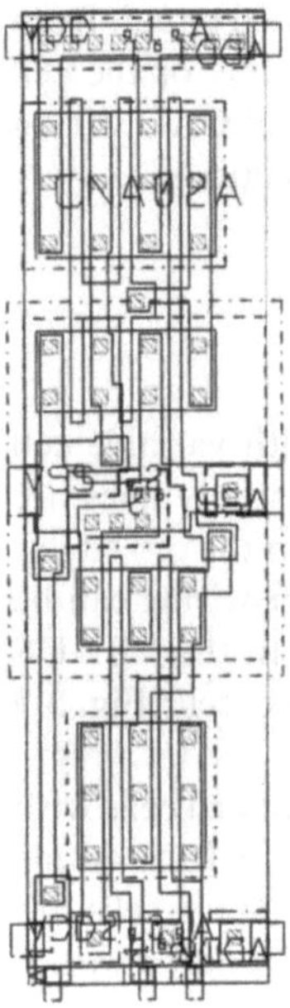

Fig. 3.8. 2-input-NAND standard cell

the number of horizontal tracks required. The blocks are combined in such a way that the adjoining cells are positioned back-to-back. Gaps in the block structure provide vertical wiring channels, for connections between blocks, which are, if possible, made on a metal layer. The adjoining blocks are normally not vertically aligned at the location of the vertical channel, a fact which inevitably leads to unutilized space on the chip. A circuit usually contains several blocks; the upper and/or the lower block may consist of one row only, including the pertinent routing channel. The padcells are set along the chip periphery.

For standard cells with terminals on two edges, no blocks are generated, and gaps for vertical wiring channels are usually not necessary. Should the exception be necessary, these gaps can be realized by inserting so-called feed-through cells at random points without taking up unutilized chip space.

There is still another alternative: standard cells which have connections only on one side are arranged without generating blocks. This variant can only be realized conveniently with a technology that supplies two metallization levels, and thus allows for vertical wiring over the cells. This makes cells with terminals on one edge equivalent to those with terminals on two edges. The VENUS design system currently offers cell libraries with about 100–150 elements for three families of standard cells, which are adjusted to various CMOS technology variants.

For the structure width of 3 μm there is a family with aluminum/polysilicon wiring and a family with two metal wiring levels. The third cell family uses also two metal levels, but with a structure width of 2 μm.

The logic elements of each family can be applied universally. The data sheets of the cell catalog contain logic symbols, function table, dimensions, time behavior, time diagram, load factors, the number of equivalent gate functions and the internal schematic of each cell.

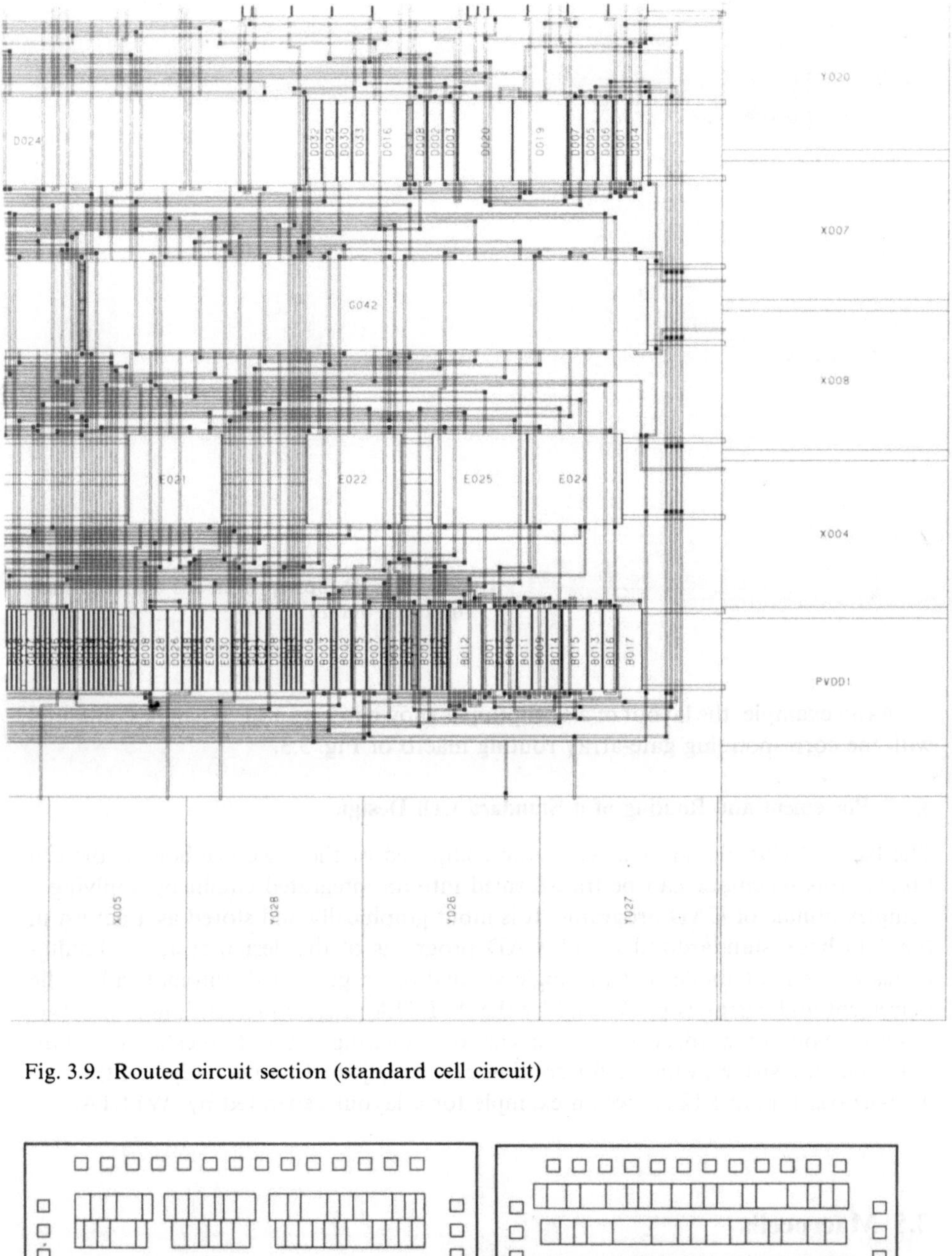

Fig. 3.9. Routed circuit section (standard cell circuit)

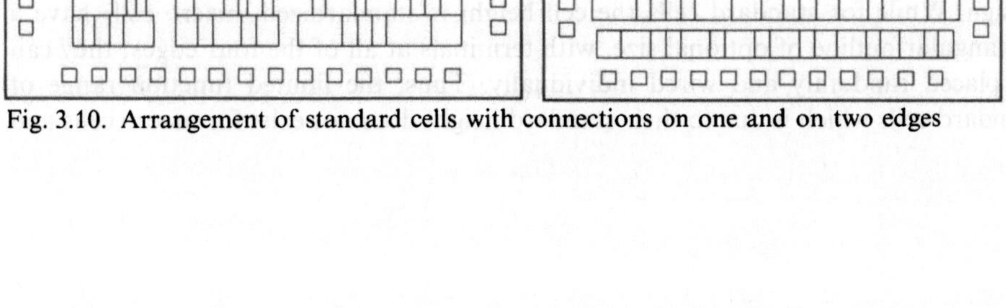

Fig. 3.10. Arrangement of standard cells with connections on one and on two edges

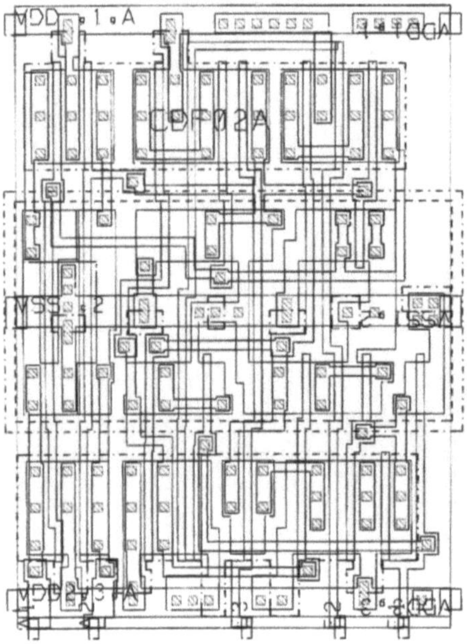

Fig. 3.11. D-flipflop from the VENUS standard cell library

As an example, the layout of a D-flipflop is shown in Fig. 3.11; it can be compared with the corresponding gate-array routing macro of Fig. 3.3.

3.4.2 Placement and Routing of a Standard Cell Design

The basis of chip design is a schematic composed of the logic symbols of the cell library; this schematic can be transformed into an integrated circuit by applying a complex bundle of CAD programs. It is input graphically and stored as a net list in the data base, standardized for all CAD programs of the design system. Simultaneously, a circuit model for the logic simulation is generated automatically. The placement and wiring is performed by the AVESTA program (*A*nordnung und *Ve*rdrahtung von *Sta*ndardzellen = arrangement and wiring of standard cells). For standard cells the space available for routing can be adjusted individually, contrary to gate-arrays. Figure 3.12 shows an example for a layout generated by AVESTA.

3.5 Macrocells

3.5.1 Design of Macrocells

The cell scheme of macro cells provides the highest degree of freedom for cell-oriented design. While for standard cells the cell height is standardized, macro cells have a rectangular outline of optional size, with terminals at all of the four edges; they can be placed randomly and wired individually. Thus, the limited function range of standard cells which is due to their standard height does not exist for macro cells and

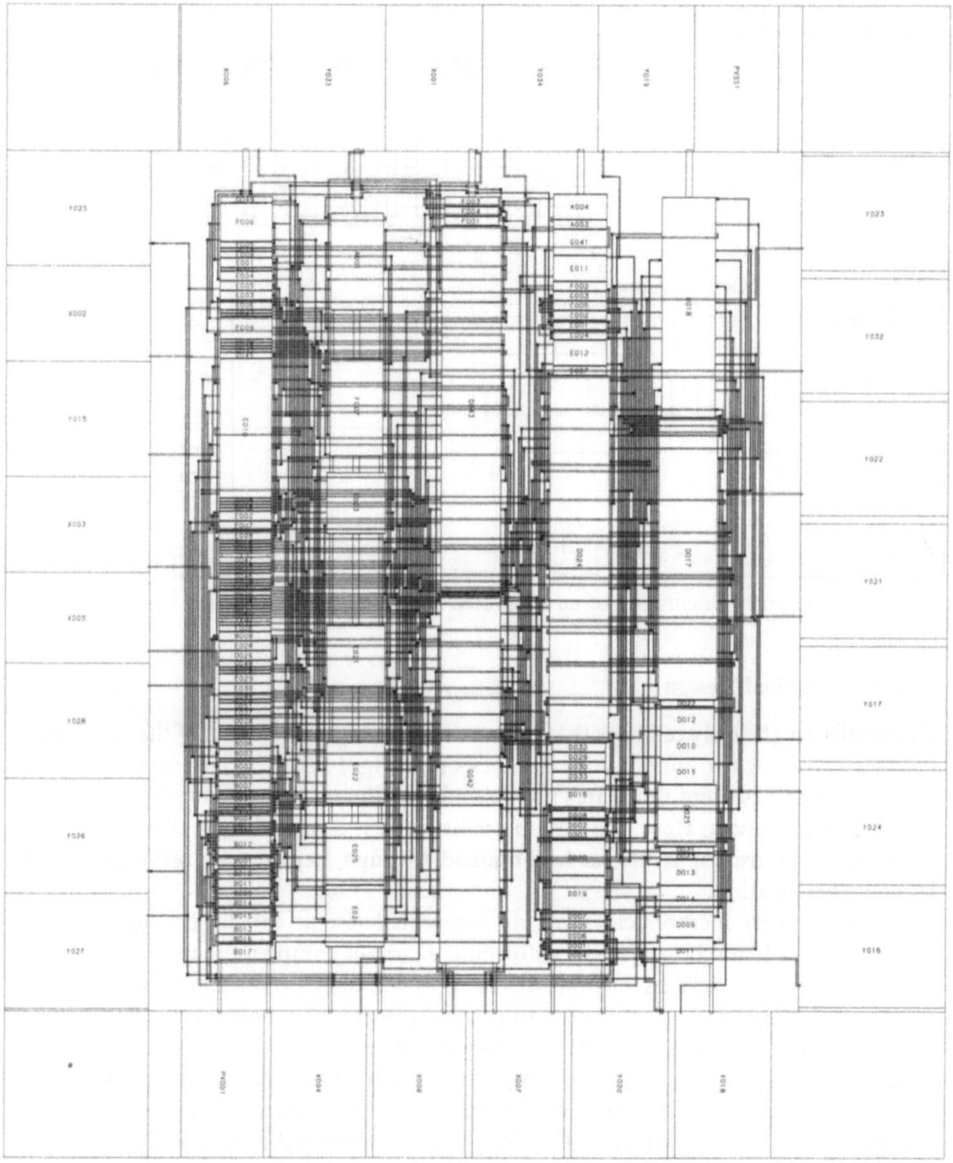

Fig. 3.12. Example for a layout generated by AVESTA

for the circuit to be realized. Simultaneously, the share required for routing can be reduced; this share may increase to more than 50% of the entire space available for designs of larger circuits. Macrocells can assume any rectangular shape; they contain memories (RAMs, ROMs), PLAs, microprocessor cores, ALUs, circuit components composed of standard cells and circuits recursively composed of macrocells (see Chap. 3.5.2). Figure 3.13 shows the schematic representation of an integrated circuit made of macrocells.

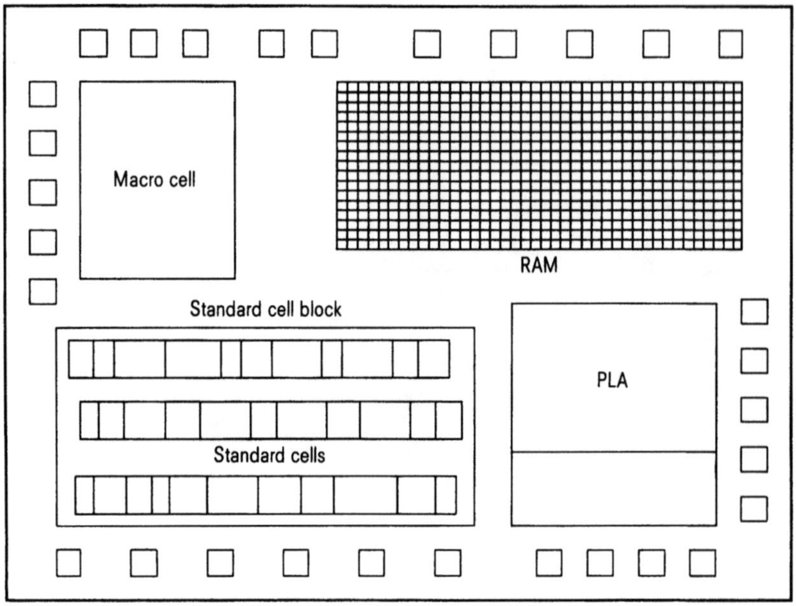

Fig. 3.13. Schematic presentation of an integrated circuit made of macrocells

3.5.2 Hierarchical Design

Macrocells are the only cell type that in turn may consist of elements of the same type and thereby allow for building of cell blocks (see Fig. 1.38). The recursive character of the macrocell concept illustrates its close connection to the structured-hierarchical circuit design concept. Here, the system architecture is designed first; then the circuit schematic at transistor level can be obtained through step-by-step refinement and detailed specification.

At the highest hierarchical level, an integrated circuit can be considered as a composition of black boxes, each performing a certain complex function, connected with the others in a defined way. At this level, block diagrams still meet the requirements for the presentation of data and control information flow, even for complex VLSI circuits; the hiding of internal details of the individual black boxes makes the presentation simple and clear. The design of so-called floor plans is also performed at the highest hierarchy level. It includes the interactive placement of the macrocells (supported by a CAD program). The floor plan is optimized in terms of complexity and length of connections; the quality of placement is a crucial factor for the success of subsequent wiring.

The hierarchic construction of macrocells made of other macrocells or standard cells makes possible a definition of a user-specific cell. Other methods for a user-specific cell generation are discussed below.

3.5.3 Placement and Routing of a Macrocell Design

For the macrocell layout process, the program CALCOS (*C*omputed *A*ided *L*ayout of *C*ell *O*riented *S*ystems) is used in the VENUS design system. This program pro-

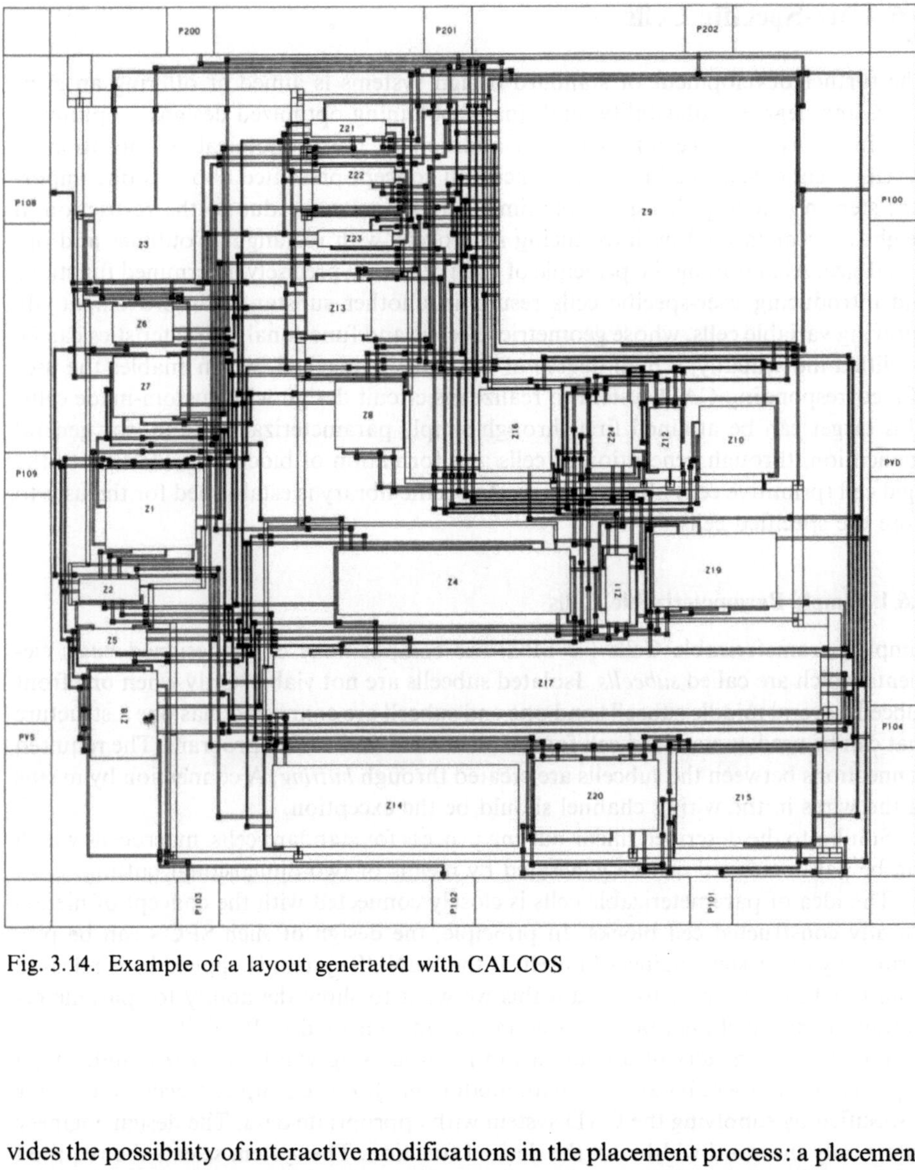

Fig. 3.14. Example of a layout generated with CALCOS

vides the possibility of interactive modifications in the placement process: a placement proposal can be improved by the design engineer by modifying the outlines of the macrocells used. This is particularly possible for cell blocks composed of standard cells; their outlines can easily be modified through appropriate arrangement of the standard cells. For macrocells, which are created by generators and which can be supplied with parameters the degree of flexibility for the cell outlines is lower. After the cell outlines have been modified the placement and wiring is repeated. For detailed wiring, a channel-router is used for macrocell designs as well as for standard cells. However, planar supply lines, too, have to be designed automatically with wire widths adapted to current density [3.5]. In Fig. 3.14, an example of a layout generated with CALCOS is shown.

3.6 User-Specific Cells

The further development of standard design systems is aimed at offering an ever-increasing degree of flexibility, and finally obtaining optimized designs – optimized also in terms of space and speed in addition to the economical optimization of development costs. The hierarchic macrocell concept presented above is one important step in reaching this goal: the limited function range due to the restriction in height is compensated by introducing macrocells with rectangular outlines and optional size. Abandoning the principle of rigid cells with precisely determined functions and introducing user-specific cells results in another substantial improvement. By applying variable cells, whose geometric, electric and functional characteristics can be modified individually, a high degree of flexibility is reached, which enables the user of a corresponding CAD system to realize his circuit design with custom-made cells. This target can be attained first through simple parameterization, and, for general application, through generation of cells and formation of blocks. In addition to the rigid cell (primitive cell) library, a project-specific library is established for the user to store the specified cells.

3.6.1 Simple Parameterizable Cells

Simple parameterizable Cells (= SPC) are compositions of predesigned basic elements which are called *subcells*. Isolated subcells are not viable. Only when one front subcell, several middle subcells and one end subcell are connected, has one a structure that can be used as standard cell for the placement and wiring program. The required connections between the subcells are created through *butting*. A connection by means of the wires in the wiring channel should be the exception.

Similar to the described linear butting process for standard cells, macrocells which can be parameterized can be generated by means of two-dimensional butting.

The idea of parameterizable cells is closely connected with the concept of hierarchically constructed cell blocks. In principle, the design of such SPC's can be performed by the design engineer himself. If the VENUS system is applied he is released from this task. In order to explain this we want to show the ability for parameterization of the functional family of counters, implemented in VENUS.

Possible parameters of a counter are its processing width and the counter type requested (up, down, binary, decimal, modulo etc.). For example, a decimal counter is specified by supplying the CAD system with appropriate data. The design engineer enters only the word width and the desired function. The system selects the required subcells from those available, and connects them automatically by butting them as needed for the realization of the circuit. The front subcell, for example, contains drivers for the internal clock or the control logic for the up/down function. The middle subcells consist of the actual counter-flipflops. The end subcell terminates the cell-internal wiring (Fig. 3.15). The part of the CAD system which performs these functions is also called counter compiler. According to the definition given above, the counter cell resulting from this process is shaped like a standard cell, and can easily be integrated in the overall design.

One can construct logic blocks either by applying the described parameterization (butting the subcells), or, automatically, by using standard cells. The structure devel-

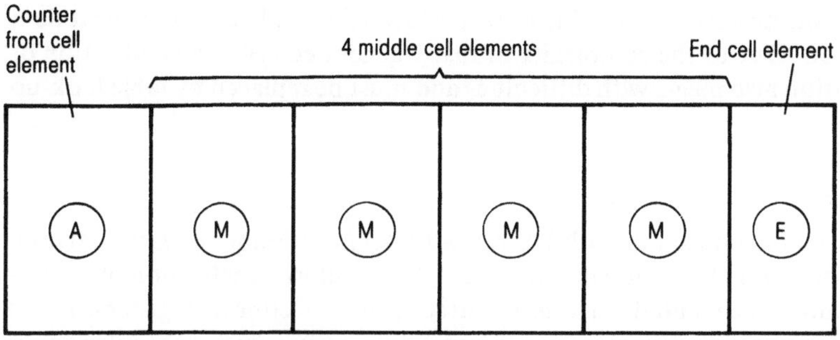

Fig. 3.15. 4-bit counter made of subcells

oped, however, does not have the characteristics of a standard cell; it is offered to the placement and routing program as a partly wired conglomerate made of several standard cells. If no pre-placement conditions are set up, it can not be guaranteed that the cells will be arranged in immediate geographic proximity.

The transition from SPC's to functionally parameterizable cells is smooth. Basically, both types can be formed by butting subcells or by prewiring of logic blocks with subsequent expansion. For those that are subject to simple parameterization the variability is, however, much more restricted. This means that the user interface of such cells is generally quite simple: for an n-bit counter, for example, one need only to give the word width n.

Characteristic representatives of cells that can be parameterized simply are n-bit counter, n-bit register or cells with variable output drivers.

3.6.2 Functionally Parameterizable Cells

Maximum flexibility is obtained through the automatic cell generation process. The user specifies the cell's behavior by means of functional descriptions: here, one number is not sufficient. One example is the data of a boolean equation in order to defined a PLA. Characteristic representatives of cells that can be parameterized functionally are PLAs, RAMs, ROMs, Finite State Machines (FSM) as well as microprogrammable microprocessors. Programs with the ability to generate such functions are called cell generators.

Cell generators predominantly, although not exclusively, apply the method of *butting* subcells. Often the subcells themselves are no longer described rigidly, but rather, by means of stick diagrams (see Chap. 3.7). They may therefore be adjusted to changing design rules. To an increasing extent, however, subcells (and the cells formed by them) are described procedurally.

One important aspect of the generator concept is that not only the cell layout is formed, but at the same time an appropriate simulation model and required test patterns are generated. In terms of cell generation, there is one basic problem which has not been solved yet: to guarantee specific cell characteristics. Rigid cells and several variants of SPC's can be subject to trial generation and measurement. This ensures the fact that the cell characteristics ascertained by means of electric simulation can also be maintained on a real chip. The high number of cell variants that can be generated does not, however, allow this to be performed completely. Statistical data is used in addition. Due to the complexity of many generated cells, the realization of electrical simulation also meets with difficulties and must be replaced by table look-up techniques.

PLA as Example

The first macrocell introduced in VENUS was a PLA (*P*rogrammable *L*ogic *A*rray). On the basis of most regular transistor structures, PLAs allow a realization of combinatorial logic and of sequential logic in connection with feedback registers. PLAs provide an opportunity to carry out an optimum mapping in compact, regular structures for complex combinational functions, which would result in large irregular structures if realized with standard cells. Combinational logic with n input and m output lines can generally be implemented by means of a memory of 2^n m-bit words. n-bit input words form the addresses of memories, the contents of which make up the m-bit output words. Because in many cases all input combinations do not occur and the total number of 2^n words are not needed a large part of the chip space would be wasted if a ROM was applied as logic element.

PLAs, on the contrary, allow for the compact implementation of a given function, for they need only one row of circuit components for those terms that actually occur. They contain two serially linked connective fields, which are termed logic AND and OR level. At the AND-level, product terms are generated from the input values. They are passed on to the OR-level, where the relevant sum terms are formed. By feeding back the outputs through memory elements to the inputs, a combinatorial PLA is turned into a sequential logic system, working as a finite-state machine.

It is relatively easy to generate a PLA because of its high degree of regularity and its simple circuit structure. Its basic structure is formed by the two matrices of input and output line bundles, which run orthogonal; at the nodal points the circuit transistors may or may not be programmed, depending on the logic function to be realized. The PLA so personalized can be "occupied" by transistors to a high or low extent. The higher the density of transistors, the better the utilization of the silicon space available. A PLA with low density can be compacted by splitting and folding. Figure 3.16 shows an example for an automatically generated PLA structure.

When the floor plan of the component is formed other advantages of the PLAs become evident. The input and output lines can be exchanged randomly; they can thus be optimally adjusted to the periphery of the PLA. Large PLAs can be partitioned in smaller ones, which makes it possible for the design engineer to carry out a flexible floor plan formation.

3.7 Free Macrocells and Manual Layout

Application-specific circuits in particular bring up the task to approach the technologic limits for reaching such goals as maximum speed or minimum power dissipation. Examples for such circuits are coder/decoder for digital wide-band communication systems, special fast signal processors for image processing etc. For such applications, too, standard design methods are appropriate if they allow for the installation of freely generated macrocells, i.e. cells designed manually on transistor level.

For that, the CAD system has to perform two functions:

- Support of the transistor-oriented design method,
- Integration of the "free cells" in the standard design process.

Since both functions are already conceptional parts of the VENUS design system, although they will be released at a later time, the transistor-oriented design process will be explained in the following. Another reason for this discussion is that the cells and masters of the VENUS design system themselves were developed using the transistor-oriented design method. This procedure is justified by the fact that cells and masters have to be highly optimized, as is explained in Chap. 5. Moreover, the majority of standard components with a high degree of complexity, such as microprocessors or memories, which justify high design effort, are currently being developed according to that method.

One generates large parts of the layout plan of these circuits manually, by placing and connecting rectangles at the various mask levels, in order to hopefully waste as little space as possible. In this context, design rules such as minimum dimensions, minimum distances etc. have to be adhered too. This process of *manual design* described in the following paragraph has a high error-susceptibility, and it is quite unflexible towards changes which are necessary due to design-rule violations or functional errors discovered in an advanced design phase. Realizing changes is difficult, as the carefully optimized structures are all interrelated with each other, according to the definitions set up by the design rules. It often occurs that such a layout plan has to be completely reorganized.

This is why computer-based processes have been developed, which allow generating an optimized layout plan on transistor basis, which is mostly error-free and easy-to-change. The basic idea of this procedure is to release the designer from doing the tedious work of graphically generating correct rectangle structures and provide him with a symbolic description method, so that he can arrange transistors, diodes, resistors and lines in any way without having to follow actual design rules and space optimization. The symbolic description of the layout is transformed by a program into the mask structures for the various production steps, in compliance with relevant construction rules. Thus, one layout may be generated according to various design-rule sets, a function which is quite important for the maintenance of complex cell libraries. As examples for a number of similar processes, the *stick-diagram method* and the *gate-matrix method* shall be examined more closely later.

3.7.1 Manual Layout

Usually the steps of the design process are as follows: starting point for the design process is the specification of the logic function required. It is often presented as a function table. This includes the selection of a certain technology, such as CMOS. The first step consists in transforming the logic function into a technology-dependent transistor circuit. Simultaneously minimization of the circuit can often be performed, based upon certain circuit criteria. This may result in several variants which are examined by a network analysis program in order to obtain an optimal circuit.

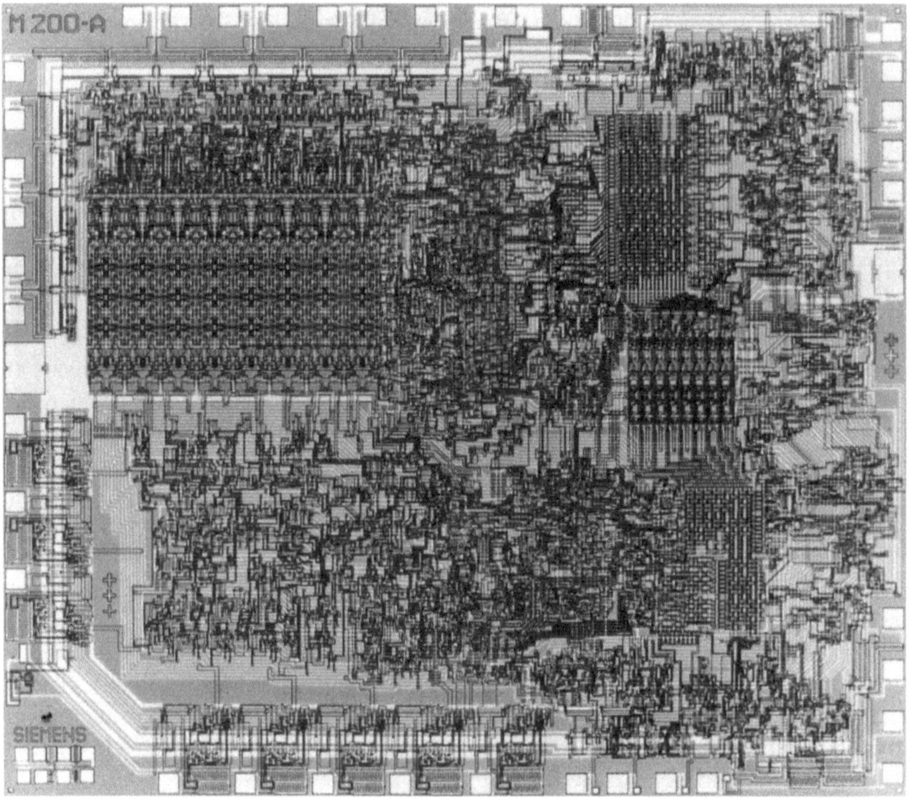

Fig. 3.17. Layout of the microprocessor Intel 8080

Then the process of geometric design can begin. The first step is to transform the transistor circuit into a symbolic layout, which is called *stick diagram*. The stick diagram is a representation on transistor level which contains, apart from the mere connection of elements, information about the relative placement of transistors, diffusion areas, metal and polysilicon lines, contact holes etc. They are represented by certain symbols in a standard color-code (blue: metal, red: polysilicon, green: diffusion, yellow: well etc.). This allows for the sketching of the rough topography of the circuit which an experienced circuit designer uses to consider the actual dimensions of the detailed layout plan. Thus, a stick diagram in practice contains almost all information about the final layout plan.

The most simple method to go from the stick diagram to the layout is to draw each transistor manually on coordinate paper; again, the aforementioned color code is used to identify the individual layers. The design rules defined by the production process must be adhered to. They are geometric rules which determine, for example, minimum width or minimum distances, thus ensuring that the chip can be generated safely and with a high degree of efficiency (see Chap. 3.7.2).

When the layout draft is available it has to be digitized for further use which can be realized by an explicit alphanumerical input of the rectangle coordinates or by means of a digitizer. The path of manual layout generation is long and susceptible to faults. The mask geometry for a LSI design of only 3000 transistors may comprise up to 50,000 rectangles. Obviously, VLSI circuits are too complex to be designed by means of this method. Figure 3.17 shows an example of a manually generated layout of the Intel 8080 microprocessor. The apparent lack or regularity is striking.

Today, CAD workstations with predefined geometric structures are available which enable one to perform all graphic manipulations interactively and make it possible to generate the layout without requiring a subsequent digitization.

Because the circuit is segmented in functional blocks, less single structures need to be designed individually. Basic patterns can be drawn at the graphic terminal and then replicated. By means of appropriate programs, one can check if the digitally stored finished layout complies with the geometric and electric design rules. The electric circuit description that is equivalent to the layout, including its actual geometry-based data, can be recovered by means of an extractor program; for verification it can again be made subject to a final electric simulation.

Figure 3.18 shows the basic steps of the described process of manual design, making use of the example of the layout design of a 2-input-AND-CMOS standard cell from the VENUS cell library. Based upon the function table the transistor circuit and the derived stick diagram come next. In the lower section of the figure, from left to right, four partial layouts are shown at a graphic terminal; in the first, only the structure of the diffusion level is realized; details are already integrated. In the second step the aluminum and polysilicon levels are added. The third step additionally includes contact holes, p-well, p^+- and n^+-implantation sections, as well as substrate and well contacts. Step 4 shows the complete standard cell, including frame and supply lines, plus names.

3.7.2 Geometric Design Rules

With the manual design, the designer must always be aware of the various technologic process steps, which he must apply to his circuit design until physical realization has been achieved. The most important interface between designer and technology is defined by a set of design rules specifying the geometric structures permissible for a certain process, as well as electric parameters. They determine the dimensions and electric characteristics that can be attained in a certain production process.

The geometric design rules are explained here in more detail with the help of the examples in Fig. 3.19. Geometric design rules describe the values permitted (in μm) for certain widths, distances and overlappings of geometric objects, which have to be adhered to on the various mask levels (see Chap. 2). These values contain important information about the technological level of a company; they are, therefore, strictly confidential. Therefore it is understandable that, for the examples shown here, the actual figures were replaced by symbolic data.

Figure 3.19 shows the design rules of five various mask levels. In the column "mask" not only the physical terms of the levels but also their characterization is given by means of a company-internal letter code and a number. This number identifies the mask levels at a CAD graphic workstation. The column structure supplies data about exposure during structure generation on the chip, by means of photo resist; we dealt with this already in Chap. 2. "Colors plot" (color/plot) allocates to each level one color in which the corresponding structures for graphic layout representations are shown graphically.

In the columns "Minimum distance", "Minimum width" and "Overlap", the actual design rules are shown which have to be interpreted as follows: structures in the LOCOS level have to have a minimum distance of a µm. Their minimum width has to be b µm. A contact hole has to be overlapped by structures on the LOCOS layer by at least c µm. The values d, e and f and k, l and m are accordingly applicable for the polysilicon and the metal level, respectively. The reference to thickoxide means that this rule is applicable outside active regions (see Chap. 2). For the contact hole mask, data is supplied about the distances between the contact holes at the LOCOS level (g) and about the polysilicon structures (h); also the dimensions (ixj) of the contact holes themselves are defined. For the p-well section, apart from minimum distance (n) and minimum width (o), there is the following rule: A p-well has to encircle an n-transistor field with a distance of q µm, and at the same time it has to have a distance of p µm from the p-transistor field.

From these few examples (the whole of geometric design rules comprises a lot of pages) one can see the complexity and error-subsceptibility of the manual design and the advantages of total or partial automation.

3.7.3 Regular Design

The introduction of regular structures results in a clear simplification of manual layout. In this context, one does not try to make the layout smaller by means of interlacing (see the layout of the microprocessor Intel 8080, Fig. 3.17), but one tries to reduce the layout effort by replicating structures several times. The regularity factor of a design is defined as

$$\frac{\text{number of structures in the layout}}{\text{number of structures drawn}}$$

Modern layouts have a regularity factor which is above 20.

Regular structures, however, are also a condition for (or a result from) the use of layout generators. Figures 3.20 and 3.21 are examples of automatically generated layouts for illustrating a high degree of regularity.

3.7.4 Symbolic layout

Since stick diagrams were first presented by Williams in 1977 [3.6, 3.7] a number of procedures have been developed which have furthered the principle of symbolic layout and of the pertinent compacting algorithms, and expanded and refined it for various technologies.

Stick Diagrams

While for manual design the stick diagram is an intermediate step of the design, which is to be transformed into the real layout manually, it serves here as input for a program. In this case, the designer has a symbol library available for transistors, contacts etc. and a number of lines of varying color or structure, which represent the connections at the various levels. This enables him to develop the symbolic layout of his circuit without having to know the actual design rules, and without having to pay attention to space optimization. The quality of the generated layout, however, is determined by the efficiency of the relative placement of the elements. The period of time required for the symbolic design of a layout is of course shorter than that required for manual positioning of the real mask structures of the same circuit. A stick diagram can usually be completed and modified easily.

For some CAD systems on stick-diagram basis, a hierarchic design is possible by inserting macros or circuit predefined segments. Changes of the design rules can be realized without having to change the symbolic layout. The more possibilities exist in the individual elements of the library for the design engineer to set up preconditions and parameters, the more powerful a system is.

Compacting

The transformation of a stick diagram into a real layout is carried out with a stick compiler. This compiler uses the relative placement information from the symbolic layout and therefore avoids complex automatic placement. The generation of the real layout is primarily carried out in two steps. First, the stick diagram is compacted in order to reach the goal of space minimization. The compacted layout follows the minimum distances given in the design rules. In a second step, the compacted stick diagram is transformed into corresponding rectangular structures. The two-dimensional problem of compacting is divided into two one-dimensional problems,

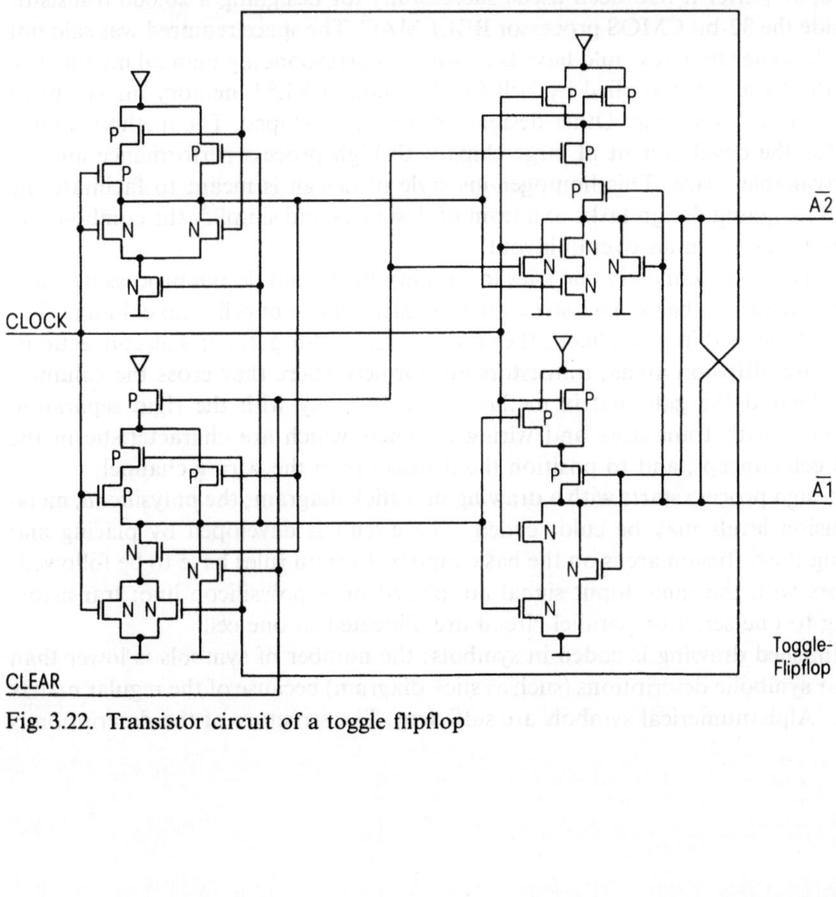

Fig. 3.22. Transistor circuit of a toggle flipflop

which are solved in sequence; a horizontal and a vertical compacting is carried through. The result is an error-free symbolic layout, which may, however, still contain unused space.

The division of the compacting process into two one-dimensional contractions is a clear simplification. Certain global situations can not be recognized, as the earlier horizontal compacting has a restrictive impact on a suitable subsequent vertical compacting. Such situations may, however, be foreseen by an experienced designer and may be removed interactively by introducing so-called "user constraints". These are, for example, used to define sections which are not meant to be subject to compaction. Distances may also be determined absolutely in advance. An important application of "constraints" is to hold the connections of one cell which must be adjusted to another cell. After the interactive compaction and the correction of the stick diagram, the real layout is automatically generated in the last step.

The entire procedure described is demonstrated in the following figures, using the toggle flipflop as an example. Figure 3.22 shows the transistor circuit of a flipflop. Figure 3.23 shows the corresponding stick diagram and its vertical and horizontal compacting.

Figure 3.24 shows an enlarged representation of the compacted stick diagram and the layout generated.

3.7.5 Layout According to the Gate-Matrix Method

The gate-matrix method was published first in 1980 by Lopez and Law, Bell Laboratories [3.8, 3.9] after it had been usede successfully for designing a 20,000 transistor block inside the 32-bit CMOS processor BELLMAC. The space required was said not to be much higher than it would have been for the corresponding manual layout. The same method was later applied by Bell for designing a VLSI-memory-management chip and a third VLSI chip. Other designs are being developed. The method is most suitable for the development of large chips with high process performance and restricted available space. This homogenous style of design is meant to facilitate the process of assigning design tasks to a team of designers and simplify the combination of the single results in an overall layout.

When using a gate-matrix one tries to simplify the layout design process by using a regular structure as basis. This structure is a matrix made of cells and columns. The columns are realized in polysilicon; they serve as transistor gates and as connections. The rows are diffusion areas; transistors are formed where they cross the columns. The idea behind the gate-matrix method is to do away with the rigid separation between cells with transistors and wiring channels which are characteristic in the standard cell concept, and to position the transistors in the wiring channel.

The design process starts with a drawing or a stick diagram; the polysilicon, metal and diffusion levels may be color coded. The circuit is developed by placing and connecting the diffusion areas on the basic matrix. Certain rules have to be followed: Transistors with the same input signal are placed on a polysilicon line; transistors belonging to one serial or parallel circuit are allocated to one cell.

The finished drawing is coded in symbols; the number of symbols is lower than for similar symbolic descriptions (such as stick diagram) because of the regular matrix structure. Alphanumerical symbols are sufficient. The structure of the description is

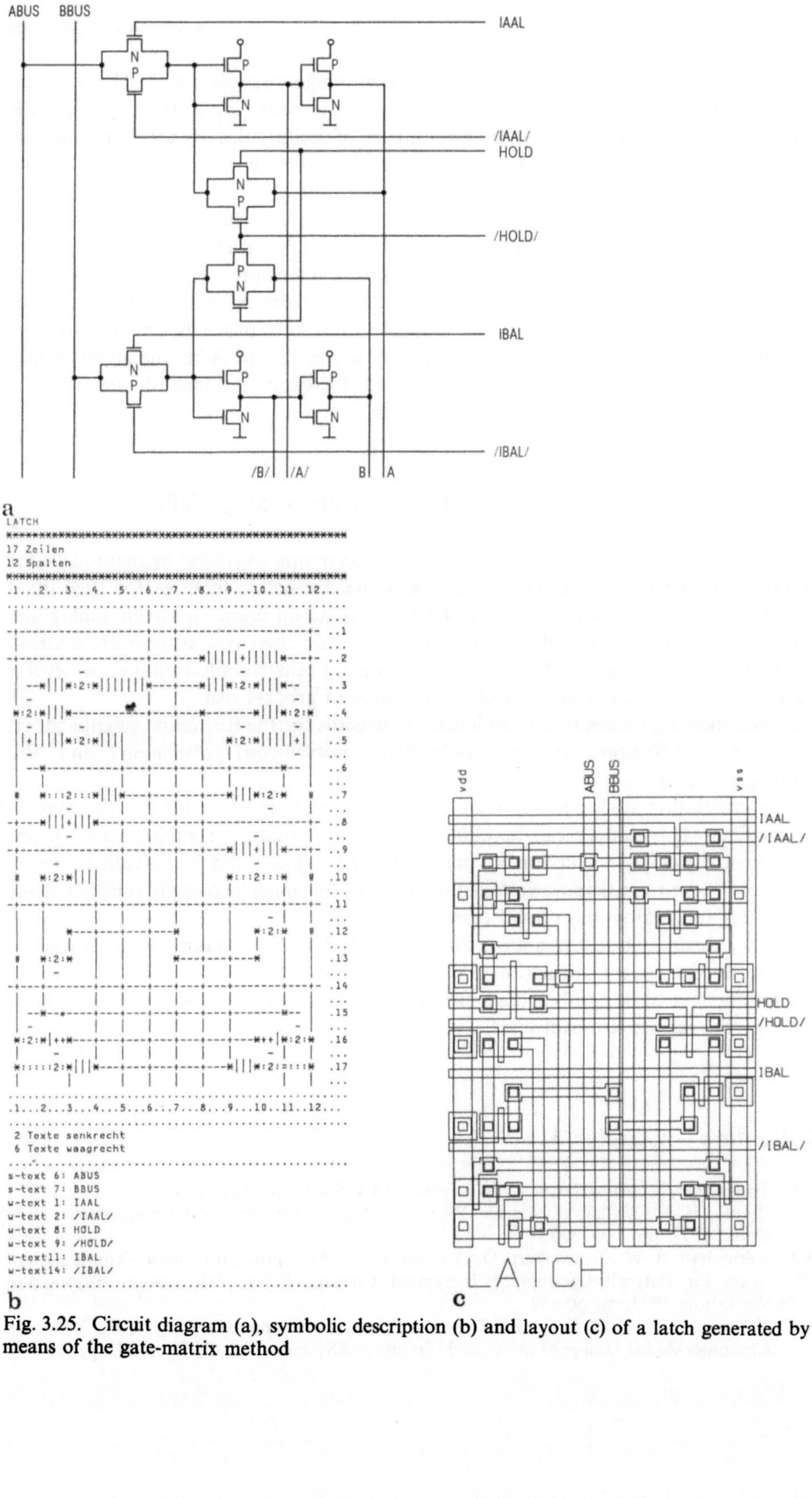

Fig. 3.25. Circuit diagram (a), symbolic description (b) and layout (c) of a latch generated by means of the gate-matrix method

similar to the stick diagram drawn, which allows a manual input check. The program translates the description into the final layout by means of a matrix design file containing the matrix parameter for a certain set of design rules. By means of that file, any design can be easily adjusted to other design rules. Figure 3.25 shows a latch generated by means of the gate-matrix method [3.10].

For this layout-generating method, it is interesting that the input of the symbolic description is not performed through a CAD terminal, but rather by means of screen oriented symbol editors at a low cost alphanumerical terminal. The symbolic input is subject to a syntax check. The faults shown can be corrected in the editor. Another program allows the representation of transistor circuits based upon the symbolic description which makes it possible to discover design errors. After the layout generation various processing programs are applied, for example those for generating a simulator input or a transistor circuit diagram.

3.8 Distinctive Features of the Layout with Analog Cells

To an ever increasing extent, analog cells are becoming available as macrocells for standard design processes. This results in some distinctive features for the circuit design. In terms of topology there is theoretically no difference between analog and digital macrocells: their outlines are rectangular, and their connections are defined. Because capacitors have to be generated for analog cells, an additional mask level is necessary, which makes the cost of the production process rise.

The following combinations of levels are feasible for the formation of capacitors: polysilicon/p-diffusion, aluminum/polysilicon, polysilicon1/polysilicon2 and aluminum1/aluminum2.

By integrating analog parts, one may induce a latch-up effect for the digital parts of a circuit. This danger can be eliminated by combining the analog parts in one section of the chip and separating them from the digital parts by so-called "*guard-rings*". In order to anticipate voltage fluctuations, one must also apply supply voltage separately for analog and digital circuit segments.

Other distinctive features arise during simulation. For analog circuit segments, only an electric circuit simulation is possible. For the simulation of the total circuit, the results have to be combined accordingly with those of the logic simulation on the gate- or the register-transfer level.

Literature – Chapter 3

3.1 Trimberger, S.: Automating Chip Layout. IEEE Spectrum, June 1982.

3.2 Breuer, M. A.: A Class of Min-Cut Placement Algorithms, Proc. 14th Design Automation Conf., 1977, pp. 284–290.

3.3 Kernighan, B. W.; Schweikert, D. G.; Persky, G.: An Optimum Channel-Routing Algorithm for Polycell Layouts of Integrated Circuits, Proc. 10th Design Automation Workshop, 1973, pp. 50–59.

3.4 Yoshimura, T.; Kuh, E. S.: Efficient Algorithms for Channel Routing. IEEE Trans. on Computer-Aided Design of Integrated Circuits and Systems, Vol. CAD-1, 1982, pp. 25–35.

3.5 Lauther, U.: Channelrouting in a General Cell Environment. Proc. VLSI Conf. 1985, Tokyo.
3.6 Williams, J. D.: STICKS – A New Approach to LSI Design. Master's Thesis, Massachusetts Institute of Technology, June 1977.
3.7 Williams, J. D.: STICKS – A Graphical Compiler for High-Level LSI Design. National Computer Conference, 1978.
3.8 Lopez, A. D.; Law H.-F. S.: A Dense Gate Matrix Layout Method for MOS VLSI. IEEE Journal of Solid-State Circuits, Vol. SC-15, No. 4, Aug. 1980.
3.9 Law, H.-F. S.: Gate Matrix: A Practical, Stylized Approach to Symbolic Layout. VLSI Design, Sept. 1983.
3.10 Lutz, W.: Softwarezellen für Datenpfadstrukturen. Diplomarbeit Technische Universität München, 1983.
3.11 Johnson, S. C.: VLSI Circuit Design Reaches the Level of Architectural Description. Electronics, May 1984, p. 121.
3.12 Wallich, P.: On the Horizon: Fast Chips Quickly. IEEE Spectrum, March 1984, pp. 28–34.
3.13 Collett, R.: Silicon Compilation: A Revolution in VLSI Design. Digital Design, Aug. 1984, pp. 88–95.
3.14 Gajski, D. D.: Silicon compilers and expert systems for VLSI. Proc. 21st Design Automation Conf, 1984, pp. 86–87.
3.15 Steinberg, L. I.; Mitchell, T. M.: A knowledge based approach to VLSI CAD. The Redesign System, Proc. 21st Design Automation Conf., 1984, pp. 412–418.

4 Test Concepts

With increased complexity, test technology has become a central field in the IC design process. Any efficient IC design system must be backed by well-founded test concepts and strategies, and it must supply the relevant tools.

This chapter first familiarizes the reader with the most important test concepts for integrated circuits. Then the required test measures are described for each phase of the IC design process. Already with system design it is important to consider a future test. Paragraph 4.1 gives an introduction to this field of problems. Paragraph 4.2 renders a survey on the methods which lead to a design for testability. In addition to methods that support a fully synchronized design, circuits which improve testability are presented, too. CAD tools for the methodical choice of an appropriate strategy leading to a design for testability do already exist in principle. They are the subject of Chap. 4.3. The features of a CAD system supporting design testability are deduced.

A large part of the design time is used for the generation of test documents (test program). This is where automation starts. The latter allows the time of test preparation to be reduced in spite of the increase in the components' complexity. A standardization by means of a design for testability is necessary.

In paragraph 4.4, the tools for producing test documents are presented. As example for the interlocking character of the single tools, test preparation is explained within the scope of the VENUS design sytem. In paragraph 4.5, the actual test of the components and the necessary test machines are described. Self-test processes are presented in paragraph 4.6 in order to illustrate future outlooks.

The terms "chip" or "element" usually describe an object to be tested.

4.1 Introduction to the Problem of Testing

While the cost of production is mostly stable, the cost of testing for highly integrated components rises when complexity increases. This is due to two facts: On the one hand the number of gates per component increases; on the other hand the number of pins per gate decreases. Because only the external pins for a component can be tested the number of internal states subject to a test sequence increases. This is why the test effort increases. This is called the increase of *logic* and *sequential* depth. Thus 2^{n+m} test patterns are required for a circuit in order to test all *m* states for *n* input variables, if there is no appropriate strategy available for reduction. One example: For a microprocessor with 25 inputs and 50 states the number of test patterns necessary for a complete test is 2^{25+50}. A tester with a cycle time of 10 MHz would need 10^8 years.

Tools suitable for reducing the number of test patterns while at the same time maintaining the quality standard are therefore absolutely necessary.

For elements of low complexity, the IC design engineer need only observe the boundary conditions of performance and space. Then, when the cell's complexity increases, he also has to pay attention to the boundary conditions of a later test. Cost of development and production as well as the product quality are effected by test time, test preparation and degree of testability.

With the normally low piece counts of semi-custom components the space minimization is a less important goal of the development process than the minimization of the design time. The design time is increasingly influenced by the effort spent for test preparation. Similar to the chip construction, test preparation also requires automation. This makes a certain degree of design standardization necessary and can be obtained by following test design rules for the components' development, and by applying testing aids to support the automatic process.

The term *design for testability* is applied if not only correct functioning but also a high degree of testability has been reached for the circuits. Generally, additional chip space is required for redundancies or modularity. This additional space has little effect on the total cost. The cost arising from an error which is only discovered during assembly or even system is considerably higher. For this reason a high rate of error detection has to be realized at the chips' level. Only then an optimum quality standard can be reached and costs minimized for the overall system.

4.1.1 Test Strategy

For a chip test, *test patterns* are required. They consist of the *input stimuli* and the output *target values*. For determining the target values at present, simulation models of the components are used, together with simulators. For testing, the actual responses of the component under test are compared with the target values. This comparison is performed automatically with the test machine (Fig. 4.1).

A fault at one of the internal circuit nodes is considered *recognizable* if the node can be set to the values 1 or 0 through the external inputs, and can be scanned in view of its effect arising at the outputs. This is termed *controllability* and *observability* of the internal node. If the logic is merely combinational each internal node can be controlled and scanned. Only circuit redundancies are exempt from this rule. The

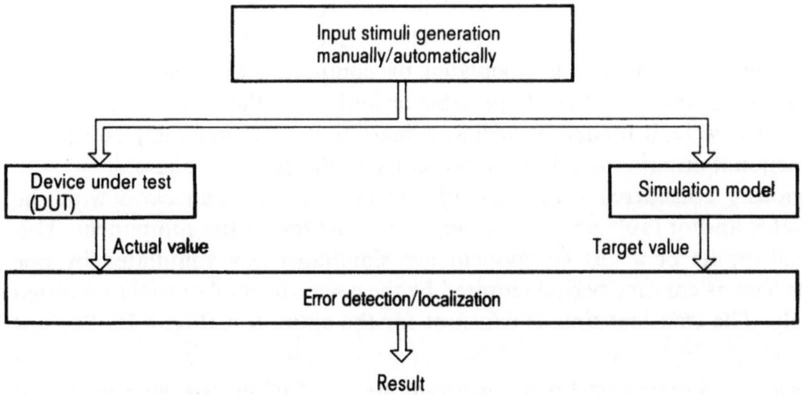

Fig. 4.1. General test principle

number of gates to be run through for control and observation determines the *logic depth* of the path that contains the node to be scanned. For a component with clocked storing elements several test patterns are required for the test (controlling and scanning) of one internal node. the number of clock steps required for the test makes up the *sequential depth* of the path. The *testability* is a result of the interaction of controllability and observability.

For improving testability additional circuits – *test aids* – have to be integrated in the component, in addition to its functional circuitry. The increase in complexity of the components leads to an increase of those cases in which a satisfying testability can no longer be guaranteed without appropriate test aids.

Input stimuli have to be generated in such a way and to such an extent that each internal node can be tested. Generally, three kinds of input stimuli can be distinguished:

- *Functional stimuli:* The IC design engineer manually generates input stimuli, based upon the functions of the test component. Each specified function can be addressed once by means of these stimuli. They are called functional stimuli. The more complex the component is, the more difficult is the manual generation.
- *Structural stimuli:* The input stimuli are automatically calculated by means of stimuli generators, according to the logic circuit structure. A number of algorithms has been developed for that purpose [4.1, 4.2, 4.3]. They are called structural stimuli. The component's complexity, above all the sequential depth, lessens the efficiency of the algorithms. By means of appropriate measures effecting the circuit's structure or by means of additional circuits, the efficiency may be improved again.
- *Functional-structural stimuli:* During the design process, functional stimuli are generated during simulation (verification); they may, however, not be sufficient for a complete test. But they can be supplemented by structural stimuli in an appropriate manner. Thus the third type is a combination of the two first types.

To control the quality standards, the *completeness proof* of the test patterns (fault coverage) is of major importance. For that purpose, *fault simulators* are applied (see Sect. 4.1.2). Calculations are performed for the internal node in order to find out if the fault condition "stuck at 0" and "stuck at 1" can be observed with the existing test patterns. This is called *"s-a-1/s-a-0" fault model*[1]. An fault model thus describes the type and number of the faults possible. The "s-a-1/s-a-0" fault model is most widely used. The fault coverage is calculated by comparing the number of faults detected with the number of all faults possible according to the basic model.

Concepts for new fault models as well as principles for proving completeness by means of functional criteria [4.4, 4.5] are presently in the research phase.

The previously considered utilization of test patterns in connection with the stimuli generator and/or fault simulator refers to a *static test* of the component. This means that all inputs of a test component are simultaneously stimulated by one pattern. After a transient time period required by the component all outputs are tested simultaneously. The transient time is constant for the entire test (Fig. 4.2). The test

[1] Instead of s-a-1/s-a-0 error model it is possible to use "s-a-1/0" or "s-a error model" as abbreviation

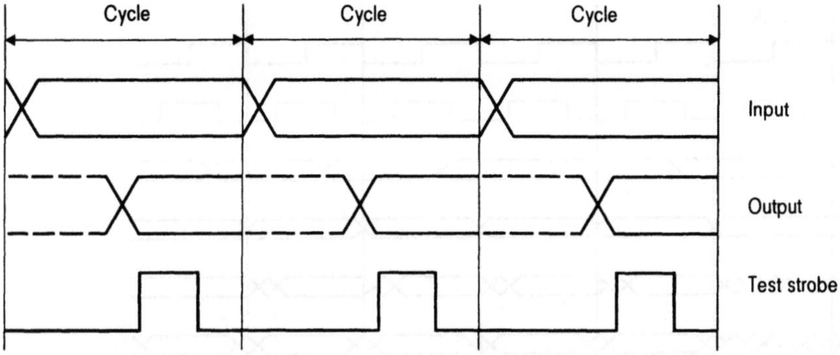

Fig. 4.2. Principle of a static test

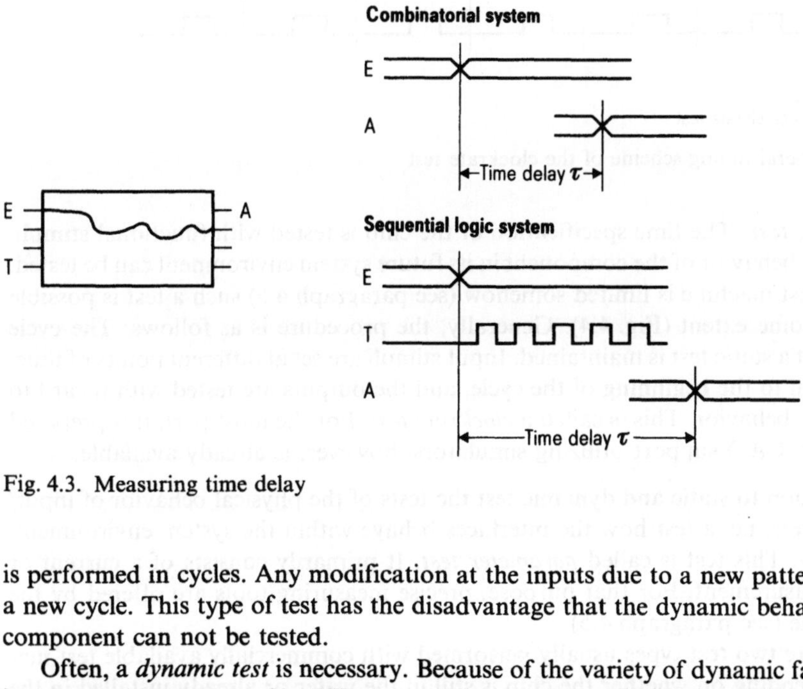

Fig. 4.3. Measuring time delay

is performed in cycles. Any modification at the inputs due to a new pattern leads to a new cycle. This type of test has the disadvantage that the dynamic behavior of the component can not be tested.

Often, a *dynamic test* is necessary. Because of the variety of dynamic fault behavior, a completeness proof is difficult to perform compared with the static test. Experience however proves that even a dynamic test which is performed at random drastically reduces the number of chips later found to be dynamically faulty in the system. Currently, two methods of dynamic test are applied in practice:

- *Time delay measurement:* The time delay from one input to one output is calculated as a target value and compared with the measured value by the test machine. The test is performed for several selected paths. If storing elements are included only the longest path (in terms of time) can be tested regarding its time behavior (Fig. 4.3). Basic methods for automatic stimuli generation with regard to sensibilization of the paths to be measured do already exist. A modification at the input of the paths induces a reaction at their output.

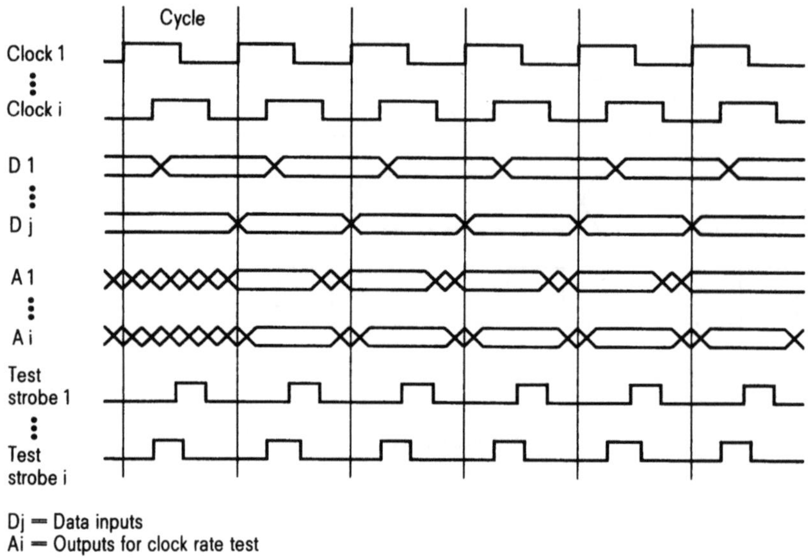

Fig. 4.4. General timing scheme of the clockrate test

- *Clockrate test:* The time specification of the chip is tested with functional stimuli. Thus, the behavior of the component in its future system environment can be tested. As any test machine is limited somehow (see paragraph 4.5) such a test is possible only to some extent (Fig. 4.4). Generally, the procedure is as follows: The cycle pattern of a static test is maintained. Input stimuli are set at different points of time, in relation to the beginning of the cycle, and the outputs are tested with regard to their time behavior. This is called a *clockrate test*. For the most part, it is prepared manually. CAD support utilizing simulators, however, is already available.

In addition to static and dynamic test the tests of the physical behavior of input/output drivers, i.e. a test how the interfaces behave within the system environment, is necessary. This test is called *parameter test*. It primarily consists of a current or voltage measurement. For that purpose, precise measuring tools are offered by the test machine (see paragraph 4.5).

There are two test types usually performed with commercially available test machines, depending on whether the chip is still in the wafer or already installed in the package. The component is tested several times. The earlier a test is performed in the life of the product, the more exact it is executed. First, the prototype is tested as "chip on the wafer" (*prototype test*). The error-free chips are cased and tested once more. If errors exist in the system, they have to be localized for a redesign. If the prototype is free of systematic errors, production and the subsequent *production test* are performed. Error localization is only required as an exception. Faulty chips due to production errors are discarded. Only design failures leading to a low yield are localized. This test is carried out manually by means of a wafer prober or an electron beam prober (see paragraph 4.5).

For the system engineer the component test is followed by a printed circuit board test and a system test.

4.1.2 Fault, Error, Failure, and Fault Model

The following example should illustrative the notions of fault, error and failure: a programmer's uptake is a fault; the consequence is an error in the written software (e.g. erroneous instruction); when this effective error produces erroneous outputs which are perceived as such, a failure occurs.

The two most important groups of faults are:

- design faults,
- production faults.

During any phase of the IC design process faults may occur, which leads to an error. This error remain latent as long as he is not activated. *Design errors* during system design and chip specification can be excluded by applying appropriate verification procedures. For a standard design system, the result of the logic design is the basis for test preparation.

It is important to remember: faults that remain undetected during the logic verification process will not be detected by means of test technology!

If logic and layout plan are generated manually, also faults in the electric circuit design and in the layout may occur. If a standard design system (such as VENUS) is used, the latter faults are impossible. The test technology need only be formed in such a way that production faults are found. Logic and layout faults are virtually impossible because experienced circuit designers have designed the elements of the cell library, have transformed them in silicon and measured them thoroughly. The cell library is subject to a strict long-term quality control. Thus, there are actually no design faults in the cells of standard design systems. Because of the strict application of the principle "correctness by construction", faults occurring during the automatic chip construction, too, are virtually impossible. The same applies to faults occurring during mask tape generation.

In the following sections, only *production faults* are considered. They result from contaminations in the process line or from other inadequacies of production technology. They are statistically distributed over the wafer. The technologic stability of the process line and the circuit complexity determines whether more or less chips on a wafer are fault-free. Their number determines the *production yield.* Faults which occur have to be allocated to different fault types. The fault types are described by means of the *fault model.*

The effects of a lot of fault types can be classified by means of the "s-a-1/s-a-0" fault model. But this is to be done with care: For CMOS components, short circuit faults between neighboring wires have the same effect as a "wired or" or a "wired and". Thus it can not be determined if the "s-a-1/0" error model is completely valid. This can mean that despite a 100% error recognition, in relation to the above mentioned error model, not every *short circuit error* is detected by means of the generated test patterns. But this is not very likely. Nevertheless, in more recently built fault simulators measures are already implemented for detecting short circuit faults.

Stuck-open faults (due to interrupts) in CMOS gates are even more problematic. While for bipolar circuits (ECL, for example) this fault type takes the same effects as s-a-faults, for CMOS a considerable part of faults can not be detected by means of fault simulation based upon this model. This type of fault which is new with CMOS

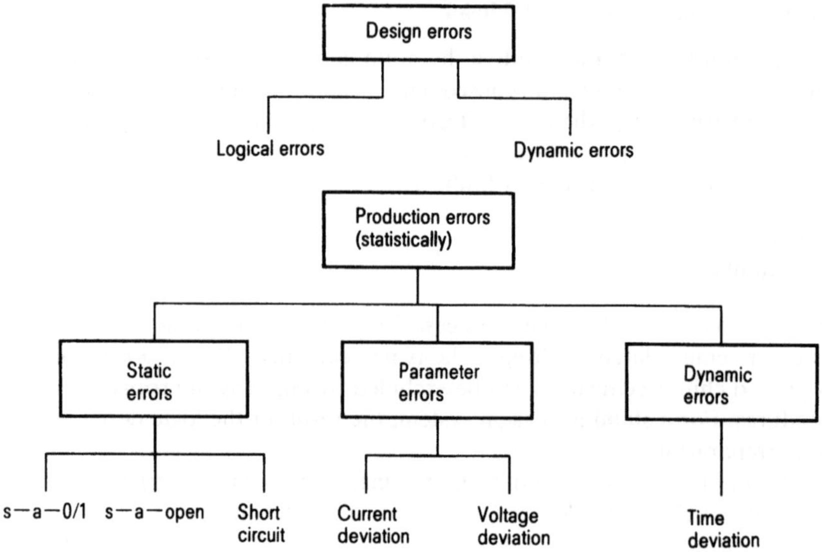

Fig. 4.5. Error classification

leads to a *specific sequential behavior* of the defective gate. That means that even for combinational logic a *sequence* of input stimuli is necessary for detecting the faults. The tests applied up to now detect this type of fault only by accident and incompletely. The derivation of suitable fault models is still in the research stage [4.6, 4.7]. To make matters worse, more complex models further raise the already high computing cost of a fault simulation. Moreover, the probability of such faults occurring has not yet been subject of a statistical survey. Another improvement in fault detection can be attained by introducing special design rules for layout generation which are designed in such a way that possible "open" errors can again be mapped on s-a-errors.

Figure 4.5 shows a fault classification. Here, design and production faults are first distinguished. By means of VENUS, components can always be tested with regard to static faults and parameter faults. A dynamic test can be performed at random in the form of a clock-rate test.

4.1.3 Phases of Test Technology in the IC Design Process

This book deals only with design technologies of integrated circuits. Nevertheless, we want to point out briefly the connections between the component's surroundings (PCB, system) for the test preparation. Then, we will describe the test technology phases in the IC design process. This aims at obtaining a closed test strategy across the levels, "system-PCB-component". If a hierarchical procedure is carried out the test aids of the components and PCB level can generally also be utilized for the system test.

If the design process is considered, the future test is prepared at each design level for the design levels which follow:

A *test strategy* must be selected during the architectural design. By determining the functional structure, modules that can be tested independently from each other can be set up by partitioning. These measures become more important as cell complexity increases. This is how one comes to the method of *hierarchical testing*: first the single modules are tested; then the test of the entire chip, based upon the results obtained in the first step, is carried out. At the architecture level, one has to determine how partitioning should be carried out and what parts of the modules should be subject to test.

During the logic design, the results determined at the architecture level are implemented. In order to reach a design for testability, some standard design systems contain a collection of *test design rules* which guide the designer, but do also restrict his freedom. These rules are available partially in formalized design rules and partially in the heuristic knowledge of the design engineer. One may, however, always define such a bundle of test design rules which guarantee that the chip is testable as a whole, and that after the logic verification not only the production-, but also the test-documents can be generated for the most part automatically. Standard design systems have this ability.

4.2 Design for Testability

The design for testability serves to guarantee the testability and to support the automation of the test preparation. Appropriate measures have to be executed for the architecture design as well as for the logic design.

All processes and methods of design for testability require additional chip space in order to improve testability. So-called test aids are installed. On the other hand, for complex chips, a reduction of design time (including the time needed for test preparation) while at the same time ensuring a high degree of quality is not possible without those test aids.

The design for testability requires standards on all design levels. For the design engineer who applies standard design systems, only the levels of the architecture and logic designs are important.

In the following paragraphs appropriate test aids for architecture and logic design are presented through examples.

4.2.1 Design Utilizing Test Design Rules

In standard design systems for integrated circuits, the basis for the automation of the test preparation is set up by means of *test design rules*. As with the concepts used for standardizing chip design which are the basis for the cell-oriented design methods a design standardization on the logic design level is recommended for test technology, too. For this purpose, mandatory test design rules are set up. Those relevant for VENUS will be examined more thoroughly here. They allow the realization of the following objectives:

- guarantee of completeness of testability,
- improvement of the dynamic noise immunity,
- restriction of the test preparation effort,
- improvement of the design's clarity,

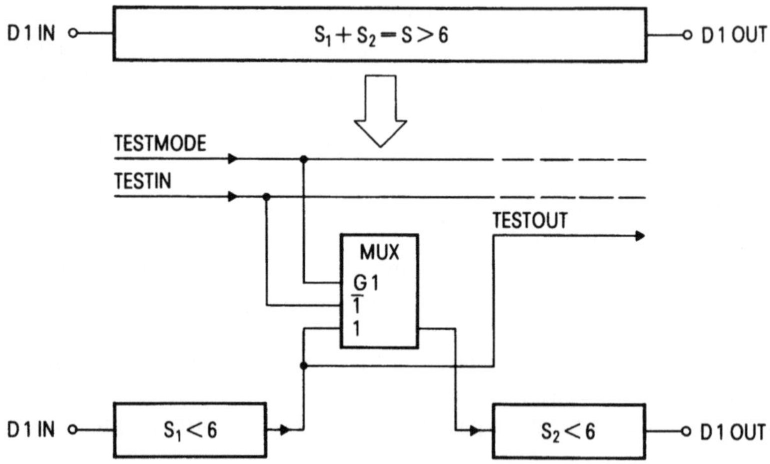

Fig. 4.6. Multiplexer used for limiting the sequential depth

- methodical design instructions for the chip designer concerning the installment of test aids,
- adaptation to the test system integrated in VENUS.

The basic principle for the test design rules is: *securing the fully synchronous design:* the internal logic states can be changed only at fixed predetermined times in accordance with a clock rate pattern.

This achieved by

- the use of edge-triggered storing elements,
- restrictions for the clock activation by data signals,
- strict separation of clock and data,
- independent controllability for internally generated clocks in test mode.

Furthermore, the automatic test pattern generation is supported by limiting the depth sequential. By applying multiplexers and additional test inputs and outputs it is always possible to adhere to the rule of the "limitation of sequential depth" (Fig. 4.6). For active TESTMODE the path S_2 is set by means of TESTIN and the path S_1 is observed by means of TESTOUT.

With regard to complex chips, it is difficult for the IC design engineer to control the adherence to the rule. This is why an *automatic rule check* is necessary as part of the logic verification. Paragraph 4.3.2 describes a software tool which checks the adherence to all rules referring to the logic plan of the circuit, and gives the design engineer advice on how to avoid violations. The test design rules listed in Chart 4.1 are established in the VENUS design system. They are *applicable for CMOS technologies as well as for ECL technologies*, as they primarily refer to the logic structure of the circuit.

In the following paragraph, the rules mentioned above are explained in more detail.

Chart 4.1. General test design rules of the design system VENUS

rule 1:	Every pad of the circuit has to be connected directly with a chip (casing's) pin
rule 2:	In the circuit design, logic and circuit redundancy is to be avoided
rule 3:	A defined initial test state of the circuit has to be externally controllable
rule 4:	Generation and deduction of pulses must not be performed by means of circuit-internal delays.
rule 5:	For the circuit test, clock are required which can be controlled externally and independently from each other in TESTMODE.
rule 6:	Circuits which may generate noise on clock lines are not permitted.
rule 7:	Clock and data have to be sent from the component's pins to the cells' inputs on separate paths.
rule 8:	In the TESTMODE, a maximum of six storing elements are allowed to be in one data path between input and output pin.
rule 9:	In the TESTMODE, the feedback part of a data path may include a maximum of two storing elements.
rule 10:	Feedback is not allowed in merely combinatorial circuit sections.

Rule 1

A cricuit pad must always be directly connected with a component pin since for wafer and component the same test program is used. In the event that this rule is not followed, a scond test program must be generated manually. An automatism, based upon a correct circuit description, can no longer be guaranteed.

Rule 2

Logic redundancy complicates automatic stimuli generation, and it lowers the degree of error detection. More severe effects are caused by circuit redundancies. If, for example, line breaks occur in elements connected in parallel, the circuit may work statically. The dynamic behavior, however, is defective which is often detected only in the system test, thus effecting an accordingly high cost.

Rule 3

In order to set a defined initial test state, either a central, externally controllable SET/RESET signal or short initialization routines are necessary. The generation of test patterns and the test itself is facilitated by a fast SET/RESET.

Rule 4

The generation of impulses by means of monoflops or their delay, shortening or extension must not be used for the realization of the circuit function. The resulting time behavior would then depend on time delay variations and the circuit's geometry. The geometric arrangement itself, however, can not be predetermined if placement and routing is carried out automatically. Moreover, the dynamic behavior of the circuit would depend on temperature and supply voltage, which could be tested only at random.

Rule 5

In order to back the dynamic behavior of a chip, it is necessary to test the SET-UP- and HOLD-times of clocked storing elements. For an automatic stimuli generation of such a test one has to distinguish between clock and data in order to apply them in proper timing relation to each other. In the TESTMODE, therefore, all chip-internal, deduced clock signals have to be directly adjustable by means of an external pin. For a test which, in the dynamic case, is limited to the measuring of SET-UP and HOLD times, supplying the same clock to all storing elements is insignificant in TESTMODE. Figure 4.7a provides an example of the procedure for a circuit with central clock generation. In TESTMODE, all storing elements are triggered by the same clock (TESTCLOCK) by means of additional multiplexers. The clock deduction outputs can be examined through external test outputs. In VENUS, the model for test pattern generation is prepared accordingly. Clock generating circuits are, therefore, arranged centrally, if possible.

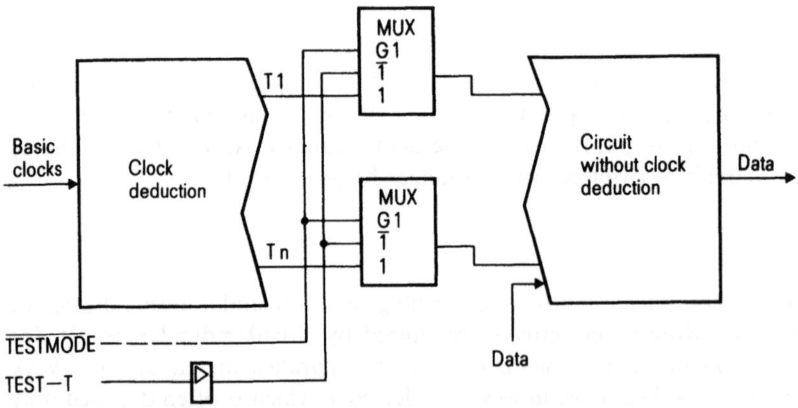

Fig. 4.7a. Separation of clock generation

Rule 6

Outside of phase generating circuits, clocks may be addressed or blocked by means of data signals only by certain circuits with combinatorial logic (see Fig. 4.7b). In this context, one must make certain that the control data is not subject to modifications during the active clock phase. For an AND combination of clocks, can no longer be controlled independently from each other. In addition, a shortening of impulses caused by time delay variations may lead to faults.

Rule 7

Only when clock and data are strictly separated from each other can a synchronous operation be realized. With the asynchronous circuit behavior which would otherwise result, unwanted dynamic effects can barely be limited. A dynamic test would become necessary for which the completeness proof could not be supplied. Figure 4.7c supplies some examples of rule 7 violations.

Rule 8

It is often impossible to generate stimuli automatically for circuits with high sequential depth. This results in a time-consuming manual correction process. Practice shows that a maximum sequential depth *S* of 6 can still be handled in a data path of stimuli generators.

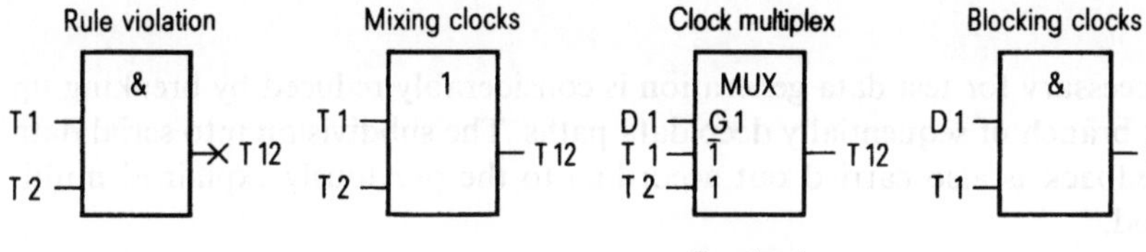

Fig. 4.7b. Combinatorial clock modification by elements

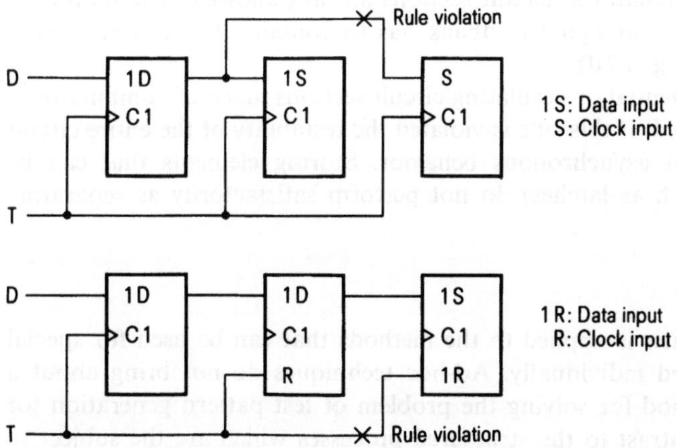

Fig. 4.7c. Examples for violations of rule 7

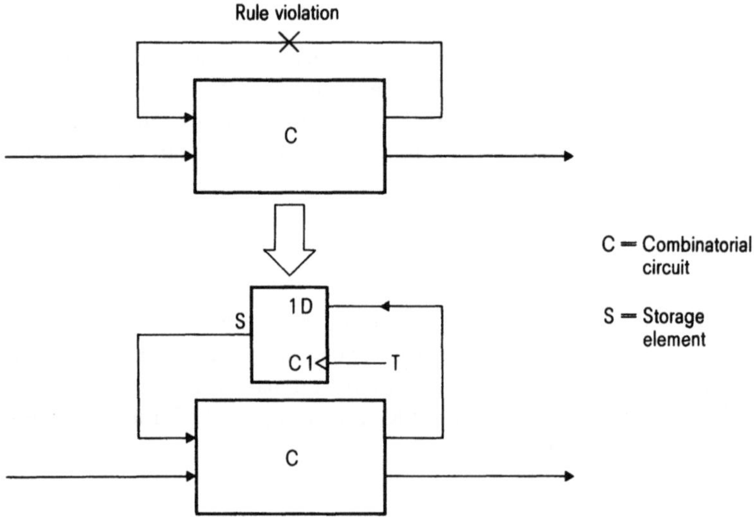

Fig. 4.7d. Break-up of combinatorial feedbacks

For that reason, data paths with $S > 6$ are to be divided into externally testable path sections where $S \leqq 6$. The principle of reducing the sequential depth by means of multiplexers has already been shown in Fig. 4.6.

Rule 9

The effort necessary for test data generation is considerably reduced by breaking up the feedback branch of sequentially deep data paths. The subdivision into serial data path and feedback is also carried out according to the previously explained multiplexer method.

Rule 10

Feedbacks in purely combinatorial circuit sections are not allowed. The loop consisting of a forward path and signal feedback has to contain at least one storing element for separation (Fig. 4.7d).

The formation of sequential or oscillating circuit sections made of combinatorial elements must be prevented. If this rule is violated the testability of the entire circuit is endangered because of asynchronous behavior. Storing elements that can be switched transparent (such as latches) do not perform satisfactorily as separators when used alone.

4.2.2 Ad-Hoc Techniques

The term ad-hoc techniques is applied to the methods that can be used for special designs and are developed individually. Ad-hoc techniques do not bring about a generally applicable method for solving the problem of test pattern generation for sequential circuits, in contrast to the structured processes which are the subject of Sect. 4.2.3. They only facilitate the process of problem solving. They are adjusted individually to the circuit in question; for that reason they require less additional silicon space. On the other hand, it is difficult to implement them in a CAD system in such a way that they are supported automatically. Moreover, they involve considerable disadvantages for CAD tools used for the automation of test preparation.

For the ad-hoc techniques now presented a circuit is always separated in such a way that the formed blocks are simple and can easily be manipulated. Processes based upon these stipulations are often grouped under the term "partitioning techniques". This partitioning process may be carried out according to the following guidelines:

- Partitioning in functional modules such as ALU, operation unit, control unit.
- Partitioning in parts that can be generated systematically such as sequential logic system, switching network.
- Partitioning according to the logic structure.

The common basis for the partitioning process is the use of multiplexers and bus systems. The details of the individual measures depend on the function plan and the objectives for the relevant chip.

The generation time for test patterns increases cubically with the number of gates. The division of a circuit in halves which can be tested individually results in the reduction of generation time to a quarter. When the systems subject to test were still being built up on a pc board, partitioning could be realized comparatively easily by means of mechanic tools. If, however, the system is completely integrated on a chip appropriate logic gates have to be used in the partitioning process. Besides this additional effort, supplementary external pins have to be provided which carry information about the actual mode (test or function a mode), or for setting and observing chip-internal circuit nodes. For integrated circuits, however, additional pins are quite expensive.

The circuit design engineer may determine the steps by which the partitioning is to be executed. He usually follows the partitioning specified by the function. Alternative partitioning may be developed by means of testability analysis (see Sect. 4.3.1) or by means of stimuli generations which have been executed experimentally (see Sect. 4.4.1), aimed at finding the circuit nodes which are most difficult to be tested.

The following examples illustrate the possibilities of partitioning. Figure 4.8 shows two modules which are integrated on one component and communicate with each other. By installation of multiplexers, both modules can be tested independently from each other. With increasing width of the interface between the two modules, the waste of space and number of test pins rises, too. Another measure of partitioning is shown in Fig. 4.6. It aims at improving the structure in order to support automatic test pattern generation. Paths with high sequential depth make generation more complex or may even prevent it under adverse conditions. Such paths can again be segmented by means of multiplexers. The model for the generation has to be modified in such a way that, instead of the multiplexer, one input and one output are inserted.

A special case for partitioning is provided by the bus architecture. The degree of testability can be raised quite easily if either a bus architecture already exists because of the function required, or if such a bus architecture can be used. A bus can be described as a bundle of signals connecting many circuit sections. A control system determines which part is used as sender, which as receiver. In order to use the bus architecture as ad-hoc technology for improving testability, the buses must first be

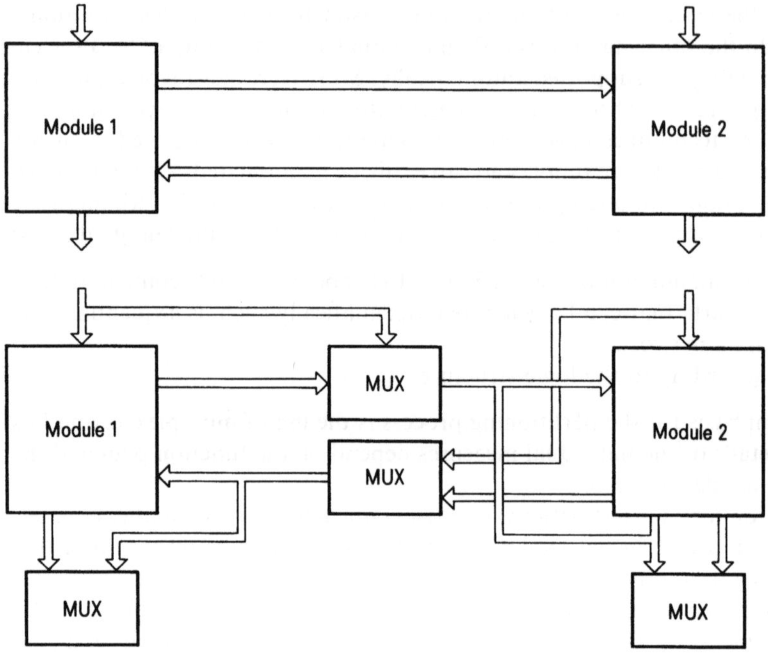

Fig. 4.8. Partitioning of two modules in one component

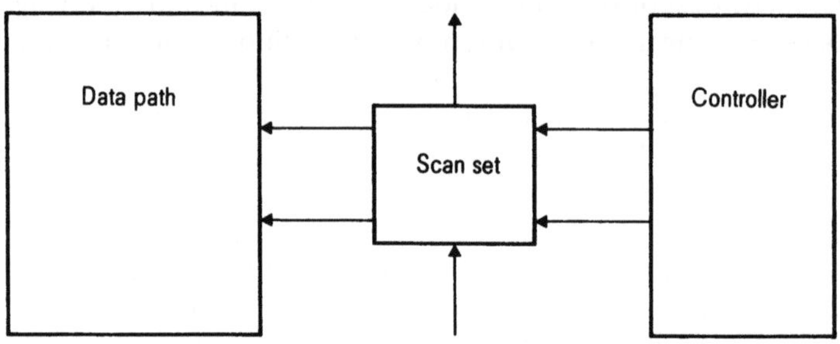

Fig. 4.9. Scan-set used for partitioning of data path and controller

externally accessible and second, by additions in the control system, one must ensure that the test machine can individually assume the function of *sender* and *receiver* for each circuit section. If these requirements are met, any circuit section can be tested independently from any other circuit section.

Another possibility of partitioning on the functional level arises through the use of shift registers (scan-set principle). Internal functional units can be set or scanned by means of serial shifting. In contrast to a scan path (see Sect. 4.2.3.1) which embeds the existing flipflops, functional flipflops are hardly used for a scan-set. A typical application of a scan-set is the partitioning of data path and controller for processor components (Fig. 4.9).

4.2.3 Structured Methods

For standard design systems the shortening of design time by means of automation is the most-desired objective. The ad-hoc technologies described are suitable only to some extent for a closed system, because of their individual character. Most appropriate are structured methods which can be applied independently from the respective function and the overall objective of the component, and which can be integrated automatically into the design system. Two important types of methods are:

- the scan-path methods,
- the random-access-scan methods.

In the following the scan-path methods is described in detail. It is supported in the VENUS design system. The random-access-scan method is only discussed briefly.

4.2.3.1 Scan-Path

The scan-path provides the opportunity of transforming a FSM into a combinatorial circuit. For that purpose, all storing elements are extended in such a way that they can be combined serially in a special operating mode and work as shift registers. Thus it is made possible to carry information to internal circuit nodes by serially inputting them, and to observe their reaction by serially shifting them out. The test of combinatorial and storing parts can be executed independently. This results in a simplified test pattern generation as only combinatorial logic has to be considered. According to the "s-a" error model, a 100% fault coverage can be obtained with automatically generated test patterns. For this method, it is important that first, by means of the CAD system, the scan path is automatically integrated into the circuit, and second, the

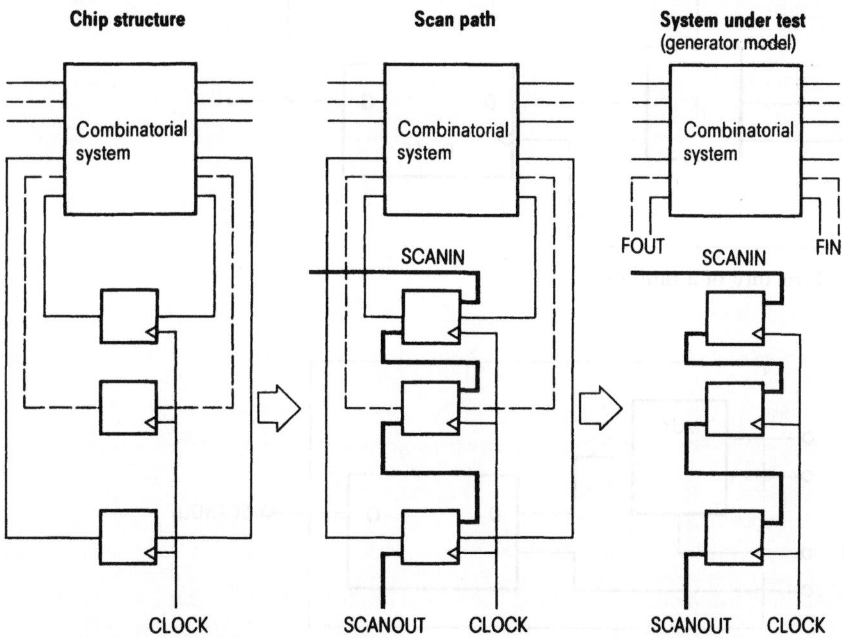

Fig. 4.10. Introduction of a scan-path

generation model for test patterns is automatically deduced. Figure 4.10 illustrates this procedure. For each storing element, a fictitious input/output is introduced for the generation process.

Scan-Path Flipflop

The scan-path concept depends on the type of storing elements (flipflops) used. One has to distinguish between level sensitive phase and edge triggered flipflops.

For state-triggered flipflops (latches), the LSSD principle ("*l*evel *s*ensitive *s*can *de*sign") is applied. In VENUS, this principle is meant for circuits with a majority of state-triggered flipflops. The different types of hardware realization have an effect only in the model preparation but not in stimuli generation itself. The concepts are independent of the technology used.

As an example for the high number of scan-path concepts [4.8, 4.9], the concept is presented, where primarily edge-triggered flipflops are used in the circuit. All flipflops of this type are built according to the same scheme (Fig. 4.11).

A master-slave flipflop is used as basic circuit. The type of stored element is determined by the supplementary combinatorial system C. From this basic circuit all types (such as RS-, JK- flipflops, counter, shift register etc.) are deduced. For extension to the scan-path flipflop, only the D-master-slave part must be considered. It is supplemented by a multiplexer which activates either the data or the SCANIN input by means of a TESTMODE signal (Fig. 4.12). The method that is described here is based upon a completely synchronous design of the circuit. The SCANOUT and

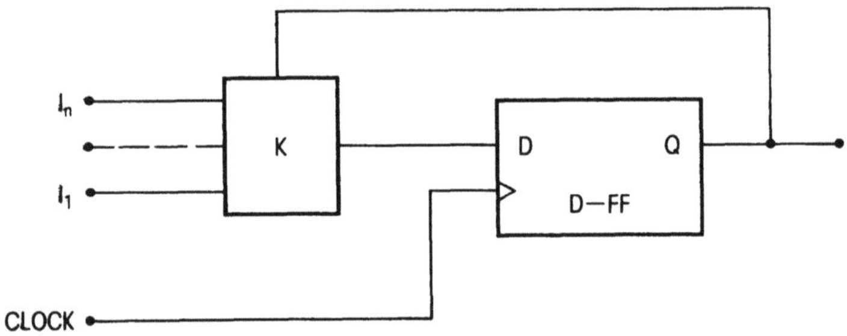

Fig. 4.11. Structure of a flipflop

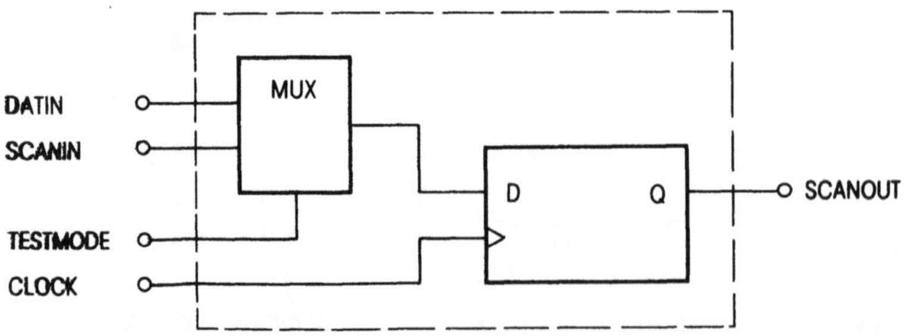

Fig. 4.12. Scan-path basic cell

Chart 4.2. Test design rules for the scan path

rule 1:	In TESTMODE, a strict separation of clock and data must be ensured.
rule 2:	In TESTMODE, each cell must be supplied with the same clock
rule 3:	All storing elements must support the scan-path and must be embedded in the scan path.
rule 4:	Feedback in combinatorial circuit sections is not allowed.

SCANIN pins of storing elements are connected and such a shift register is formed which is extended over the entire chip.

Scan-Path Design Rules

In order to secure the shift function (system clock = shift clock) and properly establish the test patterns, the design engineer has to adhere to special design rules (see also 4.2.1):

Layout Generation

The introduction of a scan path with its additional pins sets up new restrictions for the standard design systems, as far as automatic placing and routing is concerned. For optimization, the procedure is as follows:

- In the first step, the library cells are placed and wired; the wiring of the scan path is omitted.
- In the second step, the scan path is wired. Because the order of scan-path cells is random, they may be connected in compliance with the topography found in the first step. Only the expansion of some wiring channels for the TESTMODE and SCANIN/OUT lines may be necessary.

Circuit Substitute of the Scan-Path Cells for Test Pattern Generation

For the generation of test patterns the scan path (see Sect. 4.4.1) leads to a pure combinatorial circuit. By means of the scan path, the inputs of the combinatorial logic can be set and their outputs observed. For the generator, the model of a scan-path flipflop is simplified according to Fig. 4.12. Each flipflop results in one additional input/output (FIN, FOUT) which is treated like external pins by the test pattern generator. This results in short combinatorial paths and, consequently, in short generation periods.

Test Procedure at the Test Machine

The actual test with the test machine is divided into two parts:

- First, by shifting in and out various bit sequences, the storing capability and correct functioning of the scan path is tested.

- Next comes the test of combinatorial logic. For this purpose, the test patterns produced by the generator are used. A test cycle is formed as follows:
  - in the TESTMODE, a test pattern is shifted in, and the external inputs are stimulated,
  - in the operating mode, the response of the combinatorial logic in stable state is clocked into the scan path,
  - again in TESTMODE, the information is shifted out and compared with the target values.

4.2.3.2 Random-Access Scan

Another structured method to improve testability is provided by the random-access-scan principle [4.10, 4.11]. As with the scan-path principle, the circuit is divided into combinatorial system and storing elements. The generator for the test patterns again "sees" only the combinatorial logic. The substitute model is similar to the one used for the introduction of a scan path.

All storing elements are combined in a memory field in TESTMODE. Through two decoders for X and Y address lines any flipflop can be directly addressed. The setting of a flipflop is executed by means SET and RESET lines. Figure 4.13 shows the structure of a flipflop for the random-access-scan method. All SCANOUT signals are connected by means of a NAND and can be observed by means of an external output. Figure 4.14 shows the logic structure of the circuit, including the division into combinatorial system and memory field.

A test step for the combinatorial system is executed as follows:

- All external inputs are set.
- The memory field is RESET.
- Through the relevant SCAN addresses and simultaneous activation of the SET signal, the scan-information calculated by the generator for that test step is stored in the memory field.
- After the transient time of the combinatorial system, the result is stored in the memory field at a system clock.
- The result is checked at the SCANOUT output by applying the relevant scan-addresses.

For the random-access-scan method there are a number of modifications aimed at the reduction of additional space pins. Instead of the decoders, for example, one

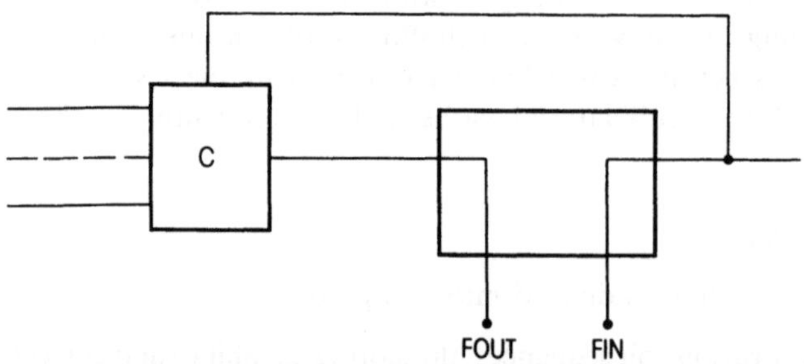

Fig. 4.13. Generator model of a scan-path flipflop

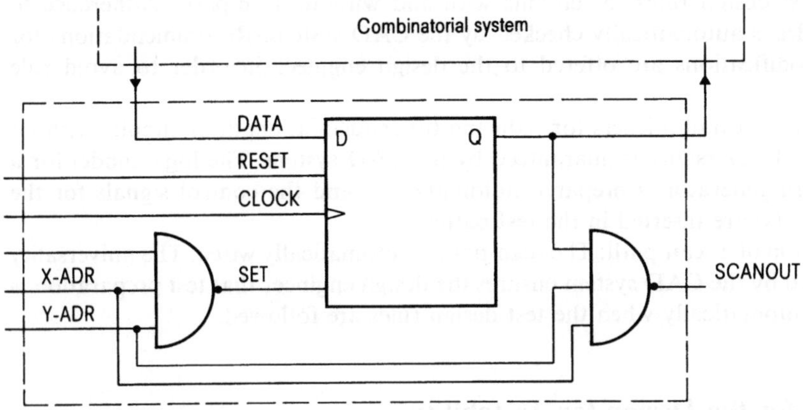

Fig. 4.14. Flipflop for random-access-scan

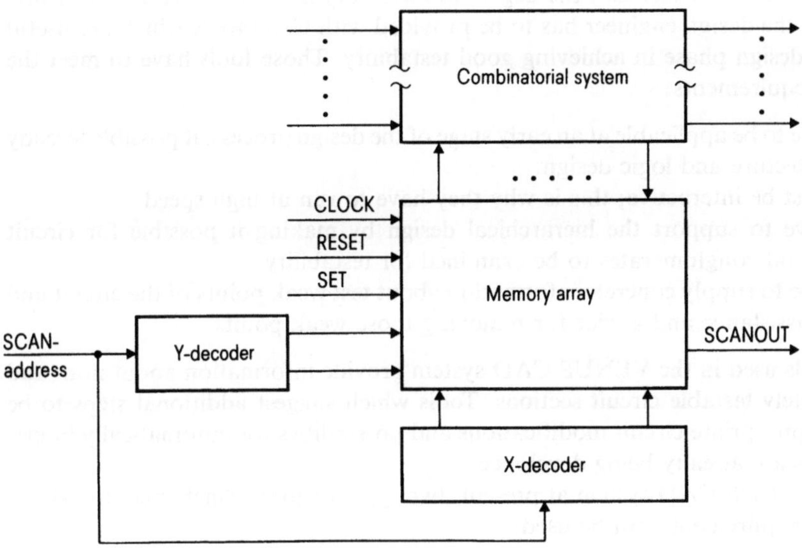

Fig. 4.15. Logic structure for random-access scan

could use shift registers, into which the previously decoded scan-address is inserted through external pins. This method seems to be especially suitable for processor chips for which the implementation of a memory field is simplified because of its regular structure. Otherwise, the scan-path method should be preferred because of its higher degree of flexibility and better space utilization.

4.2.4 Design for Testability with VENUS

With the VENUS CAD system, the following steps of design aimed at a design for testability are supported:

- Use of test design rules for circuits with and without scan-path: Adherence to design rules is automatically checked by the CAD system. Recommendations for circuit modifications are offered to the design engineer in order to avoid rule violations.
- Introduction of multiplexers for reducing the sequential depth of circuits without scan-path: Universality is guaranteed by the CAD system. The logic model for a test-pattern generator is prepared automatically, and the control signals for the multiplexer(s) are inserted in the test pattern.
- Introduction of a scan-path: The scan-path is automatically wired. The universality guaranteed by the CAD system ensures the design engineer that test preparation is realized automatically when the test design rules are followed.

4.3 Tools for the Design for Testability

The difficulty of obtaining a design for testability when circuit complexity increases is partially due to the fact that the degree of testability is more difficult to measure. This is why the design engineer has to be provided with CAD tools which are useful during the design phase in achieving good testability. Those tools have to meet the following requirements:

- They have to be applicable at an early stage of the design process, if possible already for architecture and logic design.
- They must be interactive; this is why they have to run at high speed.
- They have to support the hierarchical design by making it possible for circuit sections and conglomerates to be examined for testability.
- They have to supply concrete information about test weak points of the circuit and provide assistance and advice for removing those weak points.

The tools used in the VENUS CAD system provide information about non- and not completely testable circuit sections. Tools which suggest additional steps to be taken for appropriate circuit modifications and possibilities for automatically inserting test aids are already being developed.

In the VENUS CAD system at present, two types of tools which meet the above mentioned requirements can be used:

- tools for *testability analysis*,
- tools for *test design rule checks*.

4.3.1 Testability Analysis with VENUS

The testability analysis provides a measure for controllability and observability.

The measure for the controllability indicates the effort necessary for setting any node of the circuit to a certain logic value. It depends on the number of gates which must be run through in order to transmit a change of logic state from the external inputs to the selected node.

The measure of observability indicates the effort necessary for checking the set logic value of a node through an external output. It again depends on the number of nodes to be run through until this output is reached.

In the following, the computation algorithm is described as an example [4.12]. For each circuit node N the following values are computed:

CC0/1 (N): combinatorial controllability to 0/1,
SC0/1 (N): sequential controllability to 0/1,
CO (N): combinatorial observability,
SO (N): sequential observability,

The following equations for the controllability are for a 3-input NOR gate:

$$CC1(Y) = CC0(X1) + CC0(X2) + CC0(X3) + 1,$$
$$SC1(Y) = SC0(X1) + SC0(X2) + SC0(X3).$$

For setting the output Y to 1, the inputs X1, X2, X3 have to be set to 0. The addition of 1 is due to the combinatorial depth 1 of the NOR gate.

Use the following equations if Y is to be set to 0:

$$CC0(Y) = \min[CC1(X1), CC1(X2), CC1(X3)] + 1,$$
$$SC0(Y) = \min[SC1(X1), SC1(X2), SC1(X3)].$$

When setting Y to 0, it is sufficient to set one of the inputs X1 – X3 to 1. For this reason, the input that can be most easily set is used (minimization principle).

For testability analysis, the values for the controllability of each single node are calculated, starting from the external inputs. Sequential elements increase the sequential depth by one.

The following equations for the observability of one of the inputs (the input X1, for example) are for a 3-input NOR gate:

$$CO(X1) = C0(Y) + CC0(X2) + CC0(X3) + 1,$$
$$SO(X1) = S0(Y) + SCO(X2) + SC0(X3).$$

For observation of X1 one has to observe Y, as well as set X2, X3 to 0.

For the entire circuit, the values of each internal node are calculated according to the example given, starting from the external nodes.

The result of the testability analysis is available in a chart where the values which have been calculated for each node are indicated. If a node is difficult to access, this can be seen in the chart. Through appropriate circuit modifications, the controllability/observability can be improved.

By iterative application of the testability analysis, the efficiency of the test aids is systematically checked.

By means of a testability analysis all types of testing technology, including the ad-hoc technologies described in paragraph 4.2.2, can be supported for a design for testability.

4.3.2 Test Rule Check with VENUS

Another CAD tool useful for obtaining a design for testability is the *test rule check*. It also helps to find circuit sections which cannot be tested at all, or at least not easily. Here the steps to be taken are different from those described in paragraph 4.3.1.

The design engineer's work is based upon the test design rules described in paragraph 4.2.1. By applying these rules, he is lead to a testable circuit. If those rules are adhered to, this results in circuits for which test patterns can be generated automatically.

The measure of testability is the *number and kind* of the *rule violations* appearing in a circuit. The CAD program for rule check requires the net list of the circuit and additional information on the cells. This additional information is stored in a cell library. With this data, a structure analysis of the circuit is performed with graph-theoretical algorithms used for path search and path sensibilization.

With the help of the path structure and the cell types of the path, *rule violations* referring to the *structure* can be found. Rules referring to the *time behavior* cannot be checked directly. From the structrure of the circuit, however, one may draw some conclusions concerning time-critical paths and circuit sections (asynchronisms).

The result obtained from the rule check is a list containing the location and type of violations.

Thus, the design engineer obtains information which he can use for modification of the circuit in order to remove the rule violations, so that the testability is increased. The test rules provide assistance concerning the choice of modification.

For further development of this method, it is intended, by means of the CAD system, to have test aids automatically integrated in the existing circuit which remove rule violations.

An advantage of a method supported by rules and rule check is the support of the hierarchical design. Circuit sections can be checked for compliance with the rules. In this case, the design engineer has to see to it that he connects the individual circuit sections properly in order to form the overall circuit.

4.4 Test Data Generation

The generation of test data is carried out basically in three steps:

- Generation of the input stimuli,
- Generation and evaluation of the test patterns,
- Generation of the test program.

At the beginning of microelectronics the three steps mentioned were realized manually. With the high complexity of components, a far-reaching automation of test data generation is necessary. In the following the reader is made familiar with CAD tools which meet this requirement. The basis for this automation is the component's design for testability.

4.4.1 Generation of the Input Stimuli

Various algorithms are known [4.13, 4.14, 4.15] which more or less effectively generate input stimuli for a complete test. They primarily attempt a static test. With regards to the complete test, for purely combinatorial circuits the methods for stimuli generation are considerably more efficient than for sequentially deep circuits. For sequentially deep circuits, a sufficiently high degree of error detection can seldom be reached without manual intervention. The transformation of a circuit into a purely combina-

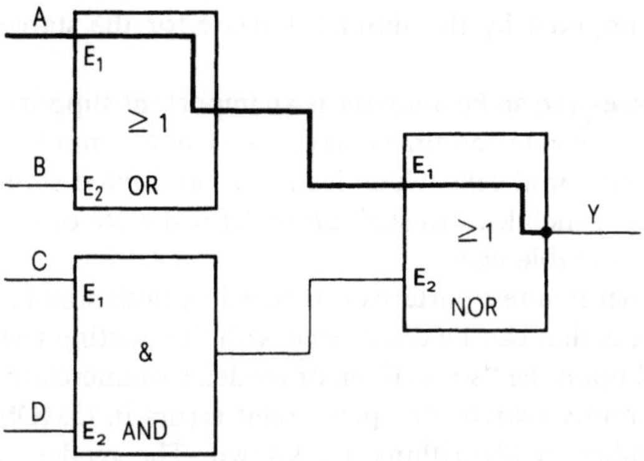

NOR	E_1	E_2	Y
	0	0	1
	1	0	0
	0	1	0
	1	1	0

OR	E_1	E_2	Y
	0	0	0
	1	0	1
	0	1	1
	1	1	1

Fig. 4.16. Example for path sensitization

torial system, with the scan-path method, for example, is useful for complete automation while simultaneously maintaining a high quality standard.

In the following, the principle of generation for combinatorial circuits is explained through an example. A common method is the D-algorithm [4.15]. The majority of the currently available stimuli generators use algorithms which are derived from the D-algorithm. A logic model described on gate level is used as the basis, i.e. only the circuit's structure is considered but not its function, as it is common for manual stimuli generation.

The generator tries to sensitize paths, starting at a circuit output and continuing towards the circuit inputs. A path is *sensitized* if a change of state at its starting point (input) can be observed at the pertinent ending point (output). To reach that goal, one has to compute stimuli for those inputs on which path sensitization depends. Every alteration of the target setting of any node on a sensitized path according to the "s-a" error model leads to a modification of the target value at the external output.

The circuit example of Fig. 4.16 illustrates path sensitization for combinatorial circuits. The bold path has to be sensitized. The generator begins at output Y of the path. In order to make a modification at input E1 of the NOR gate visible at the output of the NOR gate, E2 has to be set to 0 (see truth table NOR). To obtain this value one has to set at least one input of the AND gate to 0. The second input is optional. To further sensitize the path up to its starting point A, the OR gate at its input E2 has to be set to 0. A modification at A thus leads to a reaction at Y if the external inputs are set as follows:

B = 0, C = 0, D = 0 or 1.

With sequential deep circuits, generation is complicated because clock phases for several time levels have to be considered for the sensitization.

4.4.2 Generation and Evaluation of the Test Patterns

For manually or automatically generated input-stimuli, the pertinent output target values as well as the degrees of error detection are computed.

The output target values are ascertained by means of cycle simulations. This is based upon the same model description as for generation, but this time with time delays included. The target values computed by the simulator relate for the stable condition of the circuit.

Some cycle simulators provide a race- and spike analysis as an important support. This examines whether, because of varying gate and time delays (race), noise impulses (spikes) occur at the outputs of combinatorial cells which lead to a prohibited state change of a memory cell. Test patterns which lead to such an undefined state of the circuit have to be complemented in a suitable way.

The completeness proof for the generated test patterns is realized by fault simulation. The number and the kind of errors that can be discovered with the existing test patterns are determined. This is based upon the "s-a-0/1" error model. Commercially available fault simulators do not currently search for open circuit errors in CMOS circuits. For fault simulation, a number of algorithms are known. The reader is referred to the relevant literature [4.16, 4.17, 4.16]. All algorithms are based upon the principle, that one or several faults are installed in the circuit model in parallel. Then, one carries out the simulation process in order to see if those faults lead to a modification of the output target values. If this is true, the faults are considered detectable. As with any test pattern, several simulations are necessary until the observability can be established for all errors; fault simulation needs the highest proportion of the time used for test preparation. If the required degree of error detection has not been reached other test patterns have to be created manually or supplemented automatically. This step is also a basis for the path sensitization of the generator. The latter tries to sensitize paths only for the remaining faults.

4.4.3 Test Program Generation

A test program contains program parts for the following functions:

- *Parameter test for testing physical behavior:* In addition to a measurement of supply current, the current/voltage characteristics of the input/output drivers are primarily measured.
- *Functional tests:* For this test firstly the generated test patterns are used. For this static test an error detection, but not an error localization is performed. To reach a high quality standard it is necessary for this test to be supplemented by dynamic test patterns (see Sect. 4.1.1).
- *Statistics and shmooplots:* Statistical evaluations on the one hand assist in determining the yield, on the other hand in discovering design weak points. By means of shmooplots it is found out in which range (supply voltage, clock frequency, e.g.) the chip is functional.

The manual generation of a test program in the language of the respective test machine is quite time-consuming (see Sect. 4.5.2). By applying *postprocessors* (special program systems), automation can be achieved here, too. Depending on technological parameters and a program template which has to be generated once in the language of the respective machine, the executable test program is generated. For this purpose, test patterns are transformed into tester-specific instructions; the different parameter tests are adjusted to the chip size, and the statistics chosen by the design engineer as well as the shmooplots are integrated in the test program.

The test program generators mentioned result in a considerable time reduction for the test preparation.

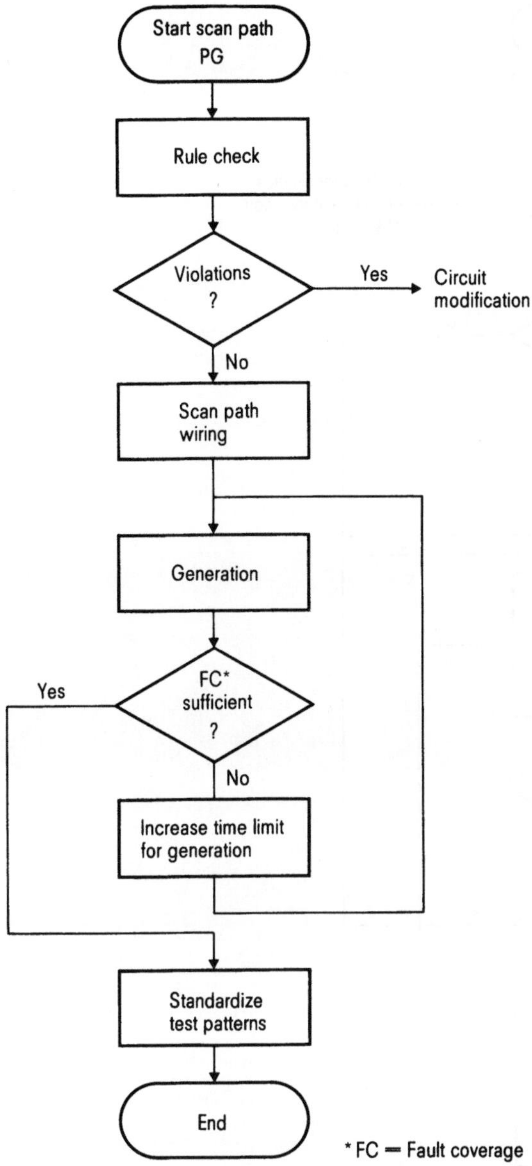

Fig. 4.17. Test pattern generation for circuits with scan-path

4.4.4 Test Preparation with VENUS

The test preparation with VENUS is automated for the most part. This is possible because a structured design of the chip is available according to the test design rules stated in Sect. 4.2.1. The test patterns as well as the test program can be generated automatically. The underlying model is also used for the verification; it is deduced from the data base. In VENUS two strategies are supported:

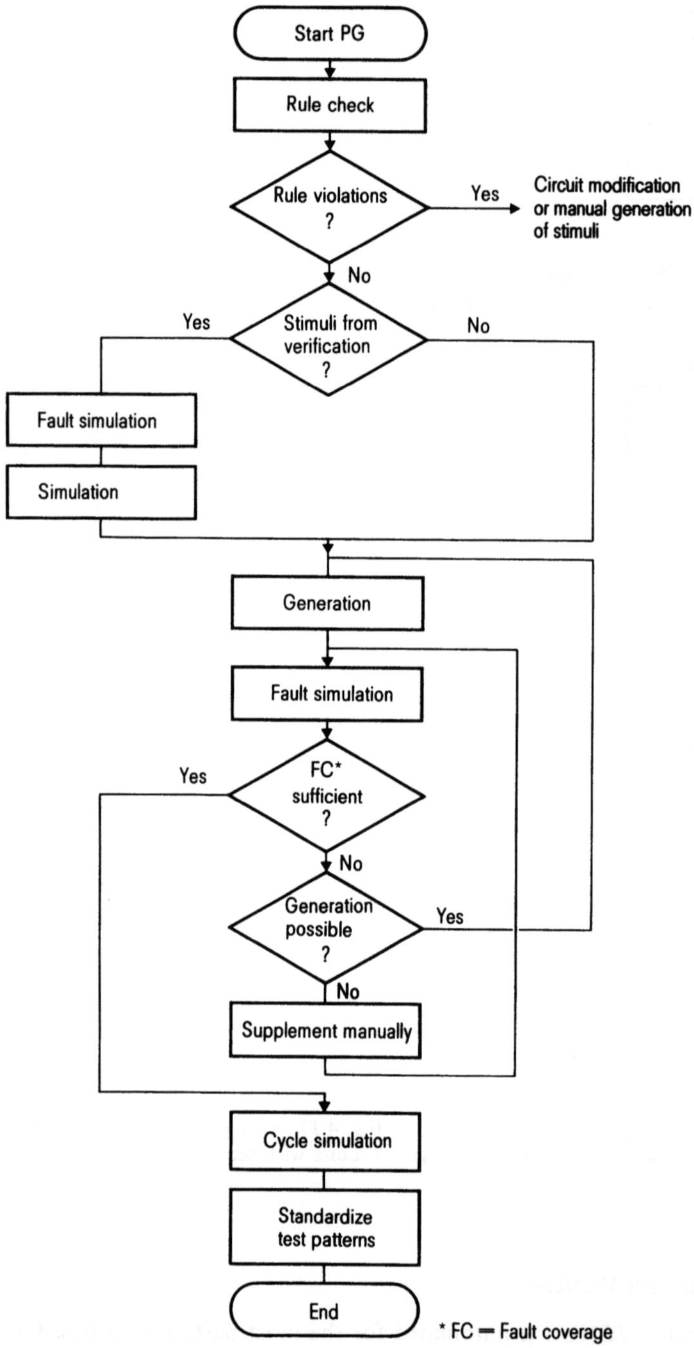

Fig. 4.18. Test pattern generation for sequentially deep circuits

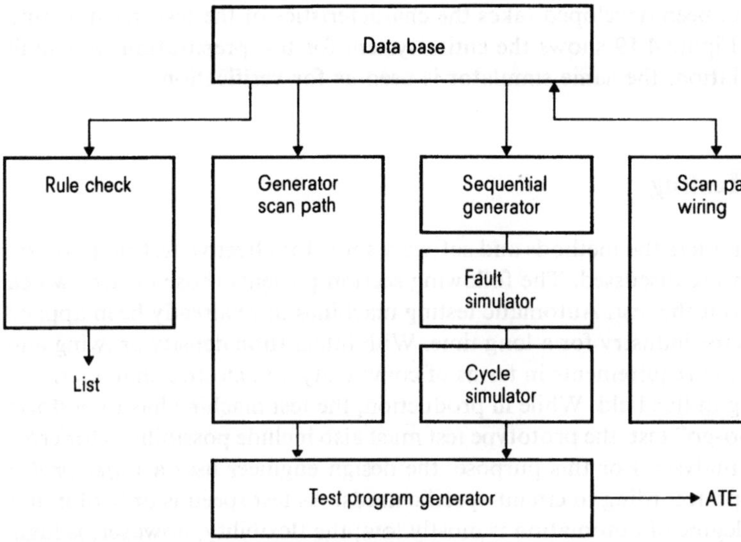

Fig. 4.19. Test preparation with VENUS

- *Circuit with scan-path:* If no test design rule violations have been discovered and the chip verification has been finished the scan path is automatically wired according to the principle discussed in Sect. 4.2.3.1. After the circuit has been completed by the scan-path, the test pattern generation comes next. It is characteristic of the generator that fault and cycle simulation is no longer necessary. The test patterns generated in parallel by the generator for fictitious inputs/outputs are transformed by means of a transformation program in serial shift instructions. The result of the generation process is the degree of fault coverage (FC) obtained and a test pattern file as standardized interface. For circuits with scan path a complete fault coverage (with redundancies excluded) is guaranteed according to the "s-a" error model without manual corrections. Figure 4.17 shows the proces from the user's viewpoint.
- *Circuit without scan-path:* Here, too, an automatic test pattern generation is possible if the test design rules are followed. Depending on the circuit structure, a rather intensive manual effort is necessary in addition. If multiplexers are used for reducing the sequential depth, the circuit model is prepared automatically for generation. For sequentially deep circuits a cycle and fault simulation is carried through in addition to generation of input stimuli. The functional stimuli from the verification process are provisionally mapped on a cycle raster; they serve as input for the fault simulator. For the errors thus found the generator need not to sensitize any paths. If the degree of fault coverage achieved by combining manually and automatically generated input stimuli is sufficient, the output target values are finaly ascertained by means of cycle simulation. Figure 4.18 shows the process from the user's view. The generated test patterns are thereafter transposed into the test machine instructions by means of a postprocessor. Furthermore, the program sections for parameter test, shmooplots and statistics are added. A dynamic test (clock rate test) is planned. Thus, functional stimuli have to be generated manually. The input lan-

guage which has been developed takes the characteristics of the test machine into consideration. Figure 4.19 shows the entire system for test preparation. For fault and cycle simulation, the same simulator is used as for verification.

4.5 Tools for Testing

In the previous chapters the methods and software tools for effective test preparation and test support were discussed. The following section presents those devices which are used to carry out the test. Automatic testing machines have already been applied in the semiconductor industry for a long time. With integration density growing and speed increasing, the requirements in terms of complexity and electric characteristics are ever increasing in this field. While in production, the test machine has to perform "only" a "go"-"no-go" test, the prototype test must also include possibilities for error localization and analysis. For this purpose, the design engineer uses a *wafer prober* which he can adapt according to circuit-specific needs. As test speed is only of minor importance, the degree of automation is mostly low; the flexibility, however, is high. The *electron beam prober* aims at analyzing circuit nodes which are not easily accessible from outside. It is the result of a development which recently emerged from the experimental stadium and has in the meanwhile lead to a generally applicable tool.

4.5.1 Wafer Prober

Wafer probers are needed for the prototype test and the continuous quality check during production. During the development of standard cells, for example, test chips on which several single cells can be measured independently from each other are manufactured. The goal here is the determination of electric behavior like delay times and driving capability, as well as the comparison with those values found by simulation and the cell specification. The test circuit is connected with the measurement system through the pins of a computer-controlled wafer prober. It consists (not only) of programmable voltage sources, pulse generators, oscilloscope and frequency counter. Positioning of the wafer's chips with respect to the measurement probes is performed automatically after the cell's dimensions have been input. This allows the realization of time-consuming measurements for whole wafers without manual interference. The obtained measurement results are stored, processed and printed. By means of a graphic display, the design engineer gets information about the local distribution of measured values on the wafer. The effect to temperature changes can be found by heating the wafer in the hot-chuck. In addition to test chips which are designed for that purpose, customer circuits, too, can be tested at the wafer prober. For the functional test there are intelligent logic analyzers now available, some with built-in pattern generators (pulse sources). These devices have a lot of the characteristics that were described in Sect. 4.5.3; they are, however, not as fast, and they have less and cheaper channels. In laboratory tests, these disadvantages are of minor importance. A more decisive criterion is the acquisition cost. For the search for systematic errors which concern all chips (logic errors, mask errors, etc.), it may often be desirable to scan voltage curves at circuit-internal nodes. This is possible to some extent if so-called micro-manipulators are used: By means of a highly precise mechanism the finest test probes available are brought into contact with metal lines on the chip. This method, of course, allows only measurements on non-passivated wafers. Moreover, only the top wiring level can be accessed. For dynamic measurements, a high capacitive load of the test probes is noted as signal distortion. This can be reduced if some test probes with built-in impedance transformers or an electron beam prober are used. By means of mechanic test probes, voltages, too, can be forced on circuit nodes which allows for further testing if, for example, a connection is missing.

4.5.2 Electron Beam Prober

The requirements of design engineers who desire highest impedance measurements at circuit internal points have been satisfied with the development of the electron beam prober (*E*lektronenstrahl*me*ß*g*erät EMG) which has been considerably advanced in the research laboratories of Siemens AG [4.19]. This method is based upon voltage-contrast technology which has been known for a long time [4.20]. The electron beam probe used as diagnostic tool is quite appropriate for laboratory application.

The structure of an EMG is similar to an electron microscope, as shown in Fig. 4.20. With the EMG, too, a bundled primary electron beam is directed towards the surface of the test object. The secondary low-energy electrons emerging from the signal line are absorbed by a collector, amplified and made visible on a monitor screen. If voltage is applied to the test circuit, lines with positive voltage are displayed darker on the screen whereas points with ground potential are displayed lighter. Positive electric fields attract the secondary electrons whereas negative fields accelerate them towards the electrode. This effect is non-linear; the logic states, however, can clearly be recognized. This comparatively simple arrangement makes it possible to carry through only static measurements; it has been in use for a long time. The big advantages of this contact-free method become evident, however, with dynamic measurement. For this purpose, the primary electron beam has to be pulsed, and the

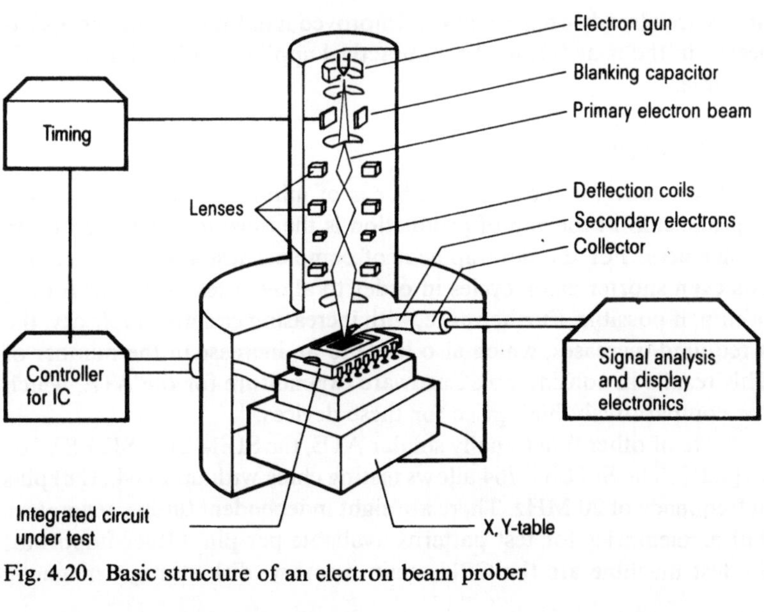

Fig. 4.20. Basic structure of an electron beam prober

secondary electrons generated have to be measured with an electron spectrometer. The voltage at the measuring point can be calculated from the energy distribution. By applying primary electron impulses (a similar method as that for the sampling oscilloscope), a high band width can be achieved for the dynamic measurement. But in this case a lower frequency limit exists, and non-recurring short impulses (such as spikes) can not be detected. The following advantages and disadvantages of the EMG at its current state of development have to be considered:

- *Advantages:*
  - High impedance measurement – for appropriate primary electron energy there is a charge balance so that no electron current flows.
  - Non-destructive scanning – the low electron energy generates no radiation damage in the cell. Mechanical damage does not occur whereas this could happen with test probes.
  - Simple positioning at any point in the circuit: because of the small beam diameter narrow lines can be accessed.
  - If reduced resolution is acceptable it is also possible to perform measurements on oxide-covered circuit nodes.
- *Disadvantages:*
  - Because of sampling technology (stroboscope effect) periodic signals are necessary.
  - Direct current measurements are not possible.
  - In contrast to mechanical testing it is not possible to force voltages on the circuit. It is therefore only suitable for measurements.
  - The measurement has to be executed in a vacuum. For this reason, it is preferably used for the analysis of chips assembled in packages (with open cover, of course).

The first systems are already on the market. Improved handling and easier operation can be expected in the near future. Then, practical application for design verification will be possible.

4.5.3 Automatic Test Equipment (ATE)

While the required test time for the wafer prober is of minor importance, the time needed for testing one chip at the end of production is the most important criterion next to test completeness. For the development of complete test systems there is a tendency towards even shorter clock cycles in order to allow a test at the operating frequency of the chip, if possible. Furthermore, with increasing circuit complexity, the number of pins required increases, which also leads to an increase in the number of test channels. This results in considerable hardware expenditure for the ATE, which is reflected in the comparatively high price for these devices.

As a representative of other functionally similar ATE, the SIEMENS SITEST 764 is described here [4.21]. The SITEST 764 allows testing chips with up to 64 (128) pins and a maximum frequency of 20 MHz. There are eight independent timing generators and four kbit buffer memories for test patterns available per pin. Other important components of a test machine are the CPU, main memory, disk storage, magnetic tape drive and the pin electronics as an interface with the test object. At a control panel, instructions can be input. The data can also be output on a printer.

By means of ATEs, chips which are assembled in the package as well as dies on the wafer can be tested. In production, wafer tests are performed first, because faulty chips should, if possible, be found before the expensive packaging is done. For both test types, the adaptation of pin electronics to the chips is carried out by means of a so-called load board which contains chip-specific test probes for the wafer test, and an appropriate test socket for a packaged chip. The test instructions can be stored as a program in a test-oriented language, or they can be automatically translated to this program by means of a post processor, as is done for VENUS.

In the first test section the dc parameters of the test object are determined and compared with the target values. For this purpose, any pin can, under program control, be connected with a precision-voltage/current measurement device combined with a programmable voltage/current source. In this case, the voltage is, for example, increased in certain steps, beginning at an initial value and ending at a certain maximum value, and the current is measured. If a given maximum voltage or current value has been reached the measurement process is stopped, and the measured values are stored. Errors made during bonding or test probes having poor contact with the contact pads of the chip can be discovered by applying negative voltage at the circuits' inputs. If the contact is good, the input protection diodes of the cell become conductive beginning at approximately – .7V, and the current flow shows the behavior of a forward biased diode. The required exactness of the values gained from the parameter test and the impossibility of measuring several pins simultaneously cause a comparatively low number of 1000 measurements per second which leads to the long test times already mentioned.

After passing the parameter test, the test with test patterns follows. Here, any tester channel can be programmed either as driver for the circuit inputs, as output comparator or as bidirectional connection. The possible driver levels and the switching thresholds of the comparators are determined by programmable reference voltage sources. The SITEST 764 facilitates the reversal of any pin between the measured

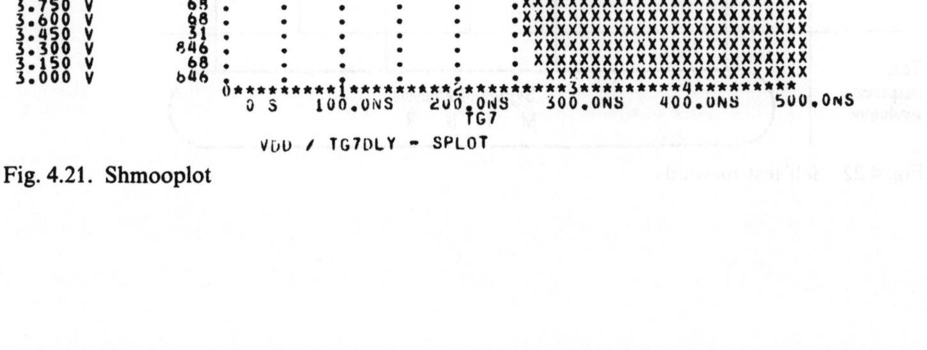

Fig. 4.21. Shmooplot

object and the reference pin. Thus, the different time delays on the test lines can be determined and adjusted to each other by means of programmable delay lines. Input and output test patterns are read-in a fast write-read memory and output serially. The response of the tested circuit to the stimuli is compared with the output test patterns; deviations may lead to a repetition of a part of a test program or to termination. Additional hardware supports the test if a scan-path exists.

At the end of each component test, the results are stored on disk or magnetic tape, or output on the printer. For the wafer test the faulty chips are provided with color marks ("inked"). After every wafer or after a certain number of chips have been tested, a list about type and number of found faults or measured chip-parameters can be printed as tables or shmooplots. Thus, information about the quality of the wafer run or about potential weak points can be obtained.

Figure 4.21 shows an example of a shmooplot. In this case, the test object is characterized in terms of circuit delay depending upon the supply voltage (VDD). An error-free function is marked by a cross (x). The good/bad limit corresponds to the maximum possible circuit speed. Column LML indicates for which test pattern the first defect occurs.

4.6 Selftest

The test strategies considered so far are based upon the assumption that the test machine stimulates the chip and also performs the comparison between target values and actual values for error detection. With increasing chip complexity it becomes more and more difficult to find suitable test machines for the dynamic requirements, and to perform a high quality functional test in an economically feasible time frame. A possible solution is the "concept of self-testing": Stimuli generation and error detection is performed directly on the chip. The resulting advantages are:

- Possibility of using simpler test machines with a reduced number of pins,
- Parallel self test of several chips at operating frequency,
- Further automation of test preparation by means of partitioning.

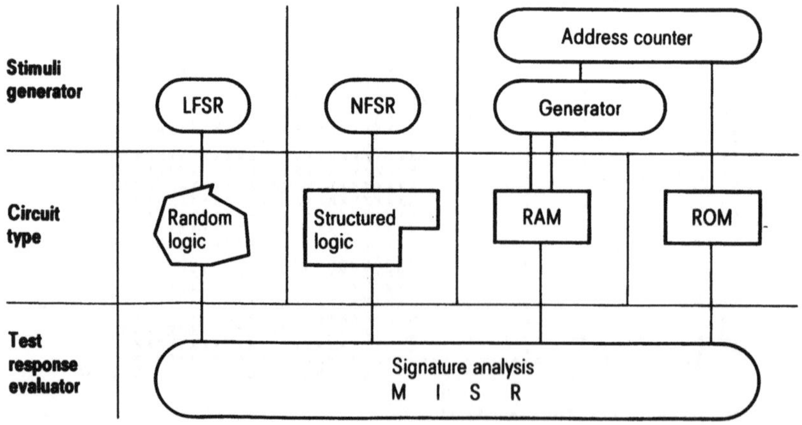

Fig. 4.22. Self-test methods

The price to be paid in exchange is:

- additional space required for the self-test circuit,
- the increased number of pins,
- a possible reduction of performance.

The self-test strategies found in technical literature at present are explained in the following by means of examples. Necessary software tools are merely referred to. Often, the circuits which are functionally necessary can be used for self-test purposes if they are provided with the necessary supplements. First, the self-test devices are stimuli generators (SG) for stimulation and test response evaluators (TRE) for error recognition. An overview can be found in Fig. 4.22. For an optimal test, the different circuit structures require programmable stimuli generators.

4.6.1 Stimuli Generators

The most simple method of providing stimuli is to store them in a ROM which, in addition, is integrated on the chip (*ROM method*). The stimuli are computed with a generator (see 4.4.1). In practice, this method which is rather space-consuming will be applied only as ad-hoc method. The stimuli generators described in the following are more suitable for a structured procedure.

For the memory test (RAM) a counter is often used as stimuli generator for addresses and data (*counter method*). For the ROM test the generation is simplified to providing addresses in ascending order.

For random logic, long test times occur if all stimuli combinations are applied and if the number of pins is high. In practice, pseudo random stimuli with an abbreviated sequence already bring about a good degree of error detection [4.21]. Those sequences can be produced with comparatively little hardware expenditure by mean of *l*inear *f*eedback *s*hift *r*egisters (LFSR). Figure 4.23 shows the principle of a four-digit random sequence. The sequence of random patterns is determined mainly by the type of feedback with EXOR gates. The X_1 through X_4 outputs are used for stimulating the test object.

This concept reaches its limits only for circuits with a lot of inputs. But even in this case, appropriate partitioning may help.

Another possibility for generating stimuli on the component is provided by the application of *n*on linear *f*eedback *s*hift *r*egisters (NFSR). They are especially suitable for a structured logic such as PLAs.

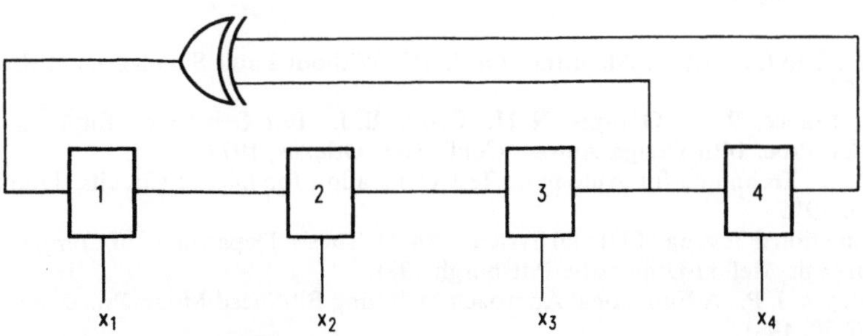

Fig. 4.23. Linear feedback shift register

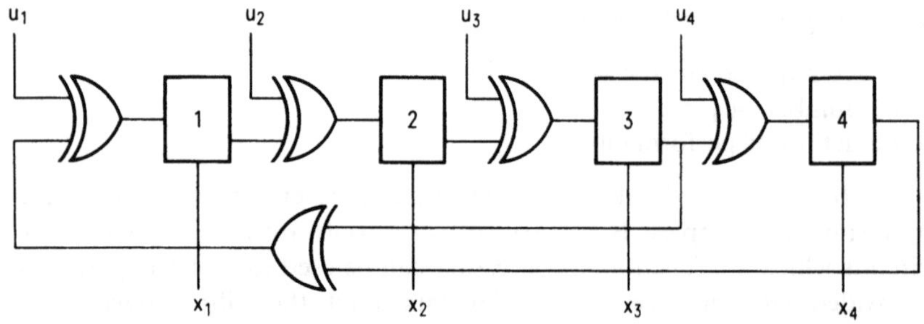

Fig. 4.24. Signature register

For these circuits, often with a high logic fan-in (= gates with a high input count), random patterns lead to a low degree of error detection. The NFSRs can be built through suitable feedback logic in such a way that they reproduce a previously computed stimuli sequence with a high degree of error detection [4.22].

4.6.2 Test Response Evaluator

After the generated stimuli have been applied to the circuit under test, the responses of the latter have to be evaluated. In simple cases, the test responses are compared with those values expected. Because a lot of space is required for storage of the expected test responses, this method is rarely used. It is used merely for memories where the comparator word is often obtained as a by-product of the test pattern generation. It is more common to evaluate the test response by means of a signature analysis. The required hardware, the signature register (*m*ultiple *i*nput *s*hift *r*egister MISR) [4.24, 4.25], is shown in Fig. 4.24. It is different from the linear feedback shift register in that the individual test responses are input in parallel through EXOR gates. These circuits safely recognize all errors which appear in one faulty test response. Multiple errors are found only with a specific probability. The target signature is computed by means of simulation. LFSRs and MISRs can be combined, forming the BILBO, the "*B*uilt-*I*n-*L*ogic-*B*lock-*O*bserver", which helps save hardware. The BILBO can be used, depending on a control input, as test pattern generator or as test response evaluator [4.26].

Literature – Chapter 4

4.1 Johansson, M.: The GENESYS-Algorithm for ATPG Without Fault Simulation. IEEE Test Conf., 1983.
4.2 Bottorf, P. S.; France, R. E.; Garages, N. H.; Orosz, E. J.: Test Generation for Large Logic Networks. Proc. 14th Design Autom. Conf., New Orleans, 1977.
4.3 Bowder, K. R.: A Technique for Automatic Test Generation for Digital Circuits. Proc. IEEE Intercon, 1975.
4.4 Lai, K.-W.: Functional Testing of Digital Systems. Ph. D. Thesis, Department of Computer Science, Carnegie-Mellon-University, Pittsburgh 1981.
4.5 Sridhar, T.; Mayes, J. P.: A Functional Approach to Testing Bit-Sliced Micro-Processors. IEEE Trans. C-30, 1981.
4.6 Bashiera, D.; Courtois, B.: Testing CMOS: A Challenge. VLSI-Design, 1984.

4.7 Wadsack, R. L.: Fault Modeling and Logic Simulation of MOS and MOS Integrated Circuits. The Bell System Technical Journal, Vol. 57, 1978.
4.8 Eichelberger, E. B.; Williams, T. W.: A Logic Design Structure for LSI Testability. Journal DAFTC, Vol. 2, 1978.
4.9 Gerner, M.; Nertinger, H.: Scan Path in CMOS-Semicustom LSI Chips? IEEE Test Conf., 1984.
4.10 Ando, H.: Testing VLSI with Random Access Scan. Digest of Papers Compcon 80, 1980.
4.11 Funatsu, S.; Wakatsuki, N.; Yamada, A.: Easily Teastable Design of Large Digital Circuits. NEC Research and Development, No. 54, 1979.
4.12 Goldstein, L. H.: Controllability/Observability Analysis of Digital Circuits. IEEE Transactions on Circuits and Systems, Vol. LAS-26, No. 9, 1979.
4.13 Thomas, J. J.: Automated Diagnostic Test Programm for Digital Networks, Computer Design, Vol. 10, No. 8, 1971.
4.14 Armstrong, D. B.: On Finding a Nearly Minimal Set of Fault Detection Tests for Combinatorial Logic Nets. IEEETC, Vol. C-21, 1972.
4.15 Roth, J. P.: Diagnosis of Automata Failures: A Calculus and a Method. IBM Journal, Vol. 10, No. 7, 1967.
4.16 Szygenda, S. A.; Thompson, E. W.: Modelling and Digital Simulation for Design Verification and Diagnosis. IEEETC, Vol. C-25, 1976.
4.17 Armstrong, D. B.: A Deductive Method for Simulating Faults in Logic Circuits, IEEETC, Vol. C-21, 1972.
4.18 Abramovici, M.; Breuer, M. A.; Kumar, K.: Concurrent Fault Simulation of Digital Circuits Described with Gate and Functional Models. Proc. STL, 1979.
4.19 Fazekas, P.; Feuerbaum, H.-P.; Wolfgang, E.: Scanning Electron Beam Probes VLSI Chips. Electronics, July 14, 1981.
4.20 Coshwait, D. L.; Ivey, F. W.: Voltage Contrast, Methods for Semiconductor Device Failure Analysis. SEM, 1974, pp. 935–940.
4.21 Short Description SITEST 764: The Computer Controlled Tester for LSI/VLSI Chips. Siemens, München.
4.22 Daehn, W.: Deterministische Testmustergenerierung für den eingegebenen Selbsttest von integrierten Schaltungen. NTG Fachberichte Nr. 82/1983, pp. 16–19.
4.23 Zuxi Sun, Laung-Terng Wang: Self-Testing of Embedded RAMs. Int. Test Conf. 1984, pp. 148–156.
4.24 Frohwerk, R. A.: Signature Analysis: A New Digital Field Service Method. Hewlett Packard Journal, May 1977, pp. 2–8.
4.25 Bhavshar, D. K.; Krishnamurthy, B.: Can We Eliminate Fault Escape in Self Testing by Polynamical Division (Signature Analysis). Int. Test Conf. 1984, pp. 134–139.
4.26 Koenemann, B.; Mucha, J.; Zwiehoff, G.: Built-in Logic Block-Observation Technique. Digest Int. Test Conf. 1979, pp. 37–41.

5 Cells and Libraries

Standard design systems are especially effective if they support several design methods (gate array, standard cells, macrocells) and several process technologies (CMOS, ECL). In this chapter some cell libraries currently offered with the VENUS design system are discussed with reference to the following aspects:

- A survey on the functional scope of the cell spectrum, the technical data of the cells and the structure of the data sheet is given.
- The similarities appearing when one goes from one design method (gate array) to another (standard cells) within one process technology are illustrated.
- The special features appearing when one goes from one process technology (3 µm CMOS) to another, higher integrated one (2 µm CMOS) are shown.
- The differences between different process technologies (CMOS and ECL) are demonstrated.

The development of the cell library by the CAD engineer is also described. This is, however, not a manual for the design of cells; it is rather intended to convince the reader that the different cell descriptions are consistent and of a high quality standard.

With this chapter it is also intended to illustrate that the user is decoupled from the complexity of library generation and the necessary effort.

The description of cells and libraries is given such that only the basic application modes can be derived. The presentation is not binding as a product description.

5.1 Application of Cell Libraries

5.1.1 Functional Scope

When determining the scope of a cell library, first a basic set of about 100 functions is considered. This is sufficient for the majority of applications. During a cell library's lifetime this basic set will inevitably be extended due to various special requirements. In order to avoid complications in terms of manageability, the spectrum of cell functions offered should not exceed approximately 200 types. This is possible if the following rules are adhered to:

- The complexity of the cells is restricted to those of discrete SSI, maybe MSI integrated circuits.
- The various electric driving capabilities of the cells are obtained by means of parameterization, not by multiplication of the cell types.
- The requirements of the target user group of design engineers are examined quite thoroughly.

The development of the VENUS cell libraries was preceded by analyses, comprising the following points:

- A cell statistics over a large range of integrated circuit projects was executed.
- A statistical analysis of SSI and MSI integrated circuits currently used by pc board developers was made as these very developers will be the developers of integrated circuits in future.
- Experienced developers of integrated circuits were involved in the planning process.
- For comparison purposes, data books of "low-integrated" IC families were evaluated.

VENUS offers cell families of three different technologies (two CMOS processes and one bipolar process). In the following, we want to take a closer look at the CMOS libraries.

5.1.1.1 CMOS Libraries

The complexity of the VENUS-CMOS cell spectrum for the 3 μm technology can be compared with the TTL 74xxx or the CMOS 4xxx series. There is, however, no 1:1 correspondence: a fully automatic 1:1 conversion from a pc board with TTL integrated circuits to a single chip, for example, does not make sense anyway because of the very different conditions of the system environment.

For a better overview, the VENUS cells have been divided into ten groups, each characterized by similar functions:

1. *Gates:* In addition to the basic gates with up to eight inputs (AND, NAND, OR, NOR) two-stage combination gates with three inputs are implemented, too. It has been shown that combination gates bring about substantial simplification and space-saving for circuit sections with "wild" logic.
2. *Drivers:* Driver cells are offered in inverted and non-inverted forms. Each driver cell has four different strengths of output driving capability. Drivers with ENABLE input (tri-state) are offered only for standard cells but not for gate-array cells.
3. *Multiplexers:* The multiplexer group includes a 2-bit and two 4-bit multiplexers. A multiplexer cell with ENABLE makes possible the construction of larger multiplexers by cascadation.
4. *Flipflops:* The implemented flipflops are edge-triggered master-slave flipflops. The common D-, RS-, JK- and T-flipflop types are offered in all variants with and without PRESET and/or CLEAR.
5. *Latches:* A simple latch is offered with and without ENABLE. The latches are realized with transfer gates.
6. *Shift registers and*
7. *Counters:* These cells are the most complex in the VENUS spectrum. Because they are offered with various additional characteristics (synchronous or asynchronous load, with backward/forward switch, to be set or reset) and since they can be connected in series, the majority of user requirements can be met.
8. *Arithmetic elements:* Half and full adders as well as comparators are implemented. Two-bit comparators can be cascaded to larger units.
9. *Decoders:* One "1 out of 4" and one "1 out of 8" decoder are offered.

10. *Padcells:* The 200 μm × 200 μm metal areas for the connection of input and output pins are called pads. Pad cells contain, in addition to those bond areas also circuits such as drivers or voltage converters and protection structures.
    Input pad cells are available in CMOS- or TTL-compatible versions, with or without Schmitt trigger characteristics. Output pad cells are built in such a way that they can drive CMOS and TTL inputs. They are available with or without ENABLE. Bidirectional padcells are available with CMOS- or TTL-compatible input.
    If a chip needs a large number of pad cells in relation to the core area, the chip area may be determined by the frame formed by the pad cells (see Chap. 6). In this case, it is better to use high and narrow pad cells.
11. *Others:* Some special cells are listed in this group. VDD and VSS taps provide compulsory H- or L-levels. Pull-up and pull-down cells are used in order to prevent tri-state lines from being at undefined potential.

3 μm-CMOS-Libraries

Currently, three different 3 μm-CMOS cell library versions are in use: Two standard cell versions (A and B) and one gate-array version (G).

The technical data of each of these 3 μm-CMOS families differ only slightly. All three are based upon a process in 3 μm-CMOS technology with p-well. Aluminum 1 and polysilicon are used as wiring levels within the cell. For the intercell wiring, aluminum 1 and polysilicon are used for the version A, and aluminum 1 and aluminum 2 are used for versions B and G. This is also important for the cell descriptions, as the cell terminals have to be set in appropriate levels. The major effect of the additional low-impedance wiring level is, apart from a certain space saving effect, the reduction of propagation delay.

The supply voltage is 5V ± 10%, a typcial gate delay is 2ns. The internal maximum clock frequencies are influenced by the cells and the way these cells are used in an actual circuit; they are around 40 MHz.

The static power dissipation of a basic cell in the core section is negligible (2 pW), for the pad cells it amounts to 0.8 mW in the L-state and 0.4 mW in the H-state.

The dynamic power dissipation is determined by capacity or charge/discharge, mainly for heavily loaded outputs. The dynamic power dissipation is proportional to the switching frequency. The typcial dynamic currents for an output driver are 300 μA/MHz for a load of 50 pF.

Currently, four different masters are supplied in the gate-array family G:

- 1000 basic cells in the core section, 42 pad cells (low and wide: G1 variant),
- 2000 basic cells in the core section, 64 pad cells (low and wide: G1 variant),
- 1000 basic cells in the core section, 82 pad cells (high and narrow: G2 variant),
- 2000 basic cells in the core section, 124 pad cells (high and narrow: G2 variant).

Chart 5.1 below gives a short description of the functions and the abbreviations used durig the VENUS design process, for all 3 μm-CMOS cells that are currently offered.

The six-character abbreviation of the cells has a homogenous structure: the first letter describes the technology used, which is C for CMOS in the cases considered

Chart 5.1. Overview of the 3 μm-CMOS families, arranged in functional groups

Short name			Function
A-family	B-family	G-family	
1.	**Gate**		
CANnnA	CANnnB	CANnnG	AND, nn inputs, realized for nn = 2 to 8
		CNAB2G	NAND, two inputs, without output driver
CNAnnA	CNAnnB	CNAnnG	NAND, nn inputs, realized for nn = 2 to 8
CORnnA	CORnnB	CORnnG	OR, nn inputs, realized for nn = 2 to 8
		CNOB2G	NOR, two inputs, without output driver
CNOnnA	CNOnnB	CNOnnG	NOR, nn inputs, realized for nn = 2 to 8
CXR02A	CXR02B	CXR02G	EXOR, two inputs
CXN02A	CXN02B	CXN02G	EXNOR, two inputs
CAO03A	CAO03B	CAO03G	AND-OR, three inputs
COA03A	COA03B	COA03G	OR-AND, three inputs
CAI03A	CAI03B	CAI03G	AND-NOR, three inputs
COI03A	COI03B	COI03G	OR-NAND, three inputs
2.	**Driver**		
CDRs1A	CDRs1B	CDRs1G	driver, inverting
CDRs2A	CDRs2B	CDRs2G	driver, non-inverting
CDRs3A	CDRs3B		driver, inverting with ENABLE
CDRs4A	CDRs4B		driver, non-inverting with ENABLE
		CDD01G	2 drivers, inverting
3.	**Multiplexer**		
CXA01A	CXA01B	CXA01G	2-bit multiplexer
CXB01A	CXB01B	CXB01G	4-bit multiplexer
CXB02A	CXB02B	CXB02G	4-bit multiplexer with ENABLE
4.	**Flipflops**		
CDF01A	CDF01B	CDF01G	D-flipflop
CDF02A	CDF02B	CDF02G	D-flipflop with CLEAR
CDF03A	CDF03B	CDF03G	D-flipflop with PRESET and CLEAR
CRS01A	CRS01B	CRS01G	RS-flipflop
CRS04A	CRS04B	CRS04G	RS-flipflop
CRS05A	CRS05B	CRS05G	RS-flipflop with CLEAR
CRS06A	CRS06B	CRS06G	RS-flipflop with PRESET and CLEAR

Short name A-family	B-family	G-family	Function
CJK05A	CJK05B	CJK05G	JK-flipflop with CLEAR
CJK06A	CJK06B	CJK06G	JK-flipflop with PRESET and CLEAR
CTF02A	CTF02B	CTF02G	T-flipflop with CLEAR
5.	**Latches**		
CLA01A	CLA01B	CLA01G	latch
CLA02A	CLA02B		latch with ENABLE
		CLA41G	four bit latch
6.	**Shift registers**		
CSS01A	CSS01B	CSS01G	shift register cell
CSP01A	CSP01B	CSP01G	shift register cell, synchronous, parallel load
CSA01A	CSA01B	CSA01G	shift register cell, asynchronous, parallel load
7.	**Counter**		
CBC02A	CBC02B	CBC02G	counter cell, synchronous, synchronous load
CUD02A	CUD02B	CUD02G	counter cell, synchronous, up/down
CCR02A	CCR02B	CCR02G	counter cell, synchronous, with RESET
CCS02A	CCS02B	CCS02G	counter cell, synchronous, with SET
8.	**Arithmetic Elements**		
CHA01A	CHA01B	CHA01G	half-adder (one bit)
CFA01A	CFA01B	CFA01G	full-adder (one bit)
CCE02A	CCE02B	CCE02G	2-bit comparator (P = Q), with control input
CCP02A	CCP02B	CCP02G	2-bit comparator (P >, <, = Q), cascadable
9.	**Decoder**		
CDX02A	CDX02B	CDX02G	decoder, binary, 1 out of 4
CDX03A	CDX03B	CDX03G	decoder, binary, 1 out of 8
10.	**Padcells**		
CIN01A	CIN01B	CIN01G1/G2	input driver, CMOS compatible
CIN02A	CIN02B(P)	CIN02G1/G2	input driver, TTL compatible
CIS01A	CIS01B	CIS01G2	input driver, Schmitt-Trigger, CMOS compatible
CIS02A	CIS02B	CIS02G2	input driver, Schmitt-Trigger, TTL compatible

Short name			Function
A-family	B-family	G-family	
10.	**Padcells**		
CIN01A	CIN01B	CIN01G1/G2	input driver, CMOS compatible
CIN02A	CIN02B(P)	CIN02G1/G2	input driver, TTL compatible
CIS01A	CIS01B	CIS01G2	input driver, Schmitt-Trigger, CMOS compatible
CIS02A	CIS02B	CIS02G2	input driver, Schmitt-Trigger, TTL compatible
COT01A	COT01B(P)	COT01G1/G2	output driver (tristate) with ENABLE
COT02A	COT02B	COT02G	output driver
	COT05B(P)		output driver (tristate) with ENABLE, 8 mA
	COT06B		output driver, 8 mA
CIO01A	CIO01B	CIO01G1/G2	input/output driver (bidirectional), CMOS compatible
CIO02A	CIO02B(P)	CIO02G1/G2	input/output driver (bidirectional), TTL compatible
		CIC01G1/G2	clock driver, CMOS compatible
	CIO06B		input/output driver (bidirectional), TTL compatible, 8 mA
	CIOD2B		input/output driver (bidirectional), TTL compatible, "open-drain", 40 mA
	CIOSDBP		input/output driver (bidirectional),Schmitt-Trigger, TTL compatible, "open-drain", 40 mA
11.	**Others**		
CVD00A	CVD00B	CVD00G	VDD tap
CVS00A	CVS00B	CVS00G	VSS tap
CPU00A	CPU00B		pull-up cell for tristate wires
CPD00A	CPD00B		pull-down cell for tristate wires

here. The sixth symbol is the letter describing the version. The letters in positions 2 and 3 are a mnemotechnic abbreviation of the function, such as AN for AND, DF for D-flipflop etc. The two symbols in positions 4 and 5 may either describe the number of inputs (nn in group 1), or they just serve for numbering within one group. If drivers are supplied with different strengths, the driving capabilities are indicated by s $(0 \leqq s \leqq 4)$.

For the pad cells, one has to distinguish between the low and wide, and the high and narrow variant. A seventh character is used for this. The seventh character of the narrow pad cells of the "B"-family is a "P". A "P" between parentheses means that the pad cell is offered in both variants. For the G-family, "G1" describes the low version, "G2" the high version.

2 µm-CMOS Technology

Currently two libraries implemented in 2 µm-CMOS technology can be used: a standard-cell family F and a gate-array family K. All functional groups of the 3 µm families exist here, too.

In order to save space, multi-bit cells have also been realized for counters and flipflops.

The CMOS technology used with characteristic structure widths of 2 µm leads to smaller chip areas and shorter delay times. A characteristic gate delay time is 1ns. For intercell wiring two aluminum levels are available.

In the standard cell family F, four types of macrocells exist at present:

- RAM (1K static, organization 256 × 4),
- ROM (4K clocked, organization 512 × 8),
- PLA (dynamic CMOS technology, ≦ 50 inputs, ≦ 50 outputs, ≦ 75 product terms),
- Standard cell block (predesigned with standard cells by the user).

In the gate-array family, two masters currently exist:

- 4000 basic cells in the core section, 120 pad cells;
- 10000 basic cells in the core section, 180 pad cells.

In the short names of the cells, the letter C indicating the technology has been omitted, which allows more character to be used for the mnemonic name. The parameter s indicates the driving capability:
s = B: Basic, s = L: Low, s = M: Medium, s = H: High, s = U: Ultrahigh.

5.1.1.2 ECL Library

In ECL technology, there is currently one gate-array family (Z-family). For wiring three metallization levels are used. The typical internal gate delay time is 0.35 ns. Two masters with 36 and 120 core cells respectively are available. In contrast to ACMOS gate-arrays, the cells in ECL technology may contain several logic functions. These logic elements may be connected by a common control (such as ENABLE), they can, however, also be completely independent from each other. There is a N:1 assignment between specific logic symbols (logic functions) and the cell placed on the IC. The design engineer uses logic symbols for schematic entry. The assignment of logic symbols to cells is realized automatically by the wiring program, trying to minimize the total wiring lengths. Furthermore, the user may define an allocation of individual logic symbols to cells (i.e. to the location where a cell is to be placed) already during schematic entry input: unused pad cell space can be occupied by core cells (logic).

The cell list of the Z-family contains in the first column the short name of the logic element and in the second column the name of the cell which contains the logic element specified. The third column describes the function of the logic element. Due to the usually high complexity of the logic functions of a logic element, it is not always

Chart 5.2. Overview of the 2 μm-CMOS families, arranged in functional groups

Name	Number of bits (n)	Driving capability (s)	Function
1.	**Gates**		
1.1	**AND/NAND**		
AN2s	-	L-H	2-input AND
AN3s	-	L-H	3-input AND
AN4s	-	L-H	4-input AND
AN5s	-	B-H	5-input AND
AN6s	-	B-H	6-input AND
AN7s	-	B-H	7-input AND
AN8s	-	B-H	8-input AND
NA2s	-	B-H	2-input NAND
NA3s	-	B-H	3-input NAND
NA4s	-	B-H	4-input NAND
NA5s	-	L-H	5-input NAND
NA6s	-	L-H	6-input NAND
NA7s	-	L-H	7-input NAND
NA8s	-	L-H	8-input NAND
1.2	**OR/NOR**		
OR2s	-	L-H	2-input OR
OR3s	-	L-H	3-input OR
OR4s	-	B-H	4-input OR
OR5s	-	B-H	5-input OR
OR6s	-	B-H	6-input OR
OR7s	-	B-H	7-input OR
OR8s	-	B-H	8-input OR
NO2s	-	B-H	2-input NOR
NO3s	-	B-H	3-input NOR
NO4s	-	L-H	4-input NOR
NO5s	-	L-H	5-input NOR
NO6s	-	L-H	6-input NOR
NO7s	-	L-H	7-input NOR
NO8s	-	L-H	8-input NOR

Name	Number of bits (n)	Driving capability (s)	Function	
1.3	**EXOR/EXNOR**			
XR2s	-	L-H	2-input EXOR	
XN2s	-	L-H	2-input EXNOR	
1.4	**AND-OR/AND-NOR**			
AO03s	-	L-H	AND-OR	2-2-3
AOI03s	-	B-H	AND-NOR	2-2-3
AOI04s	-	B-H	AND-NOR	3-2-4
AOI14s	-	B-H	AND-NOR	2-2-4
AOI24s	-	B-H	AND-NOR	2-3-4
AOI34s	-	B-H	AND-NOR	2-2-2-4
AOI0Cs	-	B-H	AND-NOR	4-3-12
1.5.	**OR-AND/OR-NAND**			
OA03s	-	L-H	OR-AND	2-2-3
OAI03S	-	B-H	OR-NAND	2-2-3
OAI04s	-	B-H	OR-NAND	3-2-4
OAI14s	-	B-H	OR-NAND	2-2-4
OAI24s	-	B-H	OR-NAND	2-3-4
OAI34s	-	B-H	OR-NAND	2-2-2-4
2.	**Drivers**			
DRIL	-	L	inverting	
DRIM	-	M	inverting	
DRIH	-	H	inverting	
DRIU	-	U	inverting	
DRM	-	M	non-inverting	
DRH	-	H	non-inverting	
DRU	-	U	non-inverting	
DREM	-	M	non-inverting with ENABLE	
DREH	-	H	non-inverting with ENABLE	
DREU	-	U	non-inverting with ENABLE	
DRGH	-	H	gated (AND)	

Name	Number of bits (n)	Driving capability (s)	Function	
3.	Multiplexer			
MXAns	1...8	L-H	multiplexer 1 out of 2	
MXBns	1...8	L-H	multiplexer 1 out of 4	
MXBEns	1...8	L-H	multiplexer 1 out of 4 with ENABLE	
MXCEns	1...4	L-H	multiplexer 1 out of 8 with ENABLE	
4.	Flipflops			
4.1	D-flipflops			
DFAns	1...8	L-H	D-flipflops	.CK
DFBns	1...8	L-H	D-flipflops	.CK,CLN
DFCns	1...8	L-H	D-flipflops	.CK,CLN,PRN
DFDns	1...8	L-H	D-flipflops	.CKN
DFEns	1...8	L-H	D-flipflops	.CKN,CLN
DFFns	1...8	L-H	D-flipflops	.CKN,CLN,PRN
DFBSns	1...8	L-H	D-flipflop with scan path	.CK,CLN
DFLSns	-	L-H	D-flipflop with LSSD	.CK,CLN
4.2	RS-flipflops			
RSBs	-	L-H	RS-flipflops	.CK,CLN,RN,SN
RSGs	-	L-H	RS-flipflops	.CK,CLN,RN,SN (event)
RSHs	-	L-H	RS-flipflops	.CK,RN,SN(one stage)
4.3	JK/T-flipflops			
JKBs	-	L-H	JK-flipflop	.CK,CLN,JN,KN
JKCs	-	L-H	JK-flipflop	.CK,CLN,PRN,JN,KN
JKBSs	-	L-H	JK-flipflop with scan path	.CK,CLN,JN,KN
JKLSs	-	L-H	JK-flipflop with LSSD	.CK,CLN
TFBs	-	L-H	T-flipflop	.CK,CLN
5.	Latches			
LATnns	1..16	L-H	latch	.LN
LATBns	1...8	L-H	latch	.LN,CLN

Name	Number of bits (n)	Driving capability (s)	Function
6.	**Shift registers**		
SSnns	1..16	L-H	serial-in,serial-out
SSSnns	1..16	L-H	as above, with scan path
SSDnns	16..64	L-H	as above, dynamic shift register
SSAnns	1..16	L-H	serial-in, serial-out, bit 1 parallel loadable
SSBnns	1..16	L-H	serial-in, parallel-out
SPns	1..8	L-H	parallel-in, parallel-out
SPSns	1..8	L-H	as above, with scan path
SPAns	1..8	L-H	parallel-in, parallel-out, asynchronous load
7.	**Counters**		
CTRns	1..8	L-H	synchronous, with RESET, with ripple carry: .CK,CLN,CIN,CON
CTLns	1..8	L-H	synchronous, synchronous load, with ENABLE, with RESET: .CK,CLN,LN,EN,CI,CO
CTLSns	1..8	L-H	as above, with scan path
CTUDns	1..8	L-H	synchronous, synchronous load, up/down with ENABLE, with ripple carry: .CK,LN,ENN,UPN,CIN,CON
CTRCs	4	L-H	like CTRns, but: 4-bit with carry-look ahead
CTLCs	4	L-H	like CTLns, but: 4-bit with carry-look ahead
CTUCs	4	L-H	like CTUDns (without CIN), but: 4-bitwith carry-look ahead, asynchronous load
8.	**Arithmetic Elements**		
HADs	-	L-H	half-adder
FADs	1...4	L-H	full-adder (ripple carry)
FADCs	4	L-H	4-bit full-adder with carry look-ahead and propagate + generate outputs
CG4s	-	L-H	carry generator for four full-adders
CEQ2s	2	L-H	comparator 2-bit equal
CMK4s	4	L-H	comparator 4-bit - magnitude (cascadable)
PGC9s	9	L-H	parity generator/checker 9 bit with inhibit

Name	Number of bits (n)	Driving capability (s)	Function	
9.	**Decoders**			
DCKBs	-	L,M	decoder, cascadable 1 out of 4 (2:4)	
DCKCs	-	L,M	decoder, cascadable 1 out of 8 (3:8)	
DXAns	1...8	L,M	demultiplexer 1 to 2	
DXBns	1...8	L,M	demultiplexer 1 to 4	
DXBEns	1...8	L,M	demultiplexer 1 to 4 with ENABLE	
ENP8s	-	L,M	priority encoder (8:3) with ENABLE	
10.	**Padcells**			
10.1	**Input pads**			
ICM		M	CMOS-compatible	
ICU		U	CMOS-compatible	
ICIU		U	CMOS-compatible,inverting	
ICPM		M	CMOS-compatible, with PULL-UP (appr. 50K)	
ITM		M	TTL-compatible	
ITU		U	TTL-compatible	
ITIU		U	TTL-compatible,inverting	
ITPM		M	TTL-compatible with PULL-UP (appr. 50K)	
ISCM		M	Schmitt-trigger, CMOS-compatible	
ISTM		M	Schmitt-trigger, TTL-compatible	
IFT		-	feed through IN	
10.2	**Output pads**			
ON3			regular	I_{out} = 3 mA
ON8			regular	I_{out} = 8 mA
OZ3			tristate	I_{out} = 3 mA
OZ8			tristate	I_{out} = 8 mA
OFT			feed through OUT	
10.3	**Bidirectional pads**			
BCM3		M	input CMOS-compatible	I_{out} = 3 mA
BCM8		M	input CMOS-compatible	I_{out} = 8 mA
BTM3		M	input TTL-compatible	I_{out} = 3 mA
BCM8		M	input TTL-compatible	I_{out} = 8 mA

Name	Number of bits (n)	Driving capability (s)	Function
10.4	**Supply pads**		
VDP			VDD-pad
VDPP			VDD-pad - for pad cells only
VDPK			VDD-pad - for core cells only
VSP			VSS-pad
VSPP			VSS-pad - for pad cells only
VSPK			VSS-pad - for core cells only

possible to render an exact description in abbreviated form. The precise function is described in the data sheet.

Those logic elements that can be used only for the larger of the two masters, the 120-cell master, are marked with a "*". All other logic elements can be used on the 36-cell master as well as on the 120-cell master.

5.1.2 Structure of the Data Sheets

The VENUS user is first supplied with the cell catalog. It helps him decide whether a certain design problem can be solved with a standard IC design system. The data sheets which make up the cell catalog must therefore provide information about all cell characteristics that might be important for application. The data sheet contains three groups of data:

- *Call Information:* This includes name, number and logic symbol with which the cell is drawn in the logic plan.
- *Function Information:* The purpose of any cell is the realization of a certain logic function, described by means of a function table. Furthermore, the design engineer requires data on cell propagation delays and the capacitive load which each cell input represents.
- *Additional Information:* Although data on the cell dimensions, complexity and internal structure are not absolutely necessary for the cell's application, they often help in deciding on one out of several circuit alternatives.

By means of an example taken from the A-family standard cell library (CDF01A) the data sheet content is explained in the following.

- *Logic symbol (Fig. 5.1):* The logic symbols correspond with the IEC standard DIN 40900; they describe the entire logic function of the cell [5.1 or 5.2]: For the experienced reader, the IEC symbols describe the cell's function sufficiently. However, more common is a function table.
- *Function table (Fig. 5.2):* The function table contains the logical dependencies between the inputs and the outputs of a cell. In the first line, the input and output names, uniform for the entire VENUS system, are listed. The second line is used

Chart 5.3. Overview of the ECL-Z-family, arranged in functional groups

Logic element	Cell name	Function
1.	**Gate**	
1.1	**OR/NOR**	
OR145	KGT01	OR/NOR, with ENABLE
OR174	KGT02	OR/NOR, with ENABLE
OR146	KGT03	OR/NOR, with ENABLE and expander input
OR147	KGT04	OR/NOR, with ENABLE
OR175	KGT05	OR/NOR, with ENABLE and expander input
OR149	KGT06	OR/NOR, with ENABLE and expander input
OR176	KGT06	NOR
OR151	KGT07	OR/NOR, with expander input
OR172	KGT08	OR/NOR, with expandable ENABLE and expander input
1.2	**EXOR/EXNOR**	
EXOR44	KEX03	EXOR/EXNOR, 2 + 2 inputs
EXOR46	KEX04	2 OR-EXOR/NOR, 2 + 2 inputs each, one expander input each
1.3	**OR-AND/OR**	
OA45	KGT22	OR-AND/NAND, 2 + 1 inputs
OA46	KGT22	OR-AND, 2 + 1 inputs
OA51	KGT20	OR-AND/NAND, 3 + 2 inputs
OA52	KGT21	OR-AND/NAND, 3 + 2 + 2 + 2 inputs
OA53	KGT23	OR-AND/NAND, 3 + 2 inputs
OA54	KGT24	OR-AND/NAND, 3 + 2 + 2 + 3 inputs
2.	**Multiplexers**	
MU60	KMX01	2 2-bit multiplexers with common select-input
MU61	KMX02	3 2-bit multiplexers with common select-input
MU62	KMX05	2 4-bit multiplexers with common select-input
MU63	KMX06	4 2-bit multiplexers
MU66	KMX07	4 2-bit multiplexers with expandable select-input

Logic element	Cell name	Function
3.	**Flipflops**	
3.1	**D-flipflops**	
FFMS27	KMS02	1 D-masterslave flipflop with RESET
FFMS21	KMS04	2 D-masterslave flipflops
FFMS28	KMS05	1 D-masterslave flipflop with SET and 2-bit input multiplexer
FFMSP2	KMS81	2 D-masterslave flipflops with SET, expandable clock input and scan path
FFMSP3	KMS83	1 D-masterslave flipflop with SET, expandable clock input, 2-bit input multiplexer and scan path
3.2	**RS-flipflops**	
FFRSP2	KRS81	2 RS-flipflops with 3 set- and reset- inputs each and scan path
4.	**Latches**	
FFD101	KLA01	2 D-flipflops with 2 common clock inputs
FFD102	KLA03	2 D-flipflops with 2-bit input multiplexers
FFD119	KLA05	2 D-flipflops with RESET and expandable clock input
FFD1PB6	KLA81	3 D-flipflops with expandable clock input and scan path
FFD1PB7	KLA83	2 D-flipflops with expandable clock input, 2-bit input multiplexer and scan path
FFD1PB8	KLA85	2 D-flipflops with expandable clock input and scan path
FFD1PB9	KLA01	1 D-flipflop with SET, RESET and scan path
5.	**Arithmetic elements**	
ADD14	KEX02	1-bit full-adder
ADD15	KVA83	2-bit adder with ripple carry
ADDP1	KAU01	half-adder
ALUP1	KAU02	half-adder
ALUP2	KAU09	decoder and driver for control signals
CPG1	KAU11	4-bit carry-look-ahead
ADDP3	KAU21	2-bit sum generator for adder
ALUP3	KAU23	2-bit sum generator for ALU
ALUP9	KAU23	binary BCD converter for ALU
CPG3	KAU81	8-bit carry-look-ahead
ALUP12	KAU84	binary BCD converter for ALU

Logic element	Cell name		Function
6.	Decoder		
DEM21	KDM03		demultiplexer 1 out of 4 with 2 data inputs
DEM20	KDM81		demultiplexer 1 out of 8 with 2 data inputs
7.	Padcells		
7.1	Input cells		
7.1.1	OR/NOR		
OR149	KGTE1		OR/NOR with expander input
OR173	KGTE2		OR/NOR
7.1.2	Input level converter		
EIN1	KITE1		input level converter, inverter
EIN2	KITE3		input level converter, inverter
7.1.3	Input level converter, combinations		
EIN 17	KKOED	*	2 OR/NOR input level converters with 2 inputs and 1 input level converter/inverter
EIN21	KKOEI	*	3 OR/NOR input level converters with 2 inputs
EIN7	KKOE1		1 OR/NOR input level converter with 2 inputs and 1 input level converter/inverter
EIN3	KKOE2		1 OR/NOR input level converter with 2 inputs
EIN3	KKOE3		1 OR/NOR input level converter with 2 inputs
EIN9	KKOE5		2 OR/NOR input level converters with 2 inputs and 1 input level converter/inverter
EIN11	KKOE6		2 OR/NOR input level converters with two inputs
EIN11	KKOE7		2 OR/NOR input level converters with two inputs
EIN13	KKOE9	*	3 OR/NOR input level converters with 2 inputs and 1 input level converter/inverter
7.1.4	Differential line receiver		
ECLDIF	KDFE1A		input level converter/inverter with differential line receiver
7.2	Output cells		
7.2.1	OR/NOR		
OR150	KGTA4		OR/NOR with expander input

Logic element	Cell name		Function
7.2.2	**Output multiplexer**		
MU64	KMXA1		2-bit multiplexer with expandable address and output level converter
MU65	KMXA2		2-bit multiplexer with expandable address and inverted output level converter
MU66	KMXA4		2-bit multiplexer with expandable address
MU67	KMXA5		2-bit multiplexer with external address and two OR-data inputs
MU73	KMXA6	*	2-bit multiplexer with expandable address and output level converter
MU74	KMXA7	*	2-bit multiplexer with expandable address and inverted output level converter
7.2.3	**Output multiplexer with delay**		
MU79	KMDA1	*	2-bit multiplexer with expandable address and output level converter
MU80	KMDA2	*	2-bit multiplexer with expandable address and inverting output level converter
MU81	KMDA6	*	2-bit multiplexer with expandable address and output level converter
MU82	KMDA7	*	2-bit multiplexer with expandable address and inverted output level converter
7.2.4	**Output level converter**		
EO5	KOTA1	*	NOR-output level converter with expandable input
EO7	KOTA2	*	OR-output level converter with expandable input
EO1	KOTA5		OR-output level converter with expandable input
EO3	KOTA6		NOR-output level converter with expandable input
OR152	KOTA7		OR/NOR with expandable input and OR-output level converter
OR152	KOTA7		OR/NOR with expandable input and NOR-output level converter
7.2.5	**Output level converter with delay**		
EO11	KODA1	*	OR-output level converter with expandable input
EO13	KODA2	*	NOR-output level converter with expandable input
EO15	KODA5	*	OR-output level converter with expandable input
EO17	KODA6	*	NOR-output level converter with expandable input
OR177	KODA7	*	OR/NOR with expandable input and OR output level converter
OR178	KODA8	*	OR/NOR with expandable input and NOR output level converter

Logic element	Cell name		Function
7.2.6	**Output level converter combinations**		
EO19	KKOAA	*	OR/NOR-output level converter with expandable input
EO19	KKOAB	*	OR/NOR-output level converter with expandable input
EO19	KKOAC	*	OR/NOR-output level converter with expandable input
OR182	KKOAD	*	OR/NOR with expandable input and OR/NOR-output level converter
OR182	KKOAE	*	OR/NOR with expandable input and OR/NOR-output level converter
OR182	KKOAF	*	OR/NOR with expandable input and OR/NOR-output level converter
MU84	KKOAG	*	2-bit multiplexer with expandable address and inverted output level converter
MU84	KKOAH	*	2-bit multiplexer with expandable address and inverted output level converter
MU84	KKOAI	*	2-bit multiplexer with expandable address and inverted output level converter
EO9	KKOA1		OR/NOR-output level converter with expander input
EO9	KKOA2		OR/NOR-output level converter with expander input
EO9	KKOA3		OR/NOR-output level converter with expander input
OR155	KKOA4		OR/NOR with expander input and OR/NOR-output level converter
OR155	KKOA5		OR/NOR with expander input and OR/NOR-output level converter
OR155	KKOA6		OR/NOR with expander input and OR/NOR-output level converter
MU75	KKOA7		2-bit multiplexer with expandable address and inverting and non-inverting output level converter
MU75	KKOA8		2-bit multiplexer with expandable address and inverting and non-inverting output level converter
MU75	KKOA9		2-bit multiplexer with expandable address and inverting and non-inverting output level converter
8.	**Extension units**		
POR11	-		expander unit with 2 inputs
POR12	-		expander unit with 4 inputs

for comments (for example: CLOCK). Positive logic is used, i.e. H = V_{DD}-level, L = V_{SS}-level.

- *Dimensions (Fig. 5.3):* The cell's dimensions (height and width) are given in µm.
- *Number of equivalent gates (Fig. 5.3):* The number of equivalent gates is a measure of a cell's complexity. It results from the total number of its transistors minus driver transitors, divided by four. (A 2-input-NAND gate contains just four logic tran-

Logic symbol

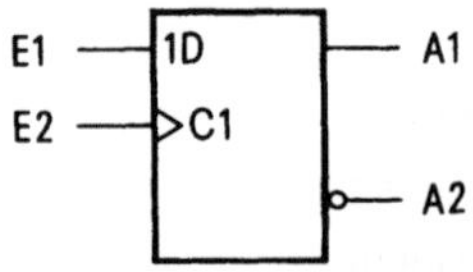

Fig. 5.1. Logic symbol DF01

Function table

Inputs		Outputs	
E1	E2	A1	A2
	Clock	Q	QN
H L	↑ ↑	H L	L H

Fig. 5.2. Function table DF01

Dimensions

[μm]	A	B
Width	156	154
Height	240	255

Gate equivalents
6.5

Fig. 5.3. Dimensions and equivalent gates DF01

Timing

[ns]	From (input)	To (output)	Fast	Typ	Slow
t_s	E1; E2*	—	3.0	5.0	13
t_h	E1; E2*	—	1.0	1.0	2.5
t_{dHL}	E2	A1, A2	3.5	7.0	20
t_{dLH}	E2	A1, A2	2.0	4.0	10
t_{CWL}	E2 @	—	5.5	8.5	18
t_{CWH}	E2 @	—	5.0	6.0	11
Type: $V_{DD} = 5$ V, T = 300 K, $C_L = 225$ fF					

* see timing diagram 1
@ see timing diagram 7

Fig. 5.4. Timing DF01

sistors.) When the numbers are added up, the result is the total number of equivalent gates of a chip; this is the number commonly used for measuring a chip's complexity.

- *Timing behavior (Fig. 5.4):* For estimating the maximum clock frequencies attainable, data about the delay times are required. Delay time means the period of time between the point when the switching threshold of the input signal is reached and the point when the switching threshold of the output signal is reached. The symbol used for the delay time is t_d. Here, two different cases have to be distinguished: a low high transition from 0 to 1 (LH) and a high-low transition from 1 to 0 (HL). Because of the different length of cell internal paths and the resulting difference of the capacitive loads, the delay times also depend on the actual input-output combination.

In addition the delay times are determined by many other values:

- The supply voltage range (operational values) permitted is 3 V to 6 V. A higher supply voltage causes shorter delay times.
- The operational temperature (environment) is 0 °C to 70 °C. An increase in temperature increases the delay times.

- Technological variations result in the fact that certain physical and geometric parameters cannot be reproduced precisely but only within certain band widths. If these values are combined such that short delay times result this is called FAST parameter set. A variation in the other direction (SLOW parameter set) leads to longer delay times. Nominal delay times are obtained if the TYP parameter set is applied.
- The higher the capacitive load C_L of an output the longer are the delay times.

The timing of the cells is indicated in the data sheet under three conditions:

- TYP: TYP parameter set, nominal geometry
  ambient temperature $T = 27\,°C$
  supply voltage $V_{DD} = 5\,V$
- FAST: FAST parameter set, minimum channel length, maximum channel width
  ambient temperature $T = 27\,°C$
  supply voltage $V_{DD} = 5\,V + 10\%$
- SLOW: SLOW parameter set, maximum channel length, minimum channel width
  ambient temperature $T = 120\,°C$ (junction)
  supply voltage $V_{DD} = 5\,V - 10\%$

A capacitive load of $C_L = 225$ fF is assumed for all core cells by default. This corresponds with the load of a typical input and a piece of aluminum wire with a length of 400 μm. For the output driver a considerably higher output load has been assumed: $C_L = 50$ pF. Approximation formulas are used to attain rough estimates; thus deviating load capacities can be taken into account.

All delay times given in the cell catalog have been determined by means of a circuit simulator and have been confirmed by measurements.

- *Logic circuit (Fig. 5.5)*: The cell concept is based upon the idea of considering cells as "*black boxes*": For the user it is irrelevant how the cell has been realized if its function is the same as is stated in the data sheet. Nevertheless, the VENUS data

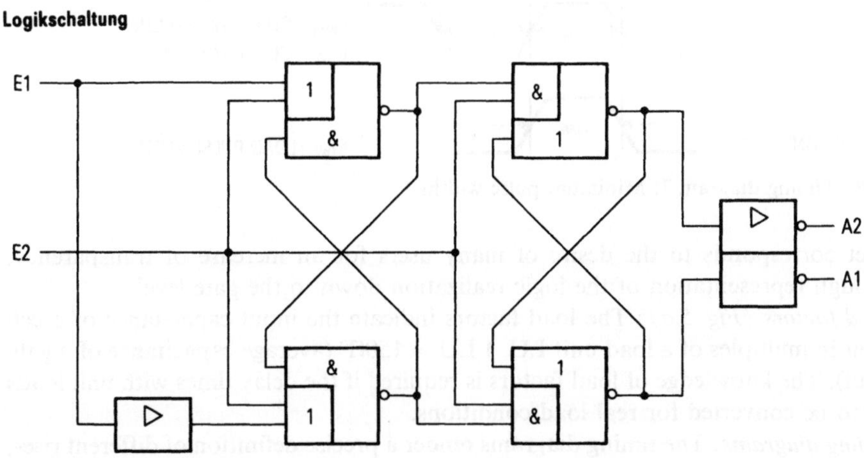

Fig. 5.5. Logic circuit DF01

Input	LU
E1	2
E2	4

Fig. 5.6. Load factors DF01

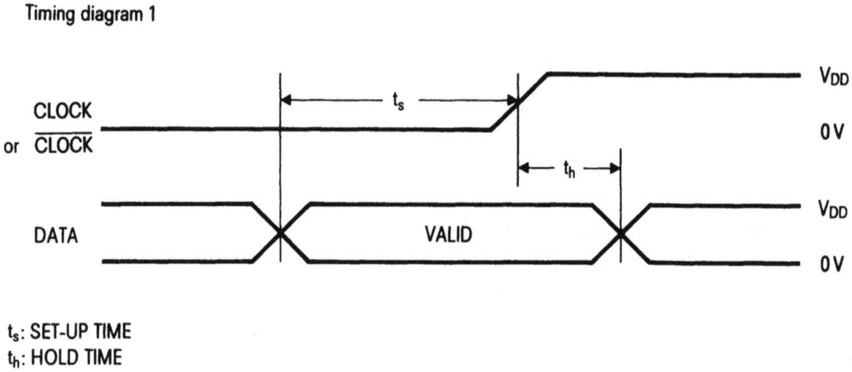

Fig. 5.7. Timing diagram 1: Set-up and hold times

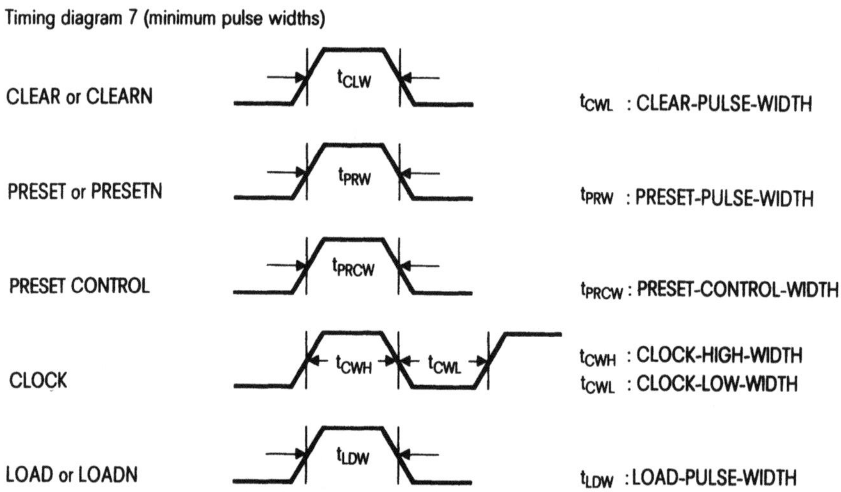

Fig. 5.8. Timing diagram 7: Minimum pulse widths

sheet corresponds to the desire of many users for an increase of transparency, through representation of the logic realization down to the gate level.

- *Load factors (Fig. 5.6):* The load factors indicate the input capacitance of a cell input in multiples of a load unit LU. 1 LU = 150fF (average capacitance of a gate input). The knowledge of load factors is required if the delay times with unit loads are to be converted for real load conditions.
- *Timing diagrams:* The timing diagrams render a precise definition of different rise-, fall- and delay-times as well as set-up- and hold-times. Figures 5.7 and 5.8 give as

examples the timing diagrams 1 (set-up- and hold-times) and 7 (minimum pulse width) for the cell DF01.

The data sheets for the gate-array cells differ only in the data concerning cell size: there the number of basic cells used is given.

For a detailed description of the data sheet please refer to the manual "SEMI-CUSTOM: Zellenorientierter Bausteinentwurf" [6.1].

The format of the ECL data sheet is (still) different from that of the CMOS families but the content is, in principle, the same. What must be considered is that several logic elements (i.e. logic cells) can be allocated to one physicall cell. This corresponds with the procedure for "discrete" component families: In one IC for the TTL series, for example, several gate functions which are independent from each other may be realized. More detailed information and a detailed description of the data sheets are included in the ECL gate-array family design manual [5.3].

5.1.3 Excerpt from the Cell Catalogs of the A/B-Families

In order to give better insight, a representative selection of 19 data sheets from the 3 μm CMOS libraries (A- and B-families) follows. These cells should be sufficient in enabling the reader to realize simple designs by himself.

N-INPUT AND, N = nn 2...8		CANnnA/CANnnB	P. 1 +
VERSION 1.3	CELL GROUP	AND/NAND	1.1

LOGIC SYMBOL

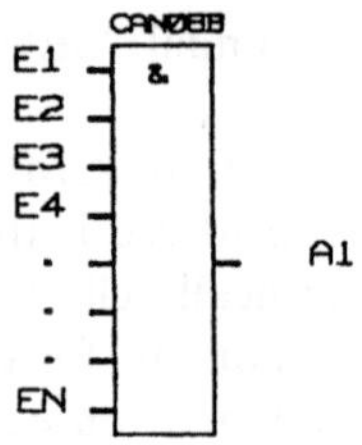

FUNCTION TABLE	
INPUT	OUTPUT
E1 E2 ... EN	A1
H H H	H
else	L

AC-CHARACTERISTICS CL = 225fF						
N	FROM	TO	2	3	4	5
[ns]	(INPUT)	(OUTPUT)	TYP	TYP	TYP	TYP
tdHL	E1,..EN	A1	2.0	2.5	2.5	3.5
tdLH	E1,..EN	A1	2.5	3.5	4.5	5.5

N	6	7	8
[ns]	TYP	TYP	TYP
tdHL	4.0	4.0	4.0
tdLH	5.5	6.5	6.5

LOAD UNITS	TYP
INPUT	[LU]
E1	1
:	:
EN	1

N-INPUT AND, N = nn = 2...8		CANnnA/CANnnB	P. 2 -
VERSION: 1.3	CELL GROUP:	AND/NAND	1.1

LOGIC CIRCUIT

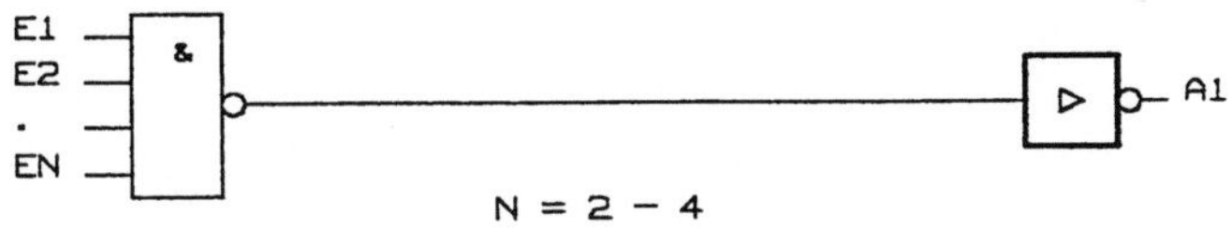

N = 2 - 4

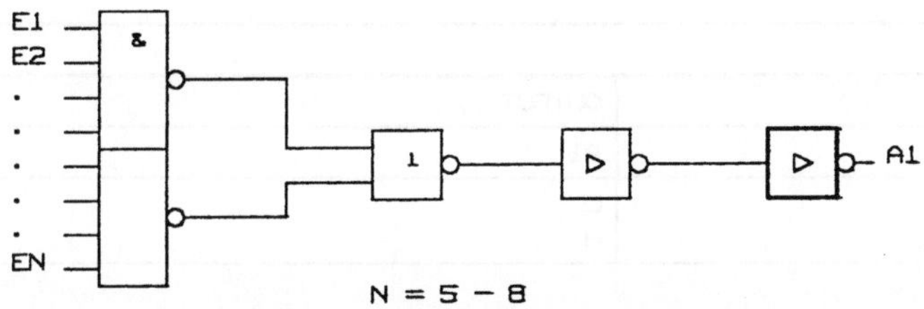

N = 5 - 8

EQUIVALENT GATES							
N	2	3	4	5	6	7	8
EQU. GATES	1.5	2	2.5	3.5	4	4.5	5

DIMENSIONS														
N	2		3		4		5		6		7		8	
[µm]	A	B	A	B	A	B	A	B	A	B	A	B	A	B
WIDTH	48	56	60	70	72	70	96	98	96	98	108	112	120	126
HEIGHT	240	255	240	255	240	255	240	255	240	255	240	255	240	255

N-INPUT NAND, N = NN = 2...8		CNAnnA/CNAnnB	P. 1 +
VERSION: 1.3	CELL GROUP:	AND/NAND	1.1

LOGIC SYMBOL

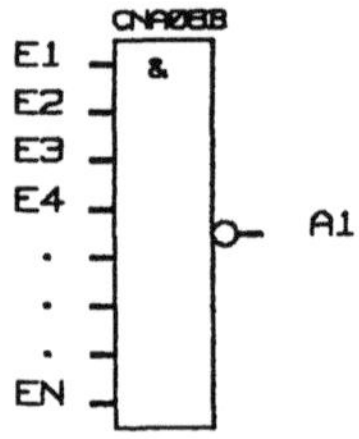

FUNCTION TABLE	
INPUT	OUTPUT
E1 E2 ... EN	A1
H H H	L
ELSE	H

AC-CHARACTERISTICS CL = 225fF							
n	2	3	4	5	6	7	8
[ns]	TYP	TYP	TYP	TYP	TYP	TYP	TYP
tdHL	3.5	4.0	5.5	5.5	5.5	6.0	6.0
tdLH	2.5	3.0	3.0	3.0	3.0	3.0	3.5

LOAD UNITS	TYP
INPUT	[LU]
E1 : EN	1 : 1

N-INPUT NAND, N = nn = 2...8		CNAnnA/CNAnnB	P. 2 -
VERSION: 1.3	CELL GROUP:	AND/NAND	1.1

LOGIC CIRCUIT

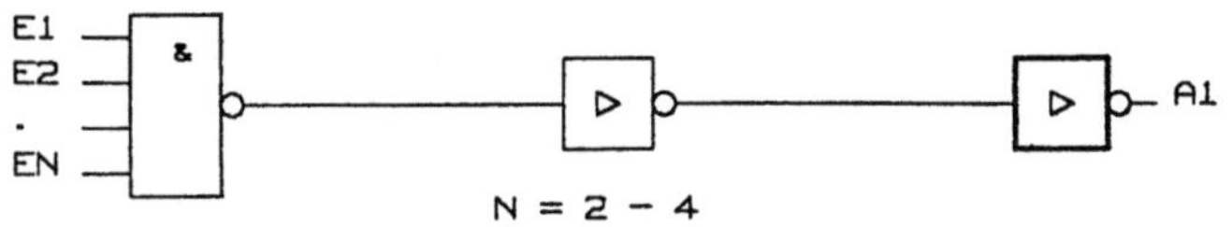

N = 2 - 4

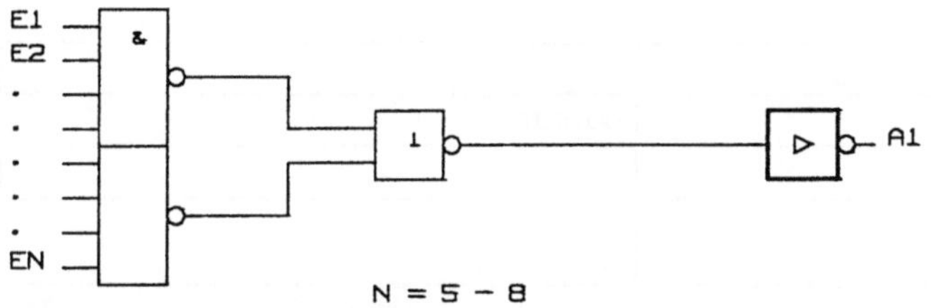

N = 5 - 8

EQUIVALENT GATES							
N	2	3	4	5	6	7	8
EQU. GATES	1	1.5	2	4	4.5	5	5.5

DIMENSION														
N	2		3		4		5		6		7		8	
[µm]	A	B	A	B	A	B	A	B	A	B	A	B	A	B
WIDTH	60	56	60	70	72	70	84	84	96	98	108	112	120	126
HEIGHT	240	255	240	255	240	255	240	255	240	255	240	255	240	255

N-INPUT NOR, N = NN = 2...8		CN0nnA/CN0nnB	P. 1 +
VERSION: 1.3	CELL GROUP:	OR/NOR	1.2

LOGIC SYMBOL

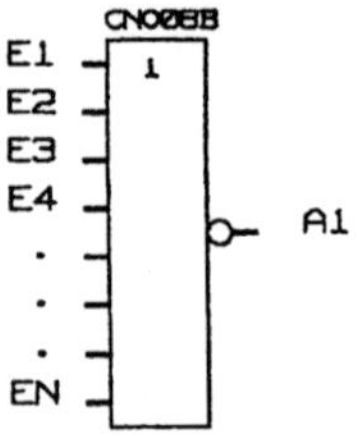

FUNCTION TABLE	
INPUT	OUTPUT
E1 E2 ... EN	A1
L L ... L OTHERWISE	H L

AC-CHARACTERISTICS CL = 225fF							
n	2	3	4	5	6	7	8
[ns]	TYP	TYP	TYP	TYP	TYP	TYP	TYP
tdHL	3.0	3.0	3.0	3.5	3.5	3.5	3.5
tdLH	3.5	4.5	6.0	5.5	5.5	6.5	7.0

LOAD UNITS	TYP
INPUT	[LU]
E1 : EN	1 : 1

N-INPUT NOR, N = nn = 2...8		CNOnnA/CNOnnB	P. 2 -
VERSION: 1.3	CELL GROUP:	OR/NOR	1.2

LOGIC CIRCUIT

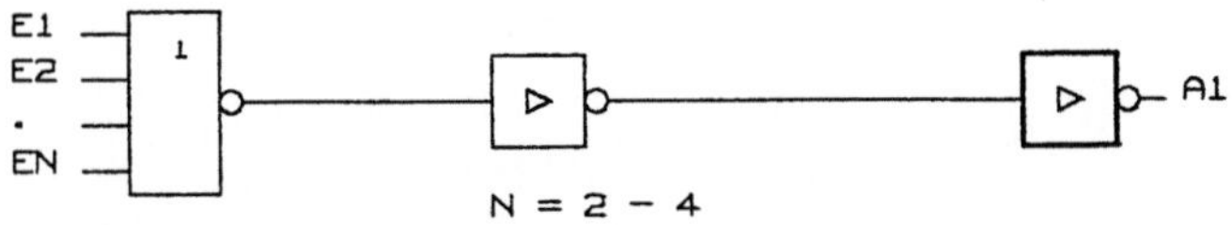

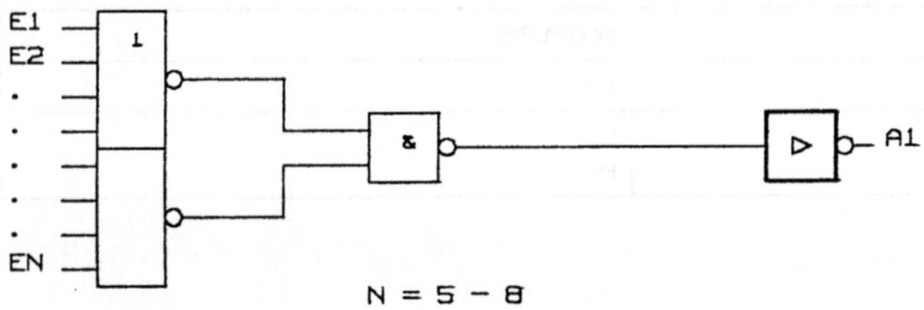

EQUIVALENT GATES							
N	2	3	4	5	6	7	8
EQU. GATES	1	1.5	2	4	4.5	5	5.5

DIMENSION														
N	2		3		4		5		6		7		8	
[µm]	A	B	A	B	A	B	A	B	A	B	A	B	A	B
WIDTH	60	56	60	70	72	70	84	84	96	98	108	112	120	126
HEIGHT	240	255	240	255	240	255	240	255	240	255	240	255	240	255

N-INPUT OR, N = nn = 2...8		CORnnA/CORnnB	P. 1 +
VERSION: 1.3	CELL GROUP:	OR/NOR	1.2

LOGIC SYMBOL

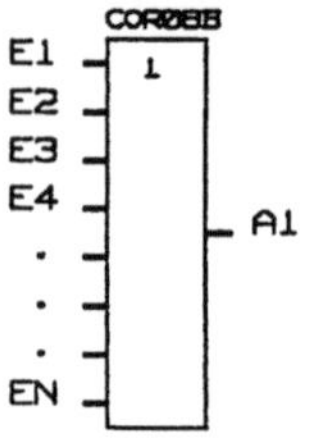

FUNCTION TABLE	
INPUTS	OUTPUTS
E1 E2 ... EN	A1
L L ... L	L
ELSE	H

AC-CHARACTERISTICS CL = 225fF							
N	2	3	4	5	6	7	8
[ns]	TYP	TYP	TYP	TYP	TYP	TYP	TYP
tdHL	3.5	4.5	6.0	6.0	6.0	7.0	7.0
tdLH	2.0	2.5	2.5	3.5	4.0	4.0	3.5

LOAD UNITS	TYP
INPUT	[LU]
E1 : EN	1 : 1

N-INPUT OR, N = nn = 2...8		CORnnA/CORnnB	P. 2 -
VERSION: 1.3	CELL GROUP:	OR/NOR	1.2

LOGIC CIRCUIT

N = 2 - 4

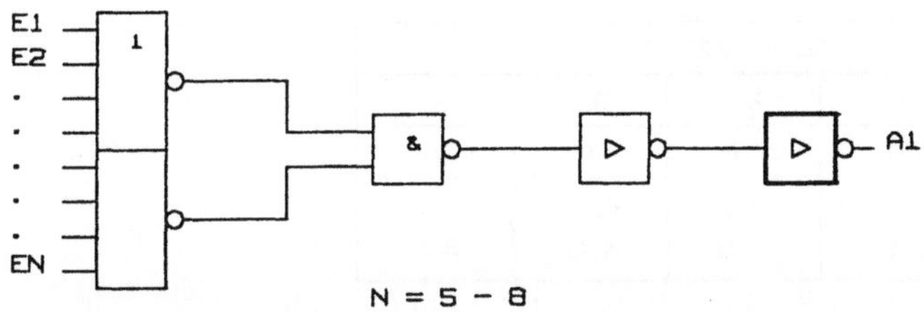

N = 5 - 8

EQUIVALENT GATES							
N	2	3	4	5	6	7	8
EQU. GATES	1.5	2	2.5	3.5	4	4.5	5

DIMENSIONS														
N	2		3		4		5		6		7		8	
[µm]	A	B	A	B	A	B	A	B	A	B	A	B	A	B
WIDTH	48	56	60	56	72	70	96	98	96	98	108	112	120	126
HEIGHT	240	255	240	255	240	255	240	255	240	255	240	255	240	255

BUFFER/DRIVER, NON-INVERTING, STRENGTH s = 0...4		CDRs2A/CDRs2B	P. 1 -
VERSION: 1.3	CELL GROUP:	BUFFER/DRIVER	1.8

LOGIC SYMBOL

FUNCTION TABLE	
INPUTS	OUTPUTS
E1	A1
H	H
L	L

AC-CHARACTERISTICS CL = 225fF					
s	0	1	2	3	4
[ns]	TYP	TYP	TYP	TYP	TYP
tdHL	2.0	3.0	3.5	4.5	4.0
tdLH	2.0	2.5	2.5	4.0	3.5

LOAD UNITS	
INPUT	[LU]
E1	1

LOGIC CIRCUIT

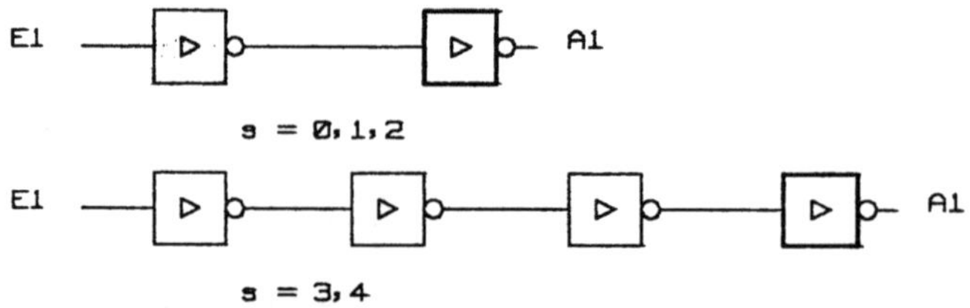

EQUIVALENT GATES
1

DIMENSION [μm]										
s	0		1		2		3		4	
	A	B	A	B	A	B	A	B	A	B
WIDTH	48	56	84	84	120	126	144	154	156	168
HEIGHT	240	255	255	240	255	240	255	240	255	240

BUFFER/DRIVER, INVERTING, STRENGTH s=0..4		CDRs1A/CDRs1B	P. 1 -
VERSION: 1.3	CELL GROUP:	BUFFER/DRIVER	1.8

LOGIC SYMBOL

FUNCTION TABLE	
INPUT	OUTPUT
E1	A1
L H	H L

AC-CHARACTERISTICS CL = 225fF					
s	0	1	2	3	4
[ns]	TYP	TYP	TYP	TYP	TYP
tdHL	3.0	3.0	3.0	3.5	3.5
tdLH	3.0	3.0	3.0	3.5	3.5

LOAD UNITS	
INPUT	[LU]
E1	1

LOGIC CIRCUIT

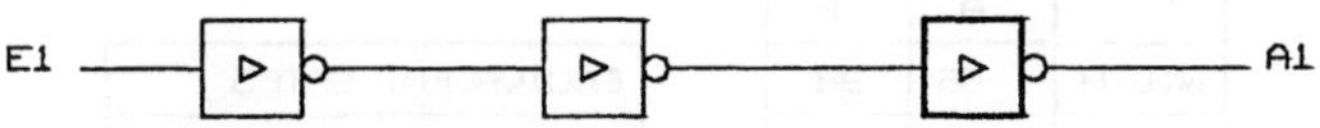

EQUIVALENT GATES
0.5

DIMENSION [µm]										
s	0		1		2		3		4	
	A	B	A	B	A	B	A	B	A	B
WIDTH	48	56	96	98	120	140	132	140	144	154
HEIGHT	240	255	255	240	255	240	255	240	255	240

LATCH		CLA01A/CLA01B	P.1 -
VERSION: 1.3	CELL GROUP:	LATCHES	1.9

LOGIC SYMBOL

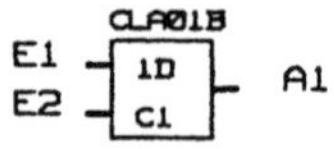

FUNCTION TABLE		
INPUT		OUTPUT
E1	E2	A1
	LOAD	
L	H	L } = E1
H	H	H } = E1
X	L	NO CHANGE

AC-CHARACTERISTICS			CL = 225fF
[ns]	FROM (INPUT)	TO (OUTPUT)	TYP
ts	E1; E2 *	-	2.0
th	E1; E2 *	-	2.0
tdHL	E1 (E2=H)	A1	3.0
tdLH	E1 (E2=H)	A1	3.0
tdHL	E2	A1	4.5
tdLH	E2	A1	4.0
tLDW	E2 ✱	-	4.0

* → TIMING DIAGRAM 11.2 ✱ → TIMING DIAGRAM 7

LOAD UNITS	
INPUT	[LU]
E1	1
E2	1

DIMENSION [μm]		
	A	B
WIDTH	96	98
HEIGHT	240	255

EQUIVALENT GATES
3

LOGIC CIRCUIT

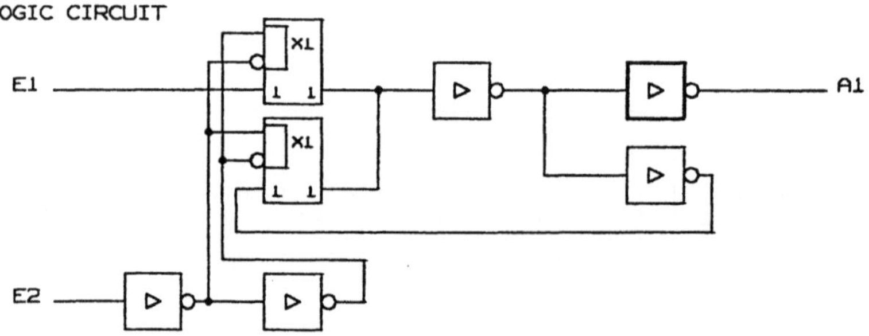

D-FLIPFLOP WITH PRESET AND CLEAR		CDFØ3A/CDFØ3B	P. 1 +
VERSION: 1.3	CELL GROUP:	D-FLIPFLOPS	1.1Ø

LOGIC SYMBOL

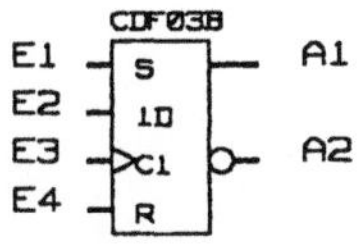

FUNCTION TABLE						
INPUT				OUTPUT		NOTE
E2	E3	E4	E1	A1	A2	
	CLOCK	CLEAR	PRESET	Q	QN	
X	X	H	H	H	H	FORBIDDEN
X	X	L	H	H	L	
X	X	H	L	L	H	
H	↑	L	L	H	L	
L	↑	L	L	L	H	

AC-CHARACTERISTICS			CL = 225fF
[ns]	FROM (INPUT)	TO (OUTPUT)	TYP
ts	E2; E3*	–	8.5
th	E2; E3*	–	Ø
th	E1, E4; E3+	–	Ø
tR	E1, E4; E3&	–	7.5
tdHL	E3	A1, A2	8.5
tdLH	E3	A1, A2	4.Ø
tdHL	E1	A2	6.5
tdLH	E1	A1	2.5
tdHL	E4	A1	7.Ø
tdLH	E4	A2	2.5
tCWL	E3 @	–	7.5
tCWH	E3 @	–	6.5
tCLW	E4 @	–	5.5
tPRW	E1 @	–	5.5

* → TIMING DIAGRAM 1 + → TIMING DIAGRAM 9
@ → TIMING DIAGRAM 7 & → TIMING DIAGRAM 12

D-FLIPFLOP WITH PRESET AND CLEAR		CDF03A/CDF03B	P. 2 -
VERSION: 1.3	CELL GROUP:	D-FLIPFLOPS	1.10

LOAD UNITS	
INPUT	[LU]
E1	2
E2	2
E3	4
E4	2

LOGIC CIRCUIT

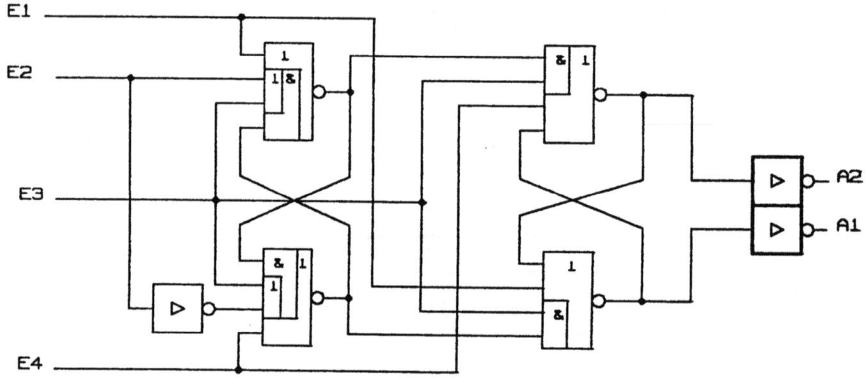

EQUIVALENT GATES
8.5

DIMENSION [µm]		
	A	B
WIDTH	180	182
HEIGHT	240	255

D-FLIPFLOP WITH CLEAR		CDF02A/CDF02B	P. 1 +
VERSION: 1.3	CELL GROUP:	D-FLIPFLOPS	1.10

LOGIC SYMBOL

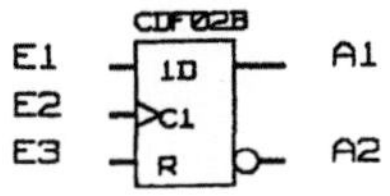

FUNCTION TABLE				
INPUT			OUTPUT	
E1	E2	E3	A1	A2
	CLOCK	CLEAR	Q	QN
X	X	H	L	H
H	↑	L	H	L
L	↑	L	L	H

AC-CHARACTERISTICS			CL = 225fF
[ns]	FROM (INPUT)	TO (OUTPUT)	TYP
ts	E1; E2 *	–	7.5
th	E1; E2 *	–	0
th	E3; E2+	–	0
tR	E3; E2 &	–	7.5
tdHL	E2	A1	7.5
tdLH	E2	A1	4.0
tdHL	E2	A2	8.5
tdLH	E2	A2	3.5
tdHL	E3	A1	4.5
tdLH	E3	A2	2.0
tCWL	E2 @	–	7.0
tCWH	E2 @	–	7.5
tCLW	E3 @	–	4.5

* → TIMING DIAGRAM 1   + → TIMING DIAGRAM 9
@ → TIMING DIAGRAM 7   & → TIMING DIAGRAM 12

LOAD UNITS	
INPUT	[LU]
E1	2
E2	4
E3	2

D-FLIPFLOP WITH CLEAR		CDF02A/CDF02B	P.2 -
VERSION: 1.3	CELL GROUP:	D-FLIPFLOPS	1.10

LOGIC CIRCUIT

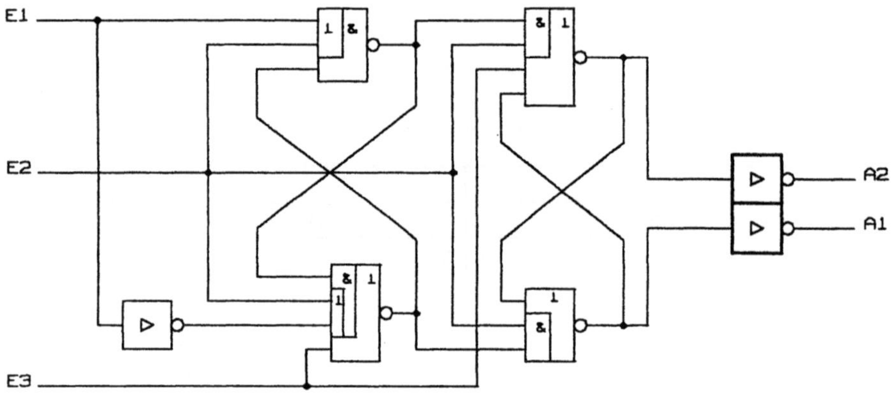

EQUIVALENT GATES
7.5

DIMENSION [µm]		
	A	B
WIDTH	168	168
HEIGHT	240	255

JK-FLIPFLOP WITH CLEAR		CJK05A/CJK05B	P. 1 +
VERSION: 1.2	CELL GROUP:	JK/T-FLIPFLOPS	1.12

LOGIC SYMBOL

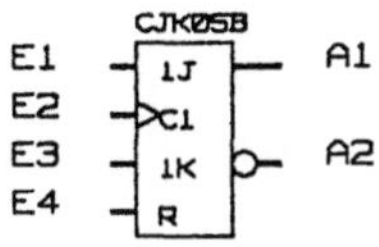

FUNCTION TABLE					
INPUT				OUTPUT	
E1	E2	E3	E4	A1	A2
	CLOCK		CLEAR	Q	QN
X	X	X	H	L	H
L	↑	L	L	NO CHANGE	
L	↑	H	L	L	H
H	↑	L	L	H	L
H	↑	H	L	TOGGLE	

AC-CHARACTERISTICS			CL = 225fF
[ns]	FROM (INPUT)	TO (OUTPUT)	TYP
ts	E1, E3; E2*	–	11
th	E1, E3; E2*	–	0
th	E4; E2 +	–	0
tR	E4; E2 &	–	7.5
tdHL	E2	A1	8.0
tdLH	E2	A1	4.0
tdHL	E2	A2	10
tdLH	E2	A2	4.5
tdHL	E4	A1	6.0
tdLH	E4	A2	2.5
tCWL	E2 @	–	7.0
tCWH	E2 @	–	8.5
tP	E2 #	–	19
tCLW	E4 @	–	4.5

* → TIMING DIAGRAM 1
@ → TIMING DIAGRAM 7
# → TIMING DIAGRAM 5
+ → TIMING DIAGRAM 9
& → TIMING DIAGRAM 12

JK-FLIPFLOP WITH CLEAR		CJKØ5A/CJKØ5B	P. 2 -
VERSION: 1.2	CELL GROUP:	JK/T-FLIPFLOPS	1.12

LOAD UNITS	
INPUT	[LU]
E1	1
E2	4
E3	1
E4	2

LOGIC CIRCUIT

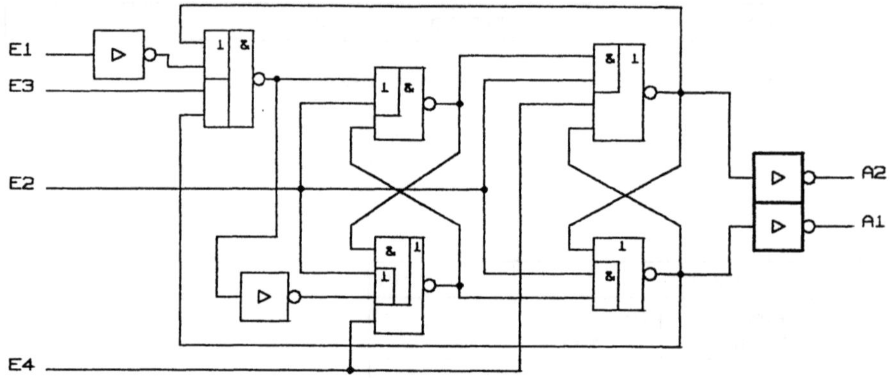

EQUIVALENT GATES
1Ø

DIMENSION [µm]		
	A	B
WIDTH	216	21Ø
HEIGHT	24Ø	255

COUNTER, SYNCHRONOUS, SYNCHRONOUS LOAD		CBC02A/CBC02B	P. 1 +
VERSION: 1.2	CELL GROUP:	COUNTER	1.14

LOGIC SYMBOL

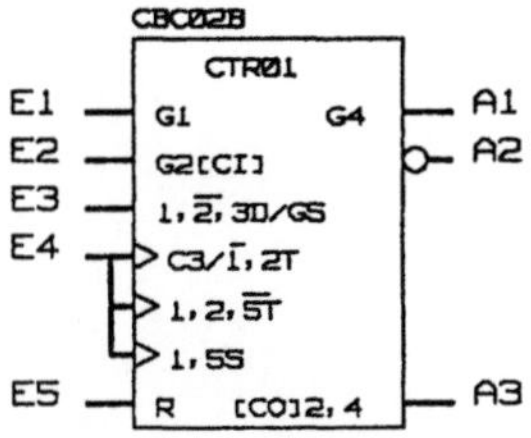

FUNCTION TABLE								
INPUT					OUTPUT			REMARK
E1	E2	E3	E4	E5	A1	A2	A3	
LOAD	CARRY-IN	DATA	CLOCK	CLEAR	Q	QN	CARRY-OUT	
X	X	X	X	H	L	H	L	
H	H	X	↑	L	EXCLUDED			REMARK 1
L	L	X	↑	L	WITHOUT CHANGE		L	
H	L	H	↑	L	H	L	L	SYNC. LOAD
H	L	L	↑	L	L	H	L	SYNC. LOAD
X	H	X	X *	L	WITHOUT CHANGE		H(A1=H) L(A1=L)	**
X	L	X	X *	L	WITHOUT CHANGE		L	**
L	H	X	↑	L	TOGGLE		H(A1=H) L(A1=L)	COUNT UP

* SWITCHING FROM E4=L TO E4=H IS NOT ALLOWED.

** CONTEXT CARRY-IN / CARRY-OUT
BOOLEAN OPERATION: A3 = (E2 AND A1)

REMARK: 1. Please note that synchronous loading is only possible with E1=H and E2=L. For cascaded counters this is achieved by a suitable stimulation of the first cell (bit 0).
→ note 1 in 4.4

2. Carry-out (A3) has no driver. The transition from carry-in (E2) to A3 needs a suitable driving capability at input E2.

COUNTER, SYNCHRONOUS, SYNCHRONOUS LOAD		CBC02A/CBC02B	P. 2 +
VERSION: 1.2	CELL GROUP:	COUNTER	1.14

AC-CHARACTERISTICS				CL=225fF
[ns]	FROM (INPUT)		TO (OUTPUT)	TYP
ts	E3; E4	*	–	10
ts	E1, E2; E4	✱=	–	10
th	E1, E2, E3; E4	✱*=	–	0
th	E5; E4	–	–	0
tR	E5; E4	&	–	7.5
td $	E2		A3	1.0
tdHL	E4		A1	9.5
tdLH	E4		A1	4.5
tdHL	E4		A2	10.5
tdLH	E4		A2	4.5
tdHL	E4		A3	10
tdLH	E4		A3	5.0
tdHL	E5		A1	7.0
tdLH	E5		A2	2.5
tdHL	E5		A3	6.0
tCWL	E4	+	–	7.0
tCWH	E4	+	–	9.0
tP	E4	|	–	22
tCLW	E5	+	–	4.5

* → TIMING DIAGRAM 1
✱ → TIMING DIAGRAM 2
\+ → TIMING DIAGRAM 7
| → TIMING DIAGRAM 5
$ td = tdHL = tdLH
= → TIMING DIAGRAM 11.1
– → TIMING DIAGRAM 9
& → TIMING DIAGRAM 12

LOAD UNITS	
INPUT	[LU]
E1	2
E2 *	3.5
E3	1
E4	4
E5	2

* → REMARK 2

COUNTER, SYNCHRONOUS, SYNCHRONOUS LOAD		CBC02A/CBC02B	P. 3 -
VERSION: 1.2	CELL GROUP:	COUNTER	1.14

LOGIC CIRCUIT

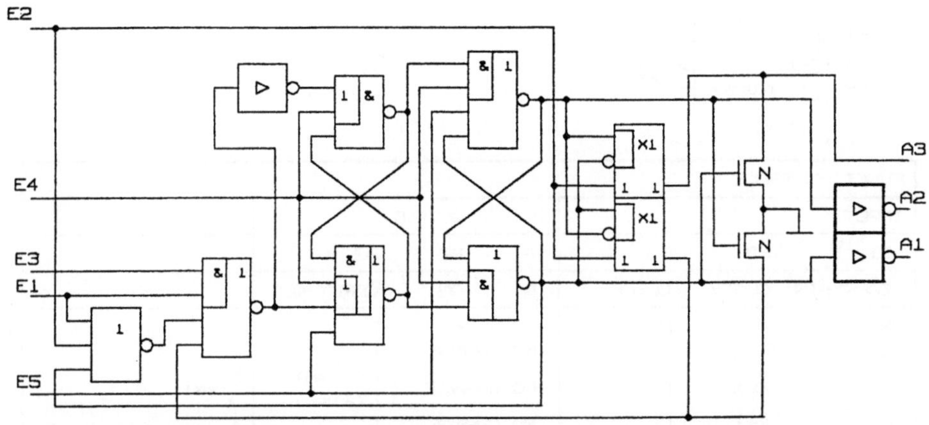

EQUIVALENT GATES
12.5

DIMENSION [µm]		
	A	B
WIDTH	264	266
HEIGHT	240	255

COUNTER, SYNCHRONOUS, WITH CLEAR		CCR02A/CCR02B	P. 1+
VERSION:	CELL GROUP:	COUNTER	1.14

LOGIC SYMBOL

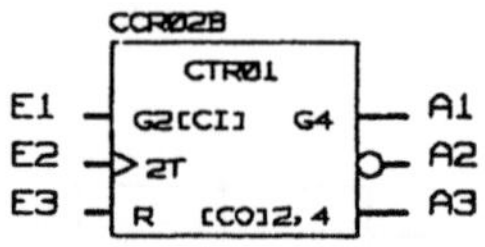

FUNCTION TABLE						
INPUT			OUTPUT			NOTE
E1	E2	E3	A1	A2	A3	
CARRY-IN	CLOCK	CLEAR	Q	QN	CARRY-OUT	
X	X	H	L	H	L	
L	↑	L	NO CHANGE		L	
H	X*)	L	NO CHANGE		H(A1=H) L(A1=L)	**)
L	X*)	L	NO CHANGE		L	**)
H	↑	L	TOGGLE		H(A1=H) L(A1=L)	COUNT UP

*) SWITCHING FROM E4=L TO E4=H IS NOT ALLOWED.
**) CONTEXT CARRY-IN / CARRY-OUT
BOOLEAN OPERATION: A3 = (E2 AND A1)

REMARK: Carry-out (A3) has no driver. The transition from carry-in (E1) to A3 needs a suitable driving capability at input E1.

COUNTER, SYNCHRONOUS, WITH CLEAR		CCR02A/CCR02B	P. 2+
VERSION 1.2	CELL GROUP:	COUNTER	1.14

AC-CHARACTERISTICS CL = 225fF			
[ns]	FROM (INPUT)	TO (OUTPUT)	TYP
tc	E1; E2 *	-	10
th	E1, E3; E2 *+	-	0
tR	E3, E2 &	-	7.5
td $	E1	A3	1.0
tdHL	E2	A1	9.0
tdLH	E2	A1	4.0
tdHL	E2	A2	11
tdLH	E2	A2	4.5
tdHL	E2	A3	9.5
tdLH	E2	A3	4.5
tdHL	E3	A1	6.0
tdLH	E3	A2	3.0
tdHL	E3	A3	6.0
tCWL	E2 @	-	7.0
tCWH	E2 @	-	10
tP	E2 #	-	21
tCLW	E3 @	-	4.5

* → TIMING DIAGRAM 2
@ → TIMING DIAGRAM 7
\# → TIMING DIAGRAM 5
$ td = tdHL = tdLH
\+ → TIMING DIAGRAM 9
& → TIMING DIAGRAM 12

LOAD UNITS	
INPUT	[LU]
E1 *)	3.5
E2	4
E3	2

*) → REMARK P.1

COUNTER, SYNCHRONOUS, WITH CLEAR		CCR02A/CCR02B	P. 3-
VERSION: 1.2	CELL GROUP:	COUNTER	1.14

LOGIC CIRCUIT

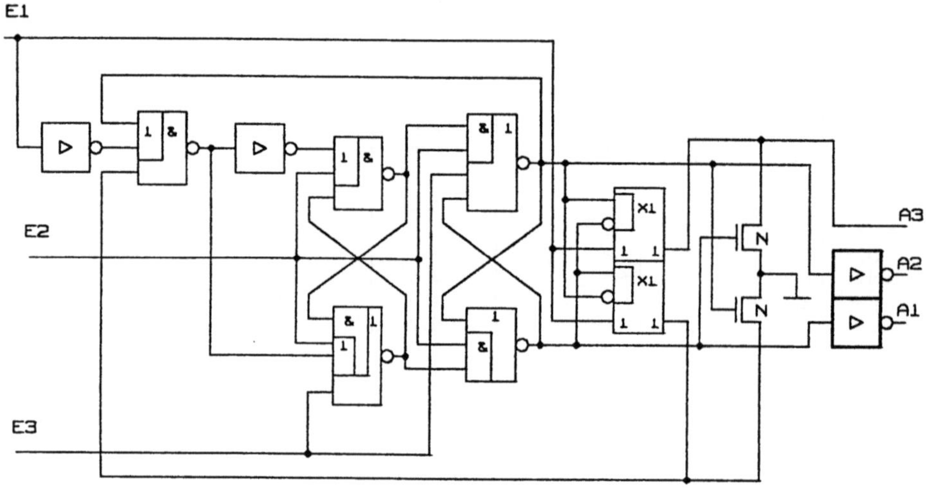

EQUIVALENT GATES
11

DIMENSION [μm]		
	A	B
WIDTH	240	238
HEIGHT	240	255

VDD-CONNECTION		CVD00A/CVD00B	P. 1 -
VERSION: 1.3	CELL GROUP:	OTHERS	1.20

LOGIC SYMBOL

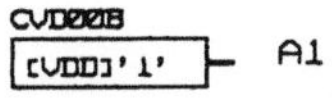

FUNCTION TABLE
OUTPUT
A1
H

LOGIC CIRCUIT

DIMENSION [um]		
	A	B
WIDTH	24	28
HEIGHT	240	255

VSS-CONNECTION		CVS00A/CVS00B	P. 1 -
VERSION: 1.3	CELL GROUP:	OTHERS	1.20

LOGIC SYMBOL

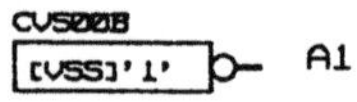

FUNCTION TABLE
OUTPUT
A1
L

LOGIC CIRCUIT

DIMENSION [µm]		
	A	B
WIDTH	24	28
HEIGHT	240	255

INPUT-BUFFER/DRIVER, CMOS COMPATIBLE		CIN01A/CIN01B	P. 1 –
VERSION: 1.3	CELL GROUP:	INPUT-PADS	2.1

LOGIC SYMBOL

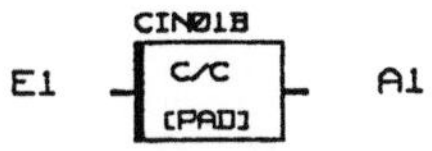

FUNCTION TABLE	
INPUT	OUTPUT
E1	A1
(PAD)	
L H	L H

AC-CHARACTERISTICS			CL = 225fF
[ns]	FROM (INPUT)	TO (OUTPUT)	TYP
tdHL	E1 *	A1	2.0
tdLH	E1 *	A1	2.0

* INPUT: CMOS-LEVEL VIL = 1.5V, VIH = 3.5
→ TIMING DIAGRAM 13.1

LOAD DEPENDENCE	TYP	
OUTPUT	kLH [ns/LU]	kHL [ns/LU]
A1	0.13	0.11

LOAD UNITS	
INPUT	[LU]
E1	6 *

* WITHOUT PACKAGE CAPACITANCE

LOGIC CIRCUIT

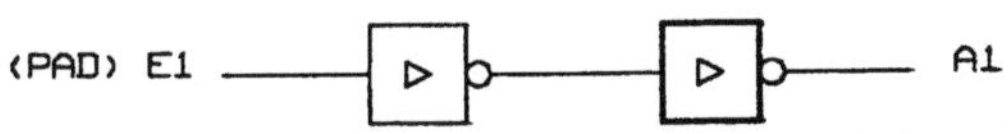

NOTE: The input must not float

DIMENSION [µm]		
	A	B
WIDTH	240	240
HEIGHT	480	480

INPUT-BUFFER/DRIVER, TTL-COMPATIBLE		CIN02A/CIN02B	P. 1 -
VERSION: 1.3	CELL GROUP: INPUT-PADS		2.1

LOGIC SYMBOL

FUNCTION TABLE	
INPUT	OUTPUT
E1	A1
(PAD)	
L H	L H

AC-CHARACTERISTICS			CL = 225fF
[ns]	FROM (INPUT)	TO (OUTPUT)	TYP
tdHL	E1 *	A1	4.5
tdLH	E1 *	A1	2.0

* INPUT: TTL-LEVEL VIL = 0.8V, VIH = 2.0V
→ TIMING DIAGRAM 13.2

LOAD DEPENDENCE	TYP	
OUTPUT	kLH [ns/LU]	kHL [ns/LU]
A1	0.13	0.11

LOAD UNITS	
INPUT	[LU]
E1	6 *

* WITHOUT PACKAGE CAPACITANCE

LOGIC CIRCUIT

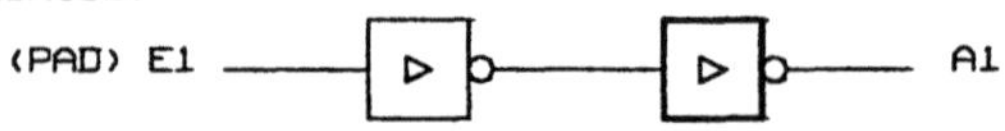

NOTE: The input must not float

DIMENSION [µm]		
	A	B
WIDTH	240	240
HEIGHT	480	480

OUTPUT-BUFFER/DRIVER		COT02A/COT02B	P. 1 +
VERSION: 1.3	CELL GROUP: OUTPUT-PADS		2.2

LOGIC SYMBOL

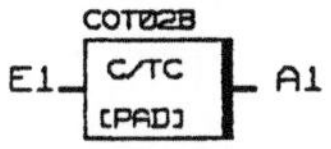

FUNCTION TABLE	
INPUT	OUTPUT
E1	A1
	(PAD)
L H	L H

AC-CHARACTERISTICS			CL = 50pF	
[ns]	FROM (INPUT)	TO (OUTPUT)	TYP	NOTE
tdHL tdLH	E1 * E1 *	A1 A1	9.0 6.5	TTL-COMP
tdHL tdLH	E1 * E1 *	A1 A1	6.5 11	CMOS-COMP

* → TIMING DIAGRAM 14

DIMENSION [µm]		
	A	B
WIDTH	240	240
HEIGHT	480	480

LOGIC CIRCUIT

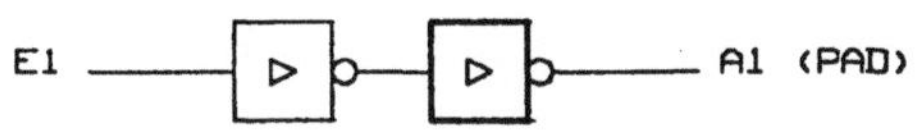

LOAD DEPENDENCE	TYP	
OUTPUT	kLH [ns/pF]	kHL [ns/pF]
A1 * A1 #	0.15 0.07	0.08 0.11

* CMOS-COMP
# TTL-COMP

LOAD UNITS	
INPUT	[LU]
E1	3

VDD-PAD		CVD01A/CVD01B	P. 1 -
VERSION: 1.3	CELL GROUP: POWER-PADS		2.4

LOGIC SYMBOL

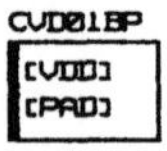

It is no signal pin, only VDD-supply for the pad cells which supply the logic cells automatically.

DIMENSION [µm]		
	A	B
WIDTH	240	240
HEIGHT	480	480

MAXIMUM SUPPLY CURRENT		
[mA]	A	B
ISS_{MAX}	110	85

For detailed information, i.e. power supply conditions, see chapter 3.6.

VSS-PAD		CVS01A/CVS01B	P. 1 -
VERSION: 1.3	CELL GROUP:	POWER-PADS	2.4

LOGIC SYMBOL

It is no signal pin, only VSS-supply for the pad cells which supply the logic cells automatically.

DIMENSION [µm]		
	A	B
WIDTH	240	240
HEIGHT	480	480

MAXIMUM SUPPLY CURRENT		
[mA]	A	B
ISS_{MAX}	110	85

For detailed information, i.e. power supply conditions, see chapter 3.6.

5.1.4 Data Sheet G-Family

D-FLIPFLOP WITH CLEARN			DFB1L1K	P. 2 -
VERSION:	1.0	CELL GROUP:	D-FLIPFLOPS	1.10

LOGIC CIRCUIT

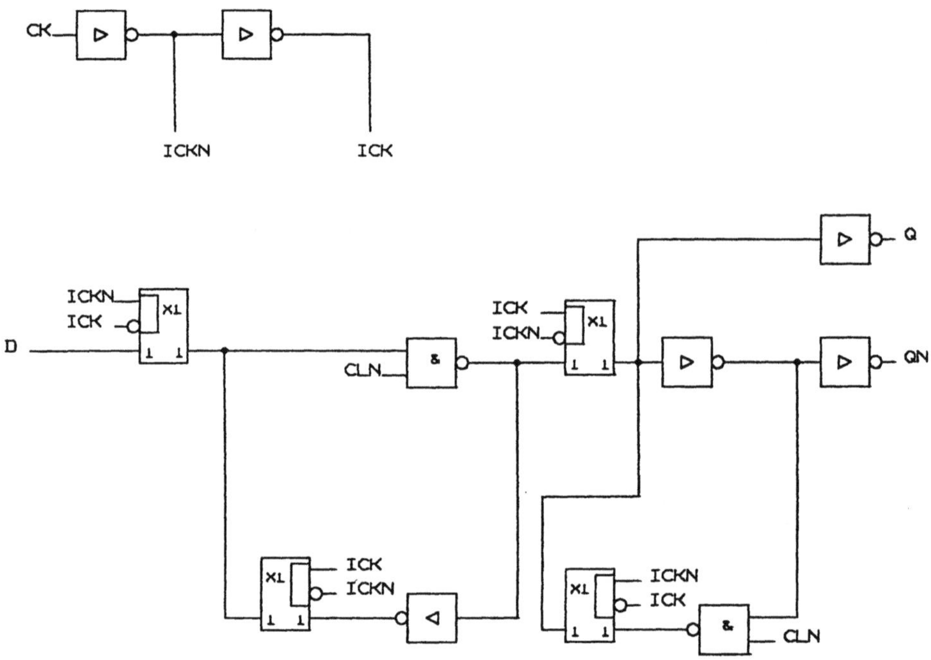

NUMBER OF CORE CELLS:	8 (DR)
EQUIVALENT GATES:	6

D-FLIPFLOP		DFA1L1F	P. 1 +
VERSION: 1.0	CELL GROUP:	D-FLIPFLOPS	1.10

LOGIC SYMBOL

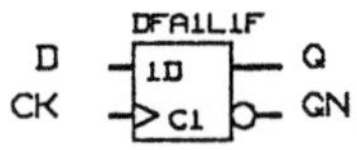

FUNCTION TABLE			
INPUT		OUTPUT	
D	CK	Q	QN
H	↑	H	L
L	↑	L	H

AC-CHARACTERISTICS			CL = 0
[ns]	FROM (INPUT)	TO (OUTPUT)	TYP
tCWL	CK *	–	3.0
tCWH	CK *	–	4.0
ts	D; CK #	–	1.5
th	D; CK #	–	2.0
tdHL	CK	Q	5.0
tdLH	CK	Q	5.0
tdHL	CK	QN	4.5
tdLH	CK	QN	4.0

* → TIMING DIAGRAM 1.11
\# → TIMING DIAGRAM 1.12

LOAD DEPENDENCE	TYP	
OUTPUT	kLH [ns/LU]	kHL [ns/LU]
Q, QN	0.18	0.18

LOAD UNITS	TYP
INPUT	[LU]
D	0.9
CK	1.1

D-FLIPFLOP		CDF01G	P. 1 -
VERSION: 1.2	CELL GROUP: D-FLIPFLOPS		1.10

LOGIC SYMBOL

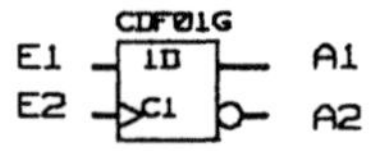

FUNCTION TABLE			
INPUT		OUTPUT	
E1	E2	A1	A2
	CLOCK	Q	QN
H	↑	H	L
L	↑	L	H

AC-CHARACTERISTICS			CL = 200fF
[ns]	FROM (INPUT)	TO (OUTPUT)	TYP
ts	E1; E2 *	-	6.5
th	E1; E2 *	-	0
tdHL	E2	A1, A2	8.0
tdLH	E2	A1, A2	4.0
tCWL	E2 $	-	5.5
tCWH	E2 $	-	6.0

* → TIMING DIAGRAM 1
$ → TIMING DIAGRAM 7

LOAD UNITS	
INPUT	[LU]
E1	2
E2	4

LOGIC CIRCUIT

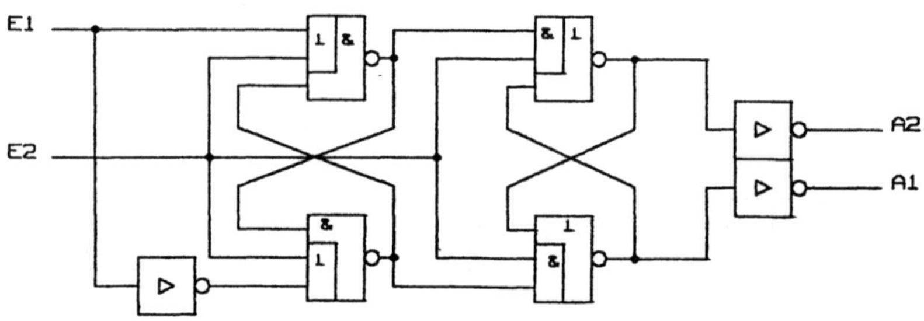

NUMBER OF CORE CELLS	
8	DOUBLE ROW

5.1.5 Data Sheets of the K- and F-Families

D-FLIPFLOP WITH CLEARN		DFB1L1K	P. 1 +
VERSION: 1.0	CELL GROUP:	D-FLIPFLOPS	1.10

LOGIC SYMBOL

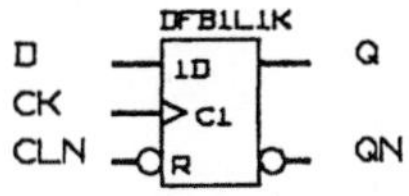

FUNCTION TABLE				
INPUT			OUTPUT	
D	CK	CLN	Q	QN
X	X	L	L	H
L	↑	H	L	H
H	↑	H	H	L

AC-CHARACTERISTICS			CL = 0
[ns]	FROM (INPUT)	TO (OUTPUT)	TYP
tCWL	CK }	–	3.0
tCWH	CK } *	–	3.0
tCLW	CLN }	–	3.0
ts	D ; CK #	–	1.5
th	D ; CK #	–	1.5
th	CLN; CK $	–	2.5
tR	CLN; CK +	–	0.5
tdHL	CK	Q	2.5
tdLH	CK	Q	3.5
tdHL	CK	QN	4.0
tdLH	CK	QN	3.5
tdHL	CLN	Q	3.0
tdLH	CLN	QN	3.5

* → TIMING DIAGRAM 1.11 $ → TIMING DIAGRAM 1.13
# → TIMING DIAGRAM 1.12 + → TIMING DIAGRAM 1.14

LOAD DEPENDENCE	TYP	
OUTPUT	kLH [ns/LU]	kHL [ns/LU]
Q, QN	0.23	0.14

LOAD UNITS	TYP
INPUT	[LU]
D	3.5
CK	1
CLN	2

D-FLIPFLOP		DFA1L1F	P. 2 -
VERSION: 1.0	CELL GROUP:	D-FLIPFLOPS	1.10

LOGIC CIRCUIT

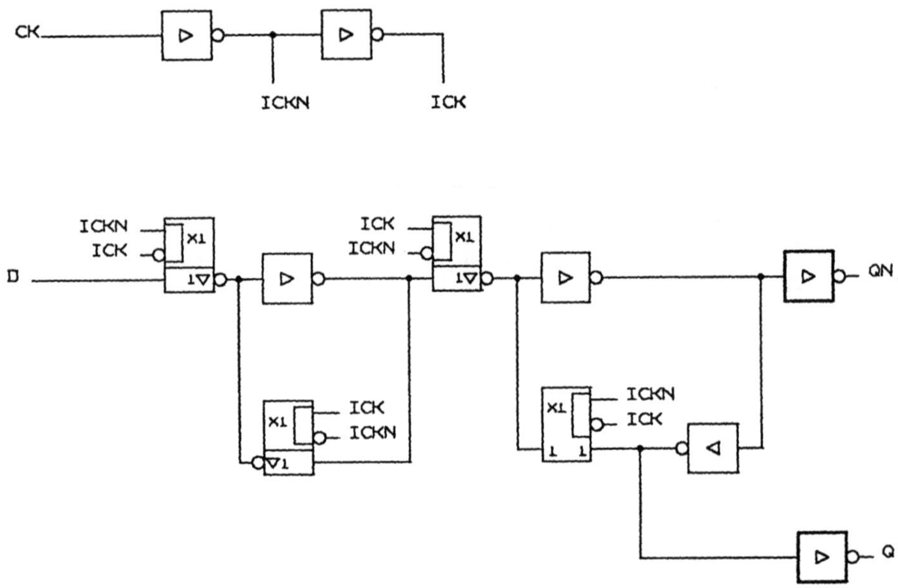

EQUIVALENT GATES
6

DIMENSION [µm]	
WIDTH	90.2
HEIGHT	134

CURRENT DISSIPATION TYP, CL=0	
STAT [µA]	DYN [µA/MHZ]
-	6

5.1.6 Data Sheet Z-Family

SIEMENS SH 100 C	2 D-MASTER-SLAVE-FLIPFLOPS WITH COMMON CLOCK	KMS04

FFDMS 21

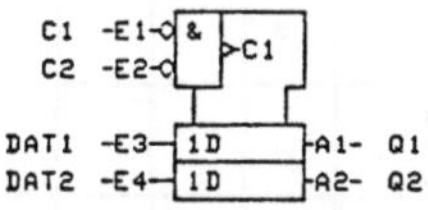

TRUTH TABLE

C1	C2	Q1	Q2
L	L	Q1	Q2
L	H	Q1	Q2
L	↓	DAT1	DAT2
H	L	Q1	Q2
↓	L	DAT1	DAT2

CELL-SITES	1 L
CURRENT CONSUMPTION (typ)	7.5 mA
POWER DISSIPATION (typ)	33.8 mW
GATE FUNCTIONS	14

SH100C1	SH100C2	SH100C3
x		x
VERS.	DATE	DES.
1	10.84	DF
PAGE : 1+		

SIEMENS SH 100 C		KMS04

CORRELATION OF PIN CONNECTIONS, FAN-IN AND FAN-OUT, EXCHANGEABLE INPUTS

SYMBOLIC CONNECTIONS		E1	E2	E3	E4	A1	A2									
ACTUAL CONNECTIONS		2	3	5	6	23	17									
INPUT OR OUTPUT LOADS	DYN.	0.45	0.45	0.4	0.4	8.5	8.5									
	STAT.	1	1	3*)	3*)	9	9									
EXCHANGEABLE INPUTS		A	A													
REF. No.																

*) ONLY FOR C1 + C2 = H ; OTHERWISE STATIC INPUT LOAD = 0.

P R O P A G A T I O N D E L A Y (TYP)

INPUT	OUTPUT	tpd [NS]
C1, C2	Q1, Q2	1,20

SIEMENS SH 100 C		KMS04

TIMING

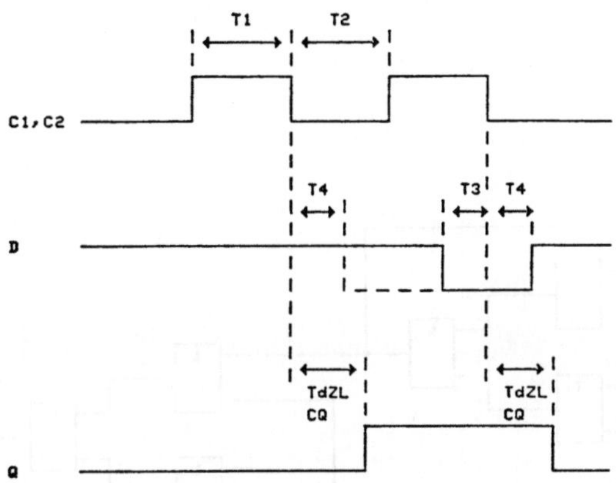

PULSE WIDTH SIGNAL C	2,0 NS <= T1
PULSE PAUSE INTERVAL SIGNAL C	2,8 NS <= T2
INTERVAL FROM TRAILING EDGE OF SIGNAL C TO SIGNAL CHANGE AT D (Tset uP)	1,2 NS <= T3
INTERVAL FROM TRAILING EDGE OF SIGNAL C TO SIGNAL CHANGE AT D (Thold)	1,2 NS <= T4 *

* WHEN MASTER-SLAVE-FLIPFLOPS WHICH ARE DRIVEN BY YHE SAME CLOCK ARE DIRECTLY CASCADED (E.G. SHIFT REGISTERS, COUNTERS), THE REQUIRED TIME T4 (Thold) IS SMALLER THAN OR EQUAL TO THE PROPAGATION DELAY C1,C2 Q.

SIEMENS SH 100 C		KMS04

OVERVIEW DIAGRAM

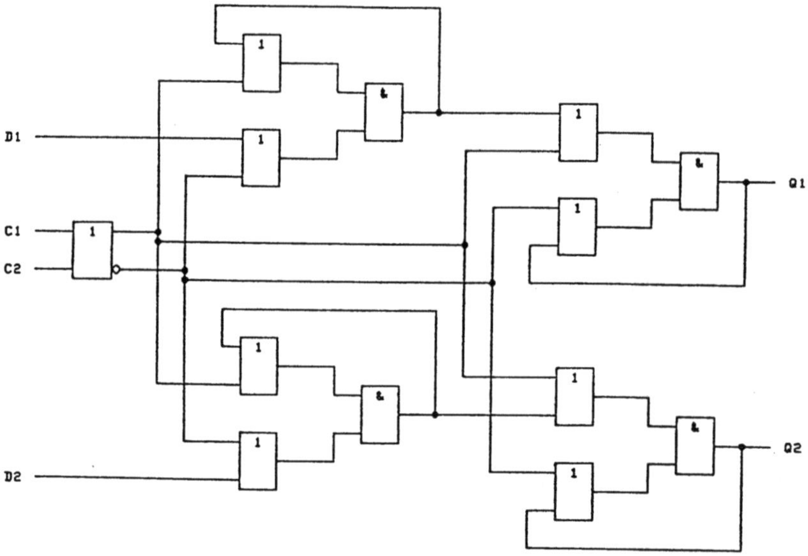

PAGE : 4-

5.2 Development of the Cell Libraries

As explained in the previous sections the simplified view of the IC design engineer is mainly due to the considerable preparatory work of the CAD engineer; during the predesign process the latter goes mainly through all the steps of general IC design that remain invisible to the user, that is the IC design engineer (see Fig. 1.30). In the following, we want to take a closer look at those steps necessary for the design of the various models and libraries presented in Sect. 1.5.2. Library development is not an easy task; this can be seen clearly in the fact that although usable placement and routing algorithms were known for years the model libraries necessary for a complete design process have not been available.

5.2.1 Objectives and Cell Concept

The most important decision for library realization is the determination of the standard design procedure. If one knows which design steps are to be supported by the CAD system and what degree of automation is to be obtained then one can also establish which model libraries are to be generated.

The main emphasis of the work in generating a cell library is put on the circuit and layout level. With the *cell concept*, the electric and geometric design is determined. A number of boundary conditions have to be observed:

- The cell libraries are integrated parts of the *CAD system* and have to be adapted appropriately.
- According to the characteristics of the CAD system the cells have to provide the *chip infrastructure*. This includes, for example, the current supply lines, the external connections (pads), special clock lines and fabrication-specific patterns.
- The planned application area of the cells determines the *production technology*.
- Cell are circuit parts which have to work correctly in varying environments. This is why they have to be designed in such way that they are electrically robust and *easy to use*.

The following takes a closer look at the individual aspects of the cell concept.

Technology

For fast gate arrays cell libraries are offered in ECL technology. The technology used for mass application, however, is CMOS. The key factors (see Chap. 2) are the low power dissipation in quiescent state, the high noise resistance, symmetric impedances and negligible residual voltages: all these characteristics prove the CMOS technology to be especially robust and easily standardized.

Ease of Use

In order not to burden the design engineer with too many rules and restrictions during the circuit design the cells have to be electrically compatible with each other. That means that at a cell output has to drive a high number of inputs as reliably as a single input. And the driver has to be designed such that rising and falling edges are of the same length even if the load varies in a wide range.

The cell input capacitance must be kept low by means of buffering. The use of transfer gates instead of logic gates saves space. But there are problems with inputs which are led directly to transfer gates without intermediate buffering: their electrical behaviour is strongly load dependent. For better manageability, all inputs are therefore buffered.

Power Supply

The fact that a cell also needs power supply should not be noticed by the user. The power supply lines which consist of wide metal bands have to be designed in such a way that not too many signal cross over are needed. This plays a major role primarily with polysilicon/aluminum wiring. A comb-shaped power supply structure, coming from the pad cells, thus results (Fig. 5.9).

The current supply lines running within the cells are integrated in the layout; the remaining supply lines are generated by the wiring program.

Basic Geometric Structure

The basic geometric structure of standard cells is adapted to the placement and routing program: a uniform cell height facilitates the wiring; terminals are only allowed at certain edges of the cell.

For complex cells, it is better to arrange two pairs of transistors above each other, and accordingly, to provide one V_{DD}-line each for the upper and the lower part of the cell and one V_{SS}-line in the middle. This structure brings about a cell height which is less favorable for simple gate cells; this, however, must be accepted, because indented cell outlines cannot in most cases be utilized by the wiring program. A restriction to uniform transistor dimensions and the resulting regularity considerably simplifies the cell layout.

Since the environment for a cell at its final position cannot be stated in advance for an automatically performed wiring the cells have to fit geometrically to each other and to the edges of the wiring channels such that no violations of the geometric design rules occur. This can be realized only if the edges are precisely specified on all mask levels.

Remarkably we again have a situation similar to software development: not all conceivable combinations can be tested. Through systematic realization of as many test chips as possible, however, a high probability of correctness can be obtained.

Padcells

The padcells connect the inner cells (which are called logic or core cells) with the environment. They include the bond area, protection structures, drivers, level converters and to some extent even logic circuits (tri-state logic). Moreover, the power supply is integrated in the padcells (see Fig. 5.9).

Cell components with a high number of logic cells and few pad cells are called "core-determined". For these components, the padcells are to be low and wide to save space. "Pad-determined" components contain a core of logic cells which does not fill the space defined by the padcells. In this case the padcells are to be narrow and high. The best solution is to offer both variants.

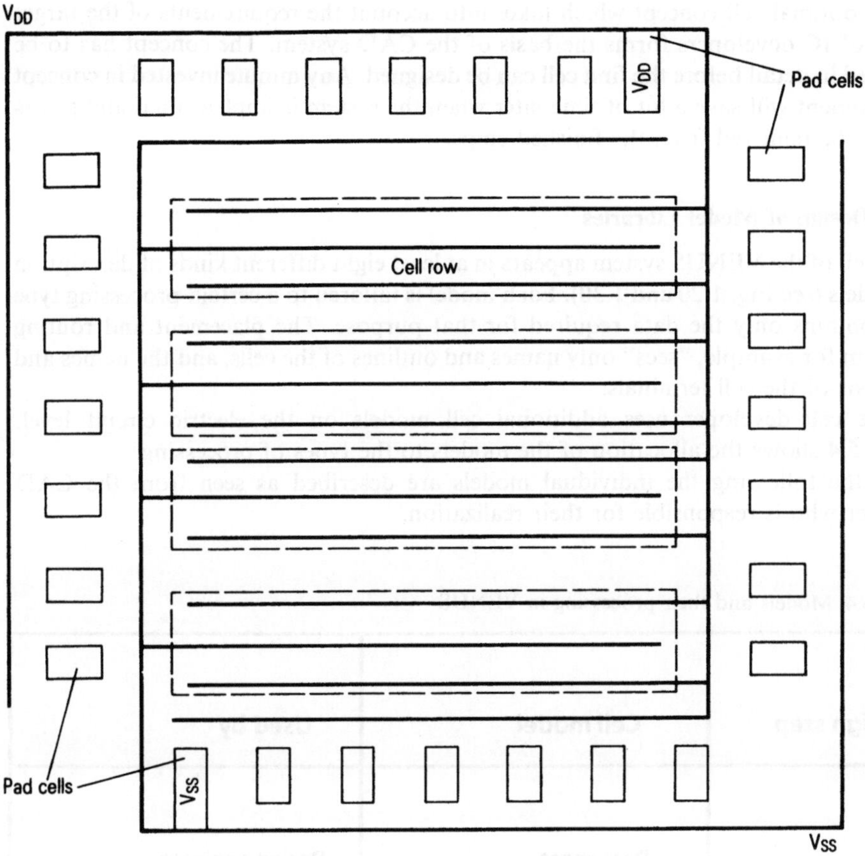

Fig. 5.9. Structure of supply lines and position of pad cells

Master Clock Lines

Clock lines have to meet special requirements concerning short delay times. This is why within the cells special lines can be provided which are preferentially led in aluminum ensuring a fast clock distribution (so-called master clock lines). A disadvantage to this procedure is the waste of space. The clock lines must also run through cells where they are not needed, for example logic gates. In addition, the majority of components needs several clocks which then are led in the wiring channel and whose delay times have to be adjusted. For this reason, no master clock lines are planned at present in VENUS.

Static Circuit Technology

The general use of static circuit technology means that one cannot make use of all possibilities of a technology but is on the safe side in terms of manageability. Dynamic circuit technology is only suitable for optimized manual designs.

An optimal cell concept which takes into account the requirements of the target group of IC developers forms the basis of the CAD system. The concept has to be specified in detail before the first cell can be designed. Any minute invested in concept development will save a lot of time later when the system is implemented and errors have to be removed from the finished chips.

5.2.2 Design of Model Libraries

Each cell of the VENUS system appears in at least eight different kinds of description or models (see Fig. 1.20 and 1.30). Each model is tailored to a certain processing type and contains only the data required for that purpose. The placement and routing program for example, "sees" only names and outlines of the cells, and the names and positions of the cell terminals.

The cell developer uses additional cell models on the electric circuit level. Chart. 5.4 shows the allocation of the models to the types of processing.

In the following the individual models are described as seen from the CAD engineer who is responsible for their realization.

Chart 5.4 Models and their processing in VENUS

Design step	Cell model	Used by
	Data sheet	Design engineer
3	Logic symbol	Schematic entry
4	Logic model	Logic simulator
5	Test model	Fault simulator
8	Layout contour	Placement and routing
9	Time delay analysis model	Time delay analyzer
10	Layout	Mask generator
11	Electric data	Test program generator
	Also during the development process:	
7	Circuit model	Circuit simulator
7,9	Switch-level model	Switch-level simulator

Data Sheet

This is the only description level whose content the design engineer has to learn explicitly and in detail. An explanation has already been given in Sect. 5.1.2.

For the design engineer the data sheet represents the development requirements (as function table) as well as a summary of the development results (delay times, e.g.).

Logic Symbols

IEC standard symbols stated in the data sheet can be found by the design engineer as a graphics menu at the workstation where he inputs the logic plan. For each input device offered in the VENUS system an individual logic symbol library has to be set up which contains the graphic representation of a symbol as well as the call information to be stored in the net list [1].

Logic Model

The basis for the logic simulator library is the function table and the timing values. First, the logical behaviour of the cells is described in models by connection of basic logic functions. Then, cell internal delay times are added in such a way that the external overall timing behaviour of the cell corresponds to the one stated in the data sheet. In this model description of a cell (an example is shown in Fig. 5.10), basic logic models (like: AND, CHFFR, EXOR) are called up. By applying appropriate stimuli,

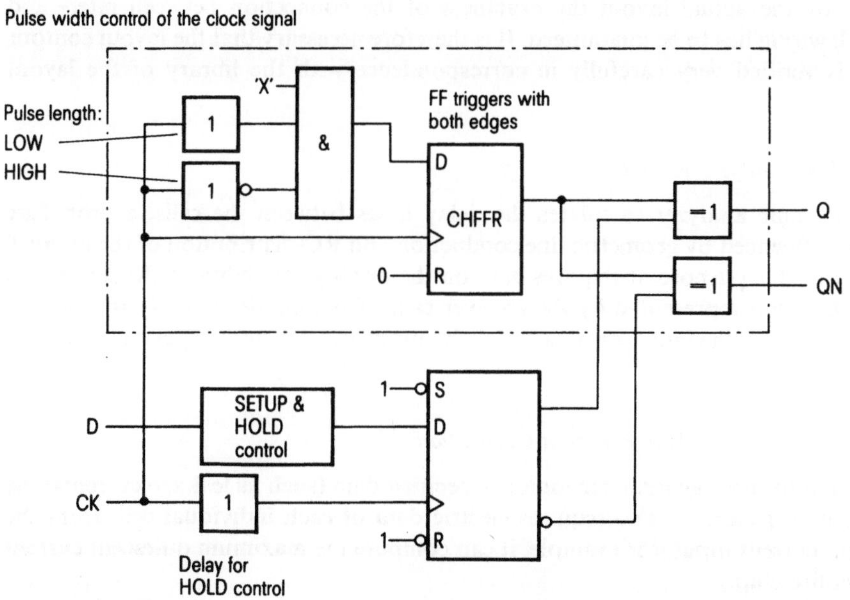

Fig. 5.10. Description of the logic model of the cell CDF01A with basic logic models

[1] The term "net list" defines a data structure which contains all cells of a logic plan and their connections.

the model is tested thoroughly by means of the logic simulator. The total of the tested model descriptions of all cells makes up the logic library.

For the modern simulator used in the current VENUS version the description of the logic function is more clearly and precisely realized in a functional description language. This is a language which is at about the same level as PASCAL, including such commands as "while", "case" etc.

Test Model

Similarly as for the logic simulator, each cell is modeled for the fault simulator. Only the degree of detail reached is higher. Because the fault simulator in principle requires the same information as the logic simulator, with the addition of information on possible error mechanisms, the use of one model for both simulators is advisable. For the fault simulator, only the fault models need to be added.

Layout Contours

In this library the outlines as well as the terminal positions of each cell are stored. Only this amount of information which is small compared with that of the complete layout is needed for the automatic chip design (placement and routing). The entire layout of the cells is only required for production data generation.

The placement and routing program runs the inter-cell wiring exactly to the position indicated by the layout-contour. For the subsequent replacement of the contour by the actual layout the exactness of the connection between inter- and intra-cell wiring has to be guaranteed. It is therefore necessary that the layout contour library is verified very carefully in correspondence with the library of the layout sections.

Delay Time Analysis Model

The delay time analyzer calculates the delay times between the cells, a procedure which is influenced by geometric line conduction and RC-distribution of the intercell wiring. For this purpose, it requires data on the driving capability of the sender cell and on the load represented by the receiver cells. The time delay analysis model of each cell contains the inner resistance of each output and the input capacitance of each input.

Electric Data for the Test Program Generation

In addition to some general, technology-specified data (such as leakage currents) the test program generator also requires electric data of each individual cell: from the quiescent current input, for example, it can compute the maximum quiescent current of the entire chip.

Layout Parts

The layout generation is the one type of cell description which requires the greatest effort. For each cell, those mask structures which will later be realized in silicon are stored in the layout library in a geometric description language.

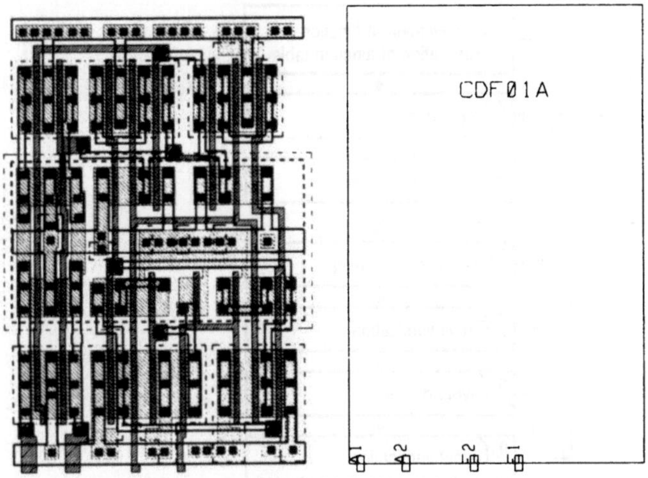

Fig. 5.11. Layout and contour of a D-flipflop

When the chip design has been finished, i.e. when it meets the requirements of the IC designer, the layout contours are replaced by the layout parts. Figure 5.11 shows the layout of the D-flipflop CDF01A and the pertinent layout contour.

The model libraries mentioned so far have been combined in the VENUS database: They are accessed by the VENUS system programs. The user may obtain information about the content of this library by read access.

For the cell designer, two more model libraries are added to the eight already existing:

Circuit Model

The electrical realization of a cell is described in the form of a transistor diagram. If this representation is translated into a circuit model this forms the basis of an electrical circuit simulation (with SPICE, e.g.). The results of any cell's circuit simulation are the general proof of correct function and the delay times for the data sheet.

Switch Level Model

If the cell designer wishes only to verify the logic function of a cell and does not need information about voltage curves or delay times he may use a simplified switch level model: each transistor is represented by controllable switch. The program run times of so-called switch-level simulators are 100 times shorter than those of circuit simulators.

5.2.3 Development Procedure

The development procedure of a cell library contains in its central section steps 3 through 9 of the general IC design process, i.e. it reaches from logic design to layout verification (Fig. 5.12). Step 4, the logic verification, can generally be omitted as the cells are relatively simple and possible logic errors will also be detected by subsequent

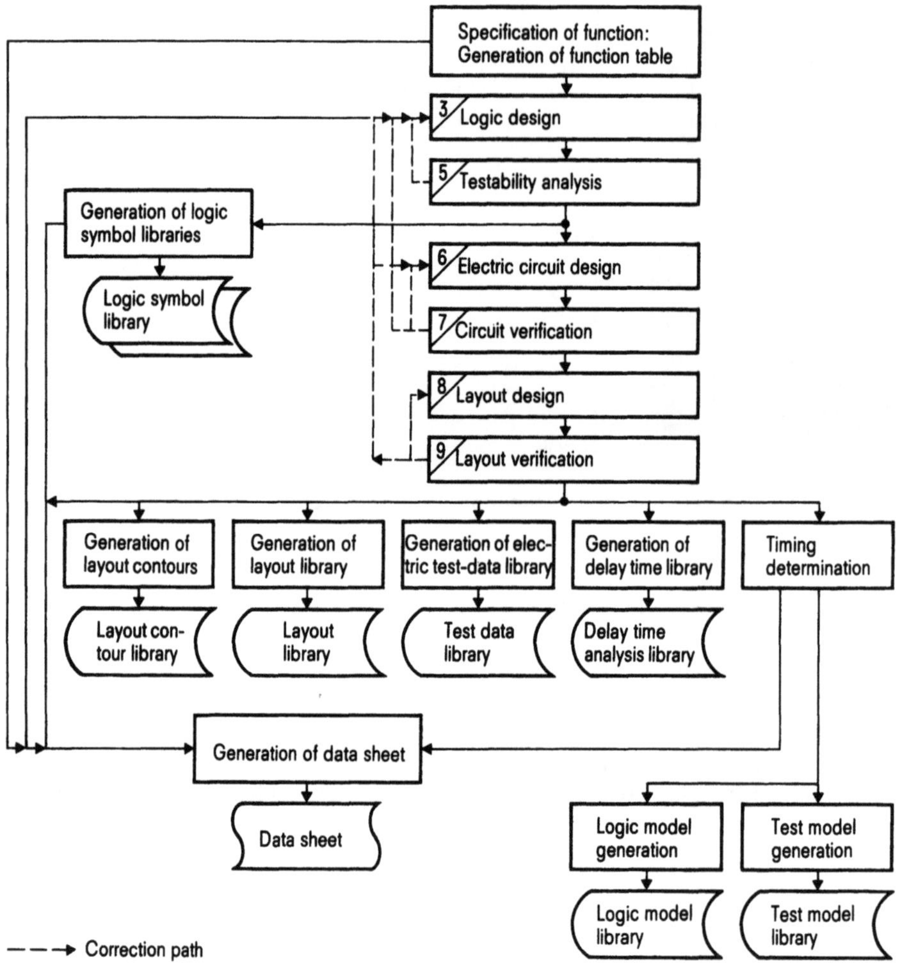

Fig. 5.12. Development procedure for a cell library

circuit verification. The verification process will be subject of a more detailed explanation in the section about quality control.

Around this central section, a number of additional procedures is arranged which are related to the generation of the eight model descriptions:

If the testability analysis has been concluded with a positive result then the generation of the logic symbol library can be initiated; for each work station one logic symbol library has to be set up.

The next important stage has been reached with a successful layout verification. Now, all data deduced from the layout can be determined and stored in the relevant library. The generation of the logic models and the test models is started already after step 5; its conclusion, however, is achieved only after the timing of the cells is known precisely under consideration of the capacities extracted from the layout.

Finally, the data sheet is formed, a task which often already is done in parallel to the development process.

Highly automatized design procedures can not yet reach optimum results with regard to space or speed if compared with a manually optimized layout. If the chip design, as for VENUS, is carried out automatically, one generally tries at least for cell design to come as close to optimum values for space utilization as possible. But in any case, procedures which are based on symbolic layout or on automatic cell generation (comparable to chip generators) tend to bring about a change.

5.2.4 Distinctive Features of the Development of User-Specified Cells

As in the case of rigid cells, the development of user-specified cells also includes layout generation and all other model descriptions required by subsequent programs. The result of the development process in this case, however, is not a cell and its model descriptions, but programs for the dynamic generation of such models.

There are two different methods for the layout generation:

- The layout is generated manually from cell elements at a graphics work station. These cell elements are connected with each other by means of a program. A layout created in this way is technology-specific.
- The layout is generated procedurally. The generation program knows the design rules and can immediately adjust the layout to technology changes. In exchange for this advantage the effort invested in program development may be considerably larger.

The generators for user-specific cells are embedded in VENUS as follows:

- User interface:
  - *Data sheet:* The information supplied by the generator is delay times, space, number of transistors and power dissipation of the cell. This data is transferred to the data sheet as actual values of the user-specific cell. It is not feasible to simulate each generated cell with a circuit simulator due to the high computing times. For this reason, the time parameters are taken from tables containing the delay times of cell elements or circuit sections. These delay times have been found by means of circuit simulation. If a cell is generated the generator computes its delay times by adding up or interpolating the values contained in the table.
  - *Cell specification:* In an input file the designer specifies the organization of the memory and the geometric form requested; for the ROM he also indicates its contents. For the PLA, for example, he defines the function by boolean equations and specifies the number of inputs and outputs. The generator then checks syntax and consistency of the description and, if necessary, issues a error list.
- *Technology interface:* For the procedural description of the layout the generator needs a technology file with the design rules.
- Interfaces with the CAD system generated by the generator:
  - *Logic model:* The central part of the model is formed by a behavioural model (*functional block description*) which is structured like PASCAL. It not only simulates the function on the logic level but also checks times (such as set-up- and hold-times) and issues appropriate messages. The necessary timing data are computed by the generator and added to the model.

- o *Test model:* The generator creates an equivalent circuit substitute for the automatic test pattern generation.
- o *Frame description and electrical data:* The program system for chip design requires data on the position and type of signal and supply terminals as well as the size of the cells (contour). The frame description includes logic connection classes for the generation of the logic symbol and the net list check. The electrical data comprise input/output capacitance and output resistance for the calculation of delay times on chip level.
- o *Logic symbols:* Based upon the frame description, a generator draws the logic symbol for schematic entry. For that purpose a PASCAL frame program generates a command file for the graphic editor which in turn generates the logic symbol.

Offering user-specific cells includes a certain risk. Because it is impossible to measure every combination on a test chip (the number of variants is much too high), one must guarantee by means of complex tests that the generator supplies correct results. To achieve this, a high number of variants is checked with the same verification procedures which are also applied for rigid cells.

5.2.5 Quality Assurance

Quality assurance begins with the exact planning of the development procedure for cells. The objective is to guarantee that all model descriptions of the cell library are:

- complete,
- internally correct and,
- consistent with each other.

There are three verification steps:

- check of data on cell level,
- check of data on chip level,
- check of manufactured cells and chips.

Check of Data on Cell Level

Just as the design procedure for cells follows the steps of the general IC design process, the verification procedure for cells corresponds with the verification procedure described in Sect. 1.5.4. Figure 5.13 shows the general verification procedure and, in the shaded area, the part which is relevant for cell design.

The logic design is followed only by a manual testability analysis. The electrical circuit design of each cell is verified by circuit simulation. If there are still errors found in the logic plan one has to restart at the logic design.

The layout plan of the cell is made subject to the most complex test. Layout verification programs report violations of the geometric design rules; after the layout extraction, the correctness of the electrical function is checked. Finally, an extraction from the layout to the logic level is carried out; on the logic level, a switch-level simulator tests whether the realized function corresponds with the one indicated in the function table.

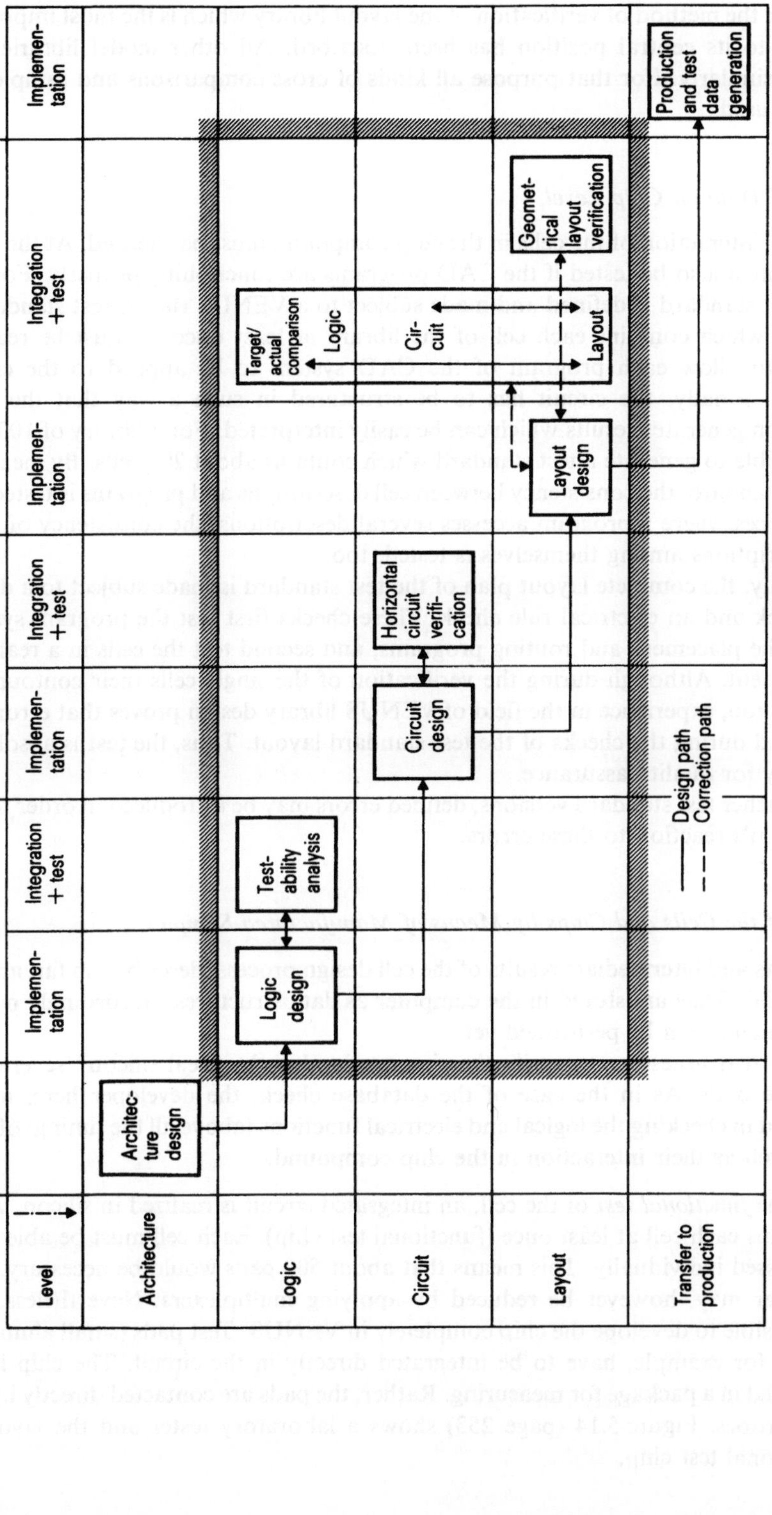

Fig. 5.13. Verification procedure for cells

So far the method of verification of the layout library which is the most important one due to its central position has been described. All other model libraries are checked similarly. For that purpose all kinds of cross comparisons and sample test runs are used.

Check of Data on Chip Level

Here, the interaction of the cells in the chip compound must be checked. At the same time, it must also be tested if the CAD programs are functioning properly. For this test a test standard is defined and made subject to a VENUS run. A test standard is a circuit which contains each cell of the library at least once; it must be realistic enough to allow each program of the CAD system to be applied to the circuit correctly. Finally, the circuit has to be structured in such a way that the logic simulation generates results which can be easily interpreted. For a library of 100 cells it is possible to generate a test standard which contains about 200 cells. By means of the test standard, the consistency between cell descriptions and programs is tested; for all instances where a program accesses several descriptions, the consistency of these cell descriptions among themselves is tested, too.

Finally, the complete layout plan of the test standard is made subject to a design rule check and an electrical rule check. These checks first test the program system, mainly the placement and routing programs, and second test the cells in a real chip environment. Although during the verification of the single cells their contours are checked, too, experience in the field of VENUS library design proves that errors are still found during the checks of the test standard layout. Thus, the test is absolutely necessary for quality assurance.

In further test standard versions, defined errors may be introduced in order to test the system's reaction to these errors.

Check of the Cells and Chips by Means of Manufactured Samples

All results and intermediate results of the cell design process, described so far, are still immaterial: They are stored in the computer as data structures. Accordingly no real measurements can be performed yet.

Since it is necessary to verify the simulated values by "real silicon" several test chips are built. As in the case of the database check, the developer here, too, is interested in checking the logical and electrical functions (above all the timing) of each cell as well as their interaction in the chip compound.

- For the *functional test* of the cell, an integrated circuit is realized in silicon, which contains each cell at least once (functional test chip). Each cell must be able to be addressed individually. This means that about 500 pads would be necessary. This number may, however be reduced by applying multiplexers. Nevertheless, it is impossible to develope the chip completely in VENUS: Test pads (small aluminum dots), for example, have to be integrated directly in the circuit. The chip is not installed in a package for measuring. Rather, the pads are contacted directly by fine test probes. Figure 5.14 (page 253) shows a laboratory tester and the layout of a functional test chip.

- If *delay times* are measured at the same functional test chip only inaccurate results will be obtained. This is because the times in the ns-range to be measured may be easily influenced by the stray capacitances of the test probes. Moreover, cell output drivers are, of course, not designed for the high capacitive loads of the test probes. If additional cells (for example a test flipflop or a pad driver) are added behind the test object, delay time uncertainties of these cells enter into the result.
  For these reasons the verification is limited to checking the correspondence of test results and simulations for some representative cells. Then it is possible, as experience has shown, to draw reliable conclusions for the remaining cells. On a *delay time test chip* correspondingly higher effort is spent for these few cells. In order to obtain times which are easier to be measured the cell type to be measured is replicated 10 to 20 times and chained. All cells of a chain must be stimulated by an edge of the same direction (rising or falling). For inverting cells this is achieved by insertion of inverters. All circuit sections introduced only for test reasons (the inserted inverters, e.g.) are once more available separately in immediate neighborhood. By measuring the two comparable chains and subtracting the results, one is sure to obtain quite precise cell delay times.
  A possibility for increasing accuracy even further is the application of the scanning electron microscope: instead of a mechanic test probe, an electron beam is used [5.4, 5.5].
- Another test chip has to be realized for the *verification of the pad cells*. Apart from delay times, electric characteristics such as input leakage currents, short circuit currents, load dependencies and voltage levels are of importance.

The result of the tests carried out with all these test chips may make corrections of the data sheets or even redesigns necessary. With VENUS cell development it was possible to avoid such corrections almost completely because with stable technology the SPICE simulation results predict the measured value with sufficient accuracy. A successful measurement which confirms the data of the data sheet, however, is an unconditional requirement for the official release of the cell.

In addition to those chips that are built for test purposes only, a high number of pilot ICs are usually developed to confirm the reliability and stability of the CAD system. The pilot ICs already correspond to real problem solutions. Thus not only the cell libraries, but also the procedures are made subject to a final, field-oriented test before being finally released.

5.2.6 Quality Standard

The quality standard of a product, especially for such a complex system as VENUS, will only be seen with practical application. The first version of VENUS has been used in production since 1983. The complexity range covers small circuits as developed by students doing term projects as well as components which actually exceeded the specified limits of the system in terms of the number of gates, size, or speed. All components designed to this point started working without problems. In no case a redesign was necessary. This proves the maturity of the cell libraries and the high quality standard of the program system.

Literature – Chapter 5

5.1 TTL Data Book. Texas Instruments.
5.2 Schaltzeichen für integrierte Schaltungen. A 49 000-C1-B-*-7435. Siemens, München.
5.3 ECL Gate Array Family Design Manual. Siemens, München.
5.4 Feuerbaum, H. P.: Electron Beam Testing: Methods and Applications. Scanning 5, (1983) pp. 14–24.
5.5 Wolfgang, E.: Electron Beam Testing: Problems in Practice. Scanning 5, (1983) pp. 71–83.

6 Application of the VENUS Design System

6.1 Summary

The following chapter presents considerations, decisions, activities and arrangements which must be considered in conjunction with the definition, design, production, test and first delivery of a semiconductor component that is developed by means of the design system VENUS.

For the presentation, it is assumed that one or several system design engineers develop such components which help in solving a specific system problem in a faster and better way than standard components or standard boards would. It is intended to enable these system engineers, when using the VENUS design system, to carry out all development steps and design processes independently. Their pertinent activities are finished when the production and test data have been transferred to the IC manufacturer. This presentation ends with information concerning the production of first samples and test program debugging.

This section places special emphasis on practical assistance for the application of VENUS, or for its preparation. This is illustrated above all by the number of check lists given.

The process steps which are explained in detail in Sect. 6.5 are illustrated by means of a VENUS run with the example HW-ADUS (see Chap. 1), in addition to an explanation given in general terms.

Unrelated to the strong points of a standard design systems already mentioned, it is advisable for the reader in each single case to examine the reasons for using such a design system for the development of integrated circuits with the following check list:

Check list 1: Reasons for the Application of VENUS:

- ☐ **Within the relevant development area, semiconductor-specific or layout-related knowledge is not available.**

- ☐ **Within the relevant development area, transistor-related circuit design knowledge is not available.**

- ☐ **The information necessary for the circuit design on the transistor level - such as design rules for the semiconductor production or electric parameters of the manufactured circuit elements - is not available.**

- If cell-oriented design is applied, one may fall utilize existing equipment which may have been used for system design, for example. It is not necessary to provide special equipment to be used only for that purpose.

- An existing problem solution at the pc-board level should be transferred to the IC level as fast and with as few problems as possible, as the functions of the former do not meet the requirements, need too much space, consume too much power or utilize expensive standard ICs only partially.

- A problem solution at the pc-board level is not possible because the system environment excludes this (environmental conditions), the error susceptibility in operation is too high or the dynamic specification can, for example, not be reached.

- The ideas of the system solution have to be kept hidden from other competitors. A pc-board built with standard chips is easy to copy, whereas the analysis and reconstruction of a cell IC requires a high degree of special knowledge and a high effort in terms of equipment and time.

- Development and production time have to be especially short. Only for cell ICs may one assume that the delivery of first samples will occur within a few months. Only for cell ICs may one exclude the necessity of a redesign with high probability.

- The risk of a design on transistor level is too high. In exchange for the design safety of cell ICs the increase of space needed is acceptable.

- The complexity of the problem to be solved is so high that the design can be managed only if it is realized by means of blocks, macros and cells. For transistor-oriented designs, function and quality assurances would be too difficult.

- The quality of the circuit is the target criterion. The cell-oriented design is based upon cells which while developed by experts, were subject to a large number of quality-assurance routines (in rest chips for the individual examination of the cells and in special pilot chips of the same design process). The quality of the designed product is thus, as opposed to the transistor level designs, assured from the start as it is part of a "serial" production.

- The chip designer wants the support of a closed design system (data base, simulator, mask data generation, fully automatic test program generation).

- ☐ Only a few samples of the devices are needed. The cell-oriented design is less expensive because of the relatively low development effort. The increase in production cost can therefore be accepted easily.

- ☐ During system development, modifications of device specification may occur. The cell-oriented design makes it possible to react quickly by supplying newly designed chips of the same quality.

- ☐ It is an important advantage that one may react flexibly to production quantity demands: The supply of samples by means of gate-arrays may be realized within a few weeks; sample series with standard-cell chips can be supplied within one or two months utilizing the same design input; high production quantities with optimized designs on transistor level can be supplied within a period of one or two years.

If one has decided to apply a standard design system, one has to get a good overview of the advertised abilities of the CAD system and its cell libraries. For this overview, one must clarify which considerations and development steps lie within the range of responsibility of the system developer, and which semi-conductor-specific or production-related problems he does not need to solve. Moreover, the system developer has to get to know which technical data – according to the data sheets of the cell catalog and its comments – are guaranteed by the design system.

The design engineer has access to the cell-oriented VENUS design system through the following documents:

- SEMICUSTOM. Zellenorientierter Baustein-Entwurf:
  - Standardzellen und Gate-Arrays,
  - Handbuch 1: Schaltungsentwicklung [6.1].
- VENUS. Arbeitsunterlagen für den Benutzer [6.2].

The VENUS version upon which this chapter is based is that of September 1985. It supports a standard process as described in Sect. 1.5.5 (Fig. 1.31) with the following steps:

- logic design,
- logic verification,
- testability analysis,
- chip design,
- layout analysis,
- mask data generation,
- test data generation.

VENUS offers the following features:

For CMOS gate-array components, masters with a complexity of 1K, 2K, 4K and 10K equivalent gate functions are offered.

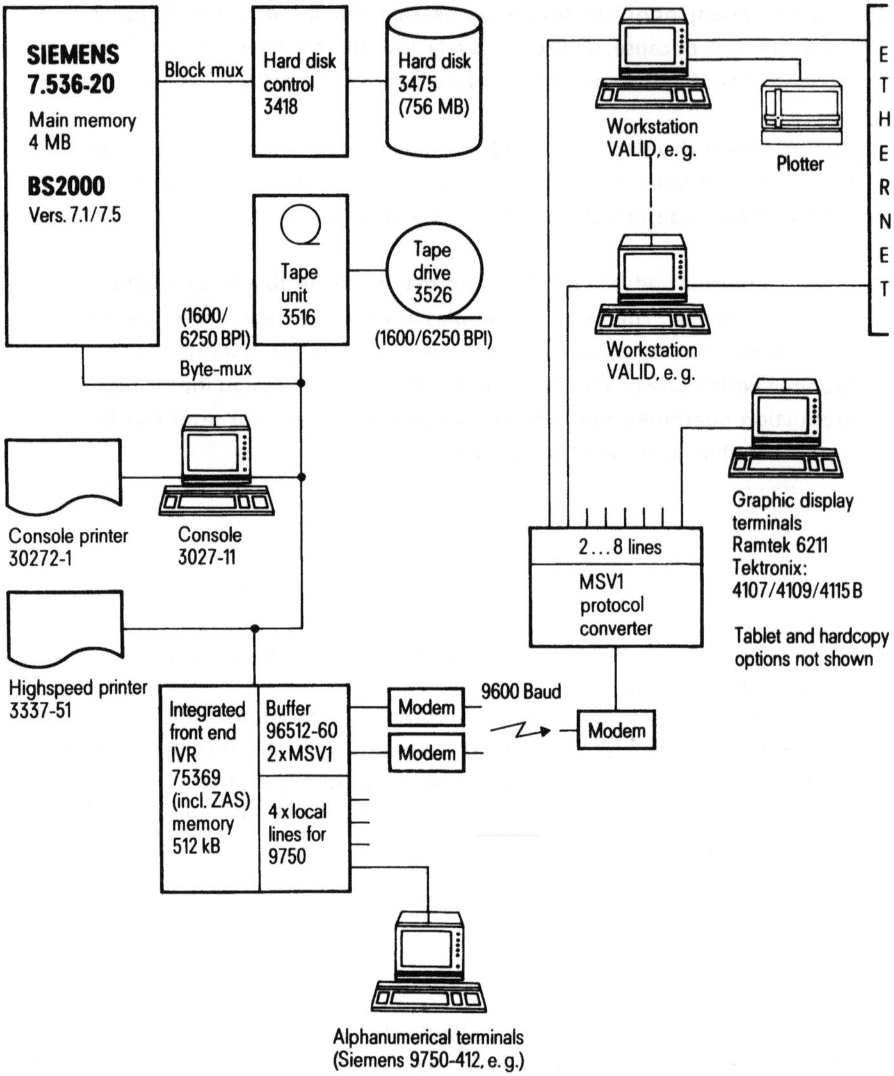

Fig. 6.1. Characteristic VENUS hardware configuration

Chips based upon CMOS standard cells can be realized if their complexities do not exceed 5.000 equivalent gate functions with up to 120 pads; chips based upon macrocells can be realized if their complexities do not exceed 20.000 equivalent gate functions with up to 256 pads.

For ECL gate-array chips, masters are offered with complexities of 960 and 2.600 gate functions.

The system VENUS runs on SIEMENS computers with the operating system BS2000. The user dialog is performed with a system manager. Generally the user is supported through menu outputs, advisory texts and messages which are easy to interpret. Schematic entry is done on a workstation. The placement and routing results are shown on color-graphic displays connected to the BS2000 system. At this terminal the interactive post-processing is also carried out. The merely alphanumerical dialog which is required for some design steps – such as verification and test preparation – can also be handled with the workstation.

Figure 6.1 shows a typical VENUS hardware configuration. The VENUS CAD-system and equipment is used by various users with quite different emphasis and intensity. According to this great variability of using the VENUS system, there is a number of possibilities concerning the division of tasks between customer, design center and manufacturer. This will be the subject of the next paragraph.

6.2 Customer/Manufacturer Interfaces

The cell-oriented VENUS design system is specified in such a way that a variety of different interfaces concerning the cooperation between customer and chip manufacturer is offered in various procedures. The basis is formed by a three-way relationship between customer, design center and manufacturer.

- *Customer:* The customer is the system designer who wishes to solve his system problem completely or to some extent by means of an application-specific IC.
- *Design center:* The associates of the design center are experts experienced in the VENUS design system and its cell libraries. Their task is to analyze the problems of the system designer during a consultation, to print out technologic boundary conditions, and, above all, to impart the use and purpose of the logic and test design rules which have to be followed for the logic design.
  Furthermore, they conduct regular training courses: first it is intended to make the system engineer familiar with the possibilities provided by the system design with application-specific components, and second, to train the design engineers on the equipment of the design center.
  The design center supplies the VENUS design system, the necessary hardware configuration and the data interfaces to the manufacturer. The cell libraries are also available within the design center system.
- *Manufacturer:* The manufacturer produces the chips that have been designed by means of VENUS. He accepts the mask tapes and test programs which have been generated with the VENUS design system and transmitted to him by the design center. The manufacturer enters into a direct contractual relationship with the customer as far as cost, guarantee and production dates are concerned.

Six different modes of cooperation are shown in the following. The consultations necessary for the system specification and for the logic development before the design system is applied are not considered; neither are production and test.

Process 1

The customer renders only a functional description of the system problem and makes known the relevant timing specifications.

The design center performs the logic design, carries out all VENUS design steps and induces production and test on the part of the manufacturer. After logic verifi-

Chart 6.1. Possibilities of dividing tasks between customer (C), manufacturer (M), and design center (DC)

Process → / Step ↓	1	2	3	4	5	6
System specification	C	C	C	C	C	C
Logic development	DC	C	C	C	C	C
VENUS logic design	DC	DC	C in DC	C in DC	C in DC	C in DC
VENUS logic verification	DC	DC	C in DC	C in DC	C in DC	C in DC
VENUS testability analysis	DC	DC	DC	C in DC	C in DC	C in DC
VENUS chip design	DC	DC	DC	DC	C in DC	C in DC
VENUS layout analysis	DC	DC	DC	DC	C in DC	C in DC
VENUS mask data generation	DC	DC	DC	DC	C in DC	C in DC
VENUS test data generation	DC	DC	DC	C in DC	C in DC	C in DC
Production	M	M	M	M	M	M
Test	M	M	M	M	M	C

cation (including layout analysis), the intermediate result of the development has to be released by the customer.

In this case the customer's know-how is not sufficient for the design of an electronic system; thus he supplies the design center with his system knowledge.

Process 2

The customer develops the system solution with the help of the semicustom manual "Circuit Design". He carries out the logic design himself and writes down the functional bit patterns. Both the logic sketch and the functional bit patterns are handed over to the design center.

Fig. 5.14. Laboratory tester (a) and functional test chip (b)

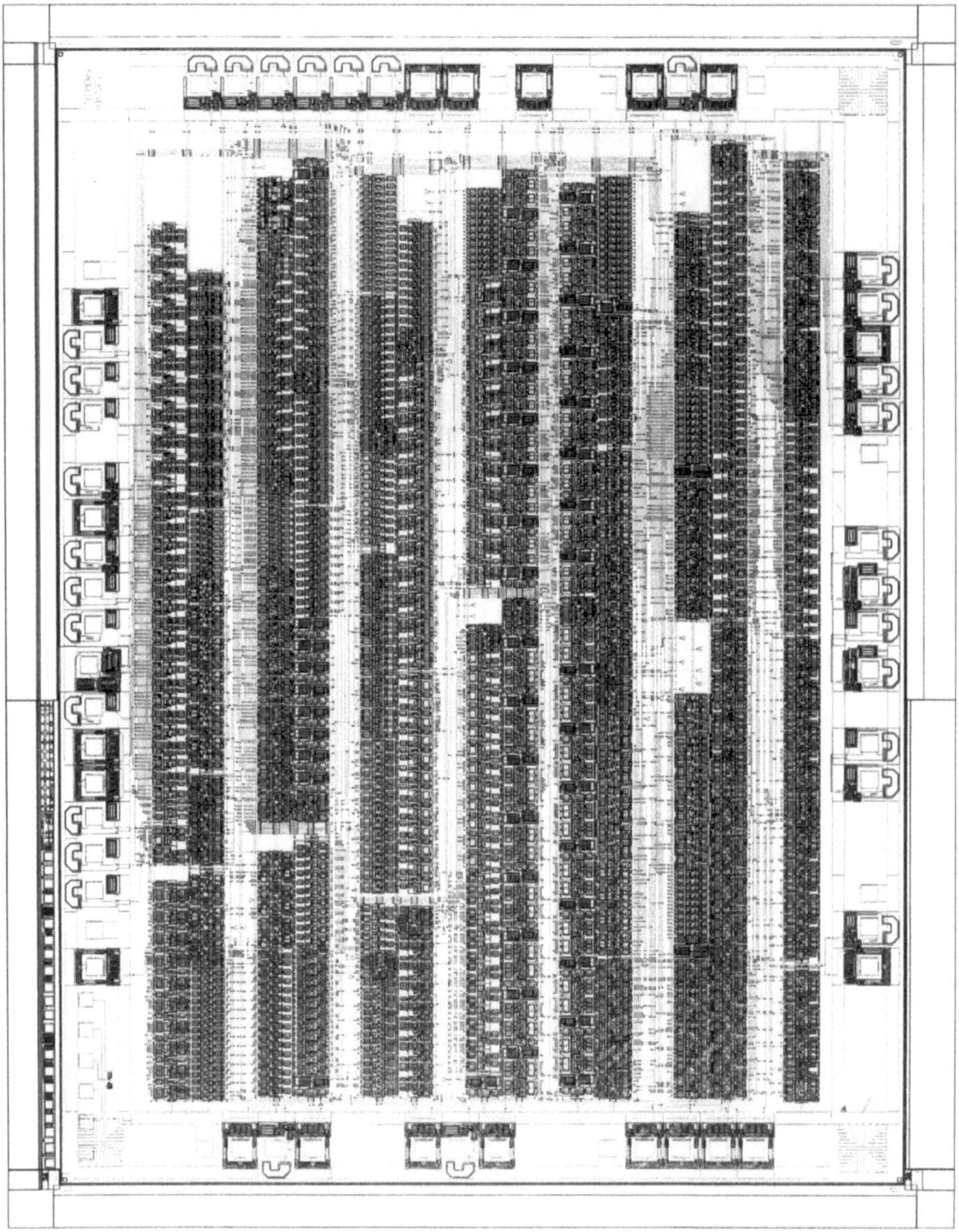

Fig. 5.14 b

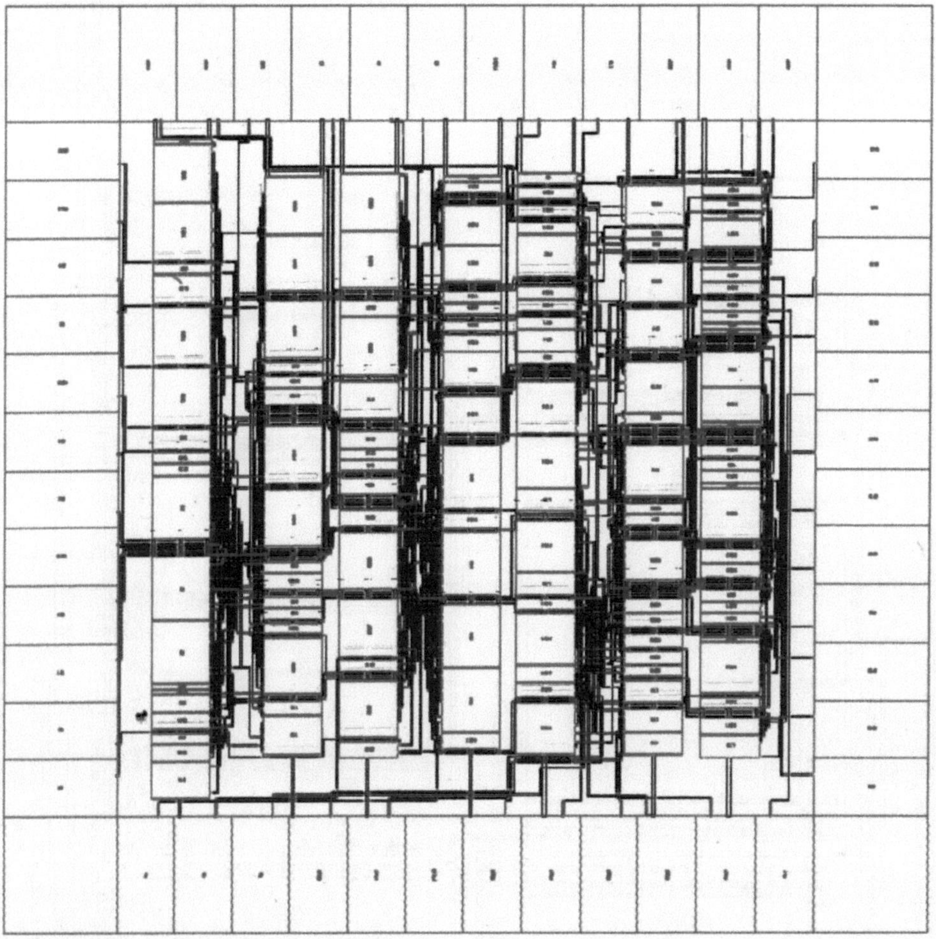

ADUS 11 a. Chip design: overall layout

```
%-----------------%
% C L U S T E R N %
%-----------------%
#CL/
CLULT = (0)/LT0,LT1,LT2,LT3,LT4,LT5/
%====================%
```

ADUS 11 b. Chip design: cluster command

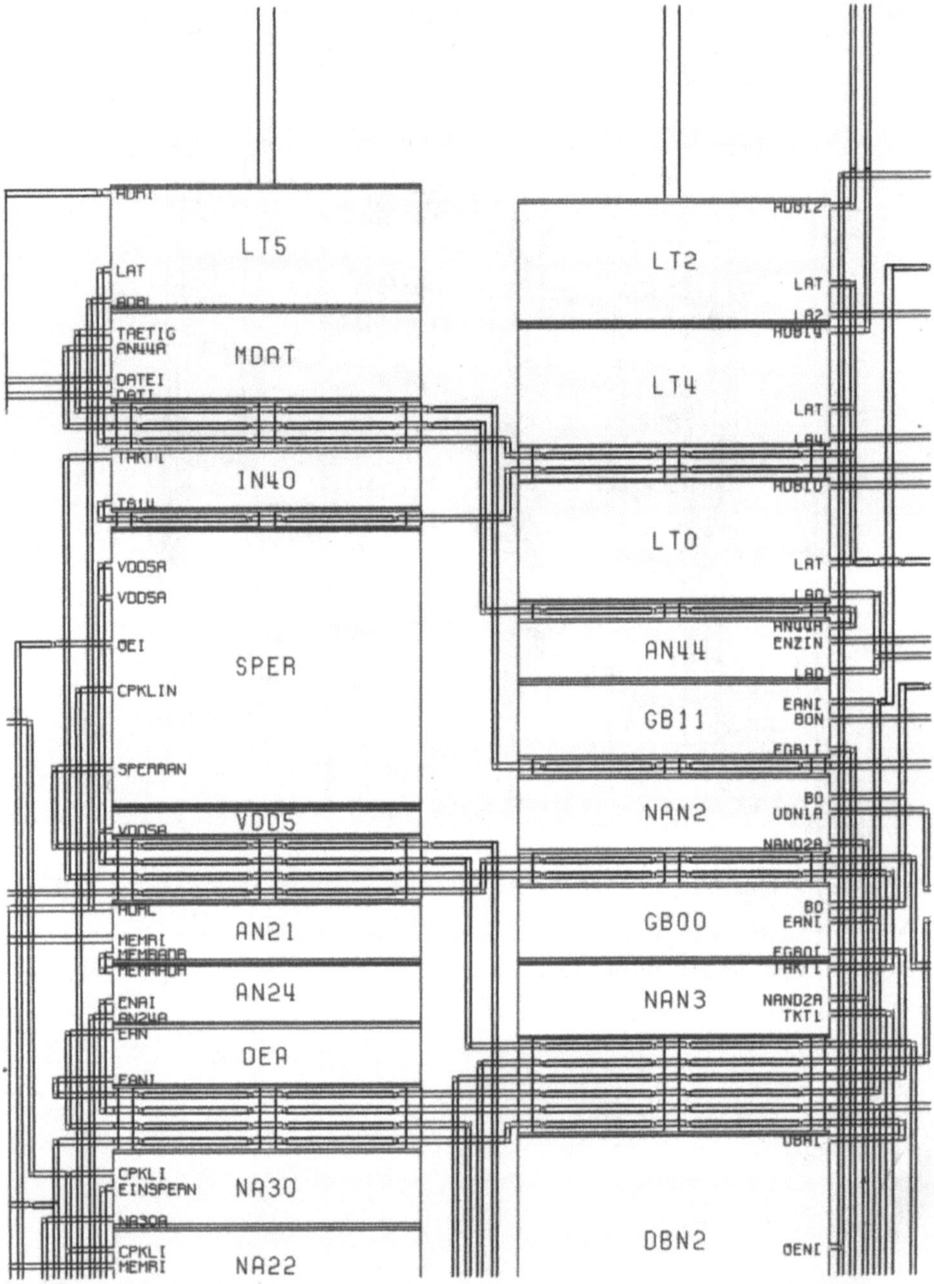

ADUS 11c. Chip design: Layout: section without cluster

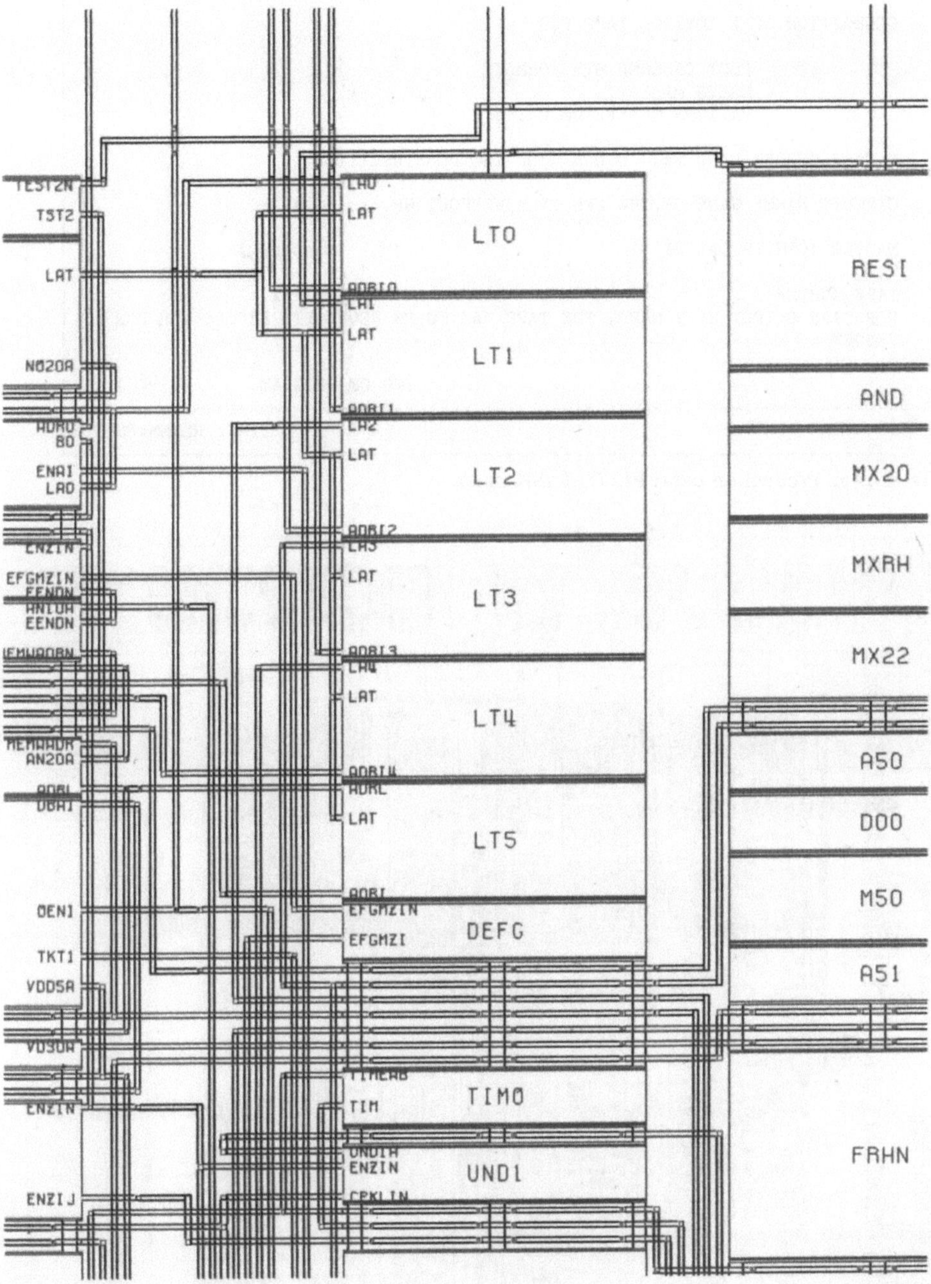

ADUS 11 d. Chip design: Layout : section with cluster

```
+------------------------------------------------------------------------------+
!    DD.MM.YY                     PLOTMASK VERSION XXX                          !
!------------------------------------------------------------------------------+
!                                                                              !
!    GENERATION OF A CONTROL TAPE FOR                                          !
!                                                                              !
!                 PLOT CALCOMP 925 FORMAT              ( )                     !
!                 MEBES MASKS                          (x)                     !
!                 PATTERN GENERATOR MASKS              ( )                     !
!                                                                              !
!    PLEASE CHOOSE MARK (x)  O N E   OF THE POSSIBILITIES                      !
!                                                                              !
!    CIRCUIT NAME (NAME OF THE HKP FILE WITHOUT HP.): (adus.1                ) !
!                                                                              !
!    MASTER IDENTIFICATION                             (masterk )              !
!                                                                              !
!    TAPE NUMBER                                       (xxxxxx)                !
!    FOR TAPE OUTPUT WITH MAREN THE TAPE HAS TO BE REQUESTED BEFORE THE RUN    !
!    SYSOUT                                            (                     ) !
!                                                                              !
!                                            END OF PROGRAM         ( )        !
+------------------------------------------------------------------------------+
!                                                          TIME: HH:MM:SS      !
+------------------------------------------------------------------------------+
```

ADUS 14a. Production data: PLOTMASK mask

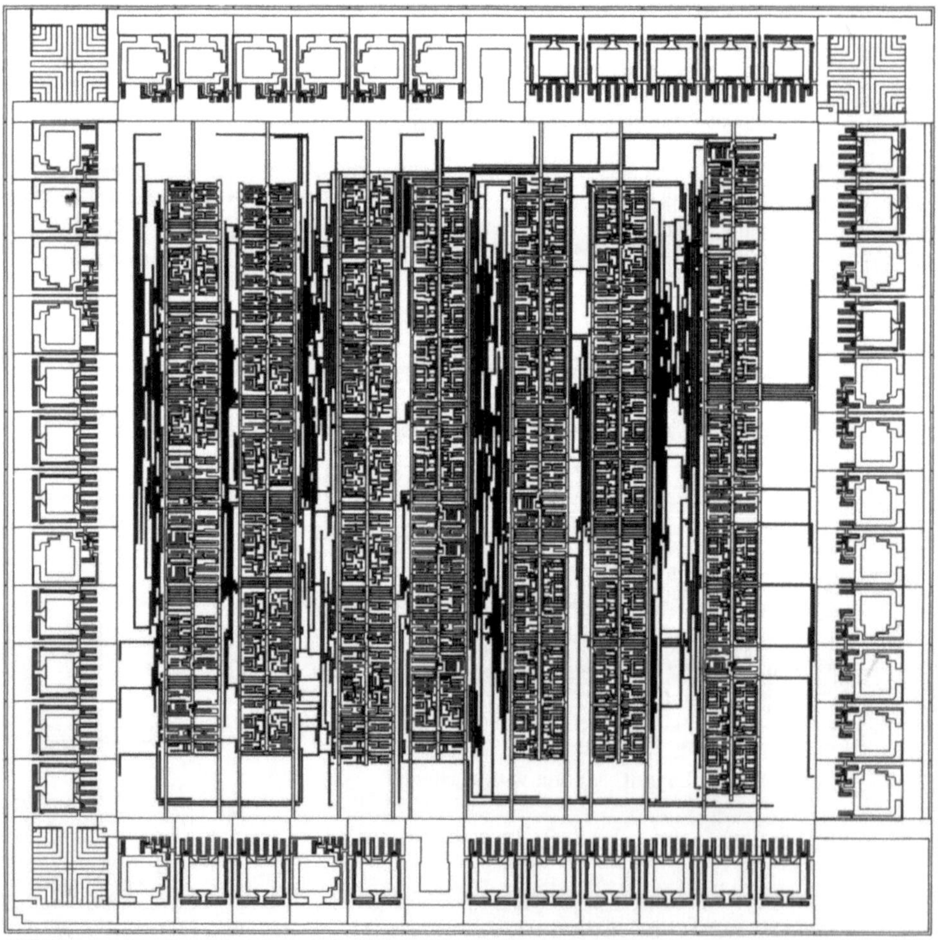

ADUS 14b. Production data: CALCOMP Plot (aluminum level)

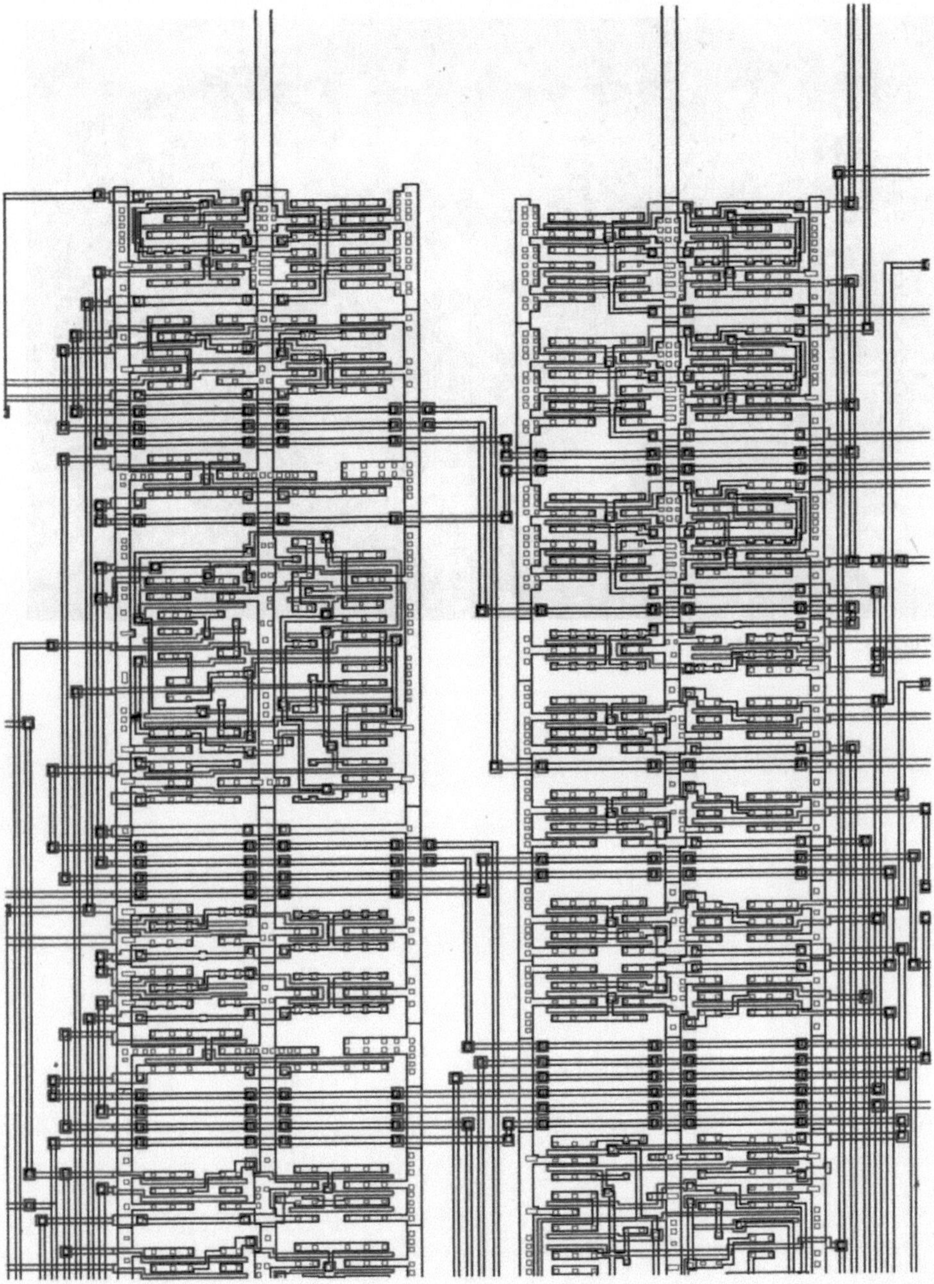

ADUS 14c. Production data: Layout (3 levels)

Fig. 6.3. Test assembly

The design center carries out all other steps in the same way as it has done for process 1; for this no detailed knowledge about the customer's system is necessary.

The customer has no knowledge of the design system; he is, however, able to provide his system-specific circuit as rough logic plan and to convert it into a VENUS logic plan sketch with the help of the user documents. The test preparation makes it necessary for the customer and the design center to make clear arrangements.

Process 3

The customer himself works in a VENUS design center and carries out the design steps up to logic verification: The testability analysis, chip construction, layout analysis, mask and test data generation steps are done by the design center.

A variant: the customer has access to a CAD work station including a part of the VENUS software and the libraries required for the customer's own design steps.

The customer does not have sufficient knowledge or hardware for solving test problems or carrying out chip construction. Concerning test preparation, the customer and design center have to make arrangements with each other.

Process 4

The customer, working in the design center, additionally takes charge of the testability analysis and test data generation design steps. The design center only carries out the chip design and mask data generation steps.

The customer does not have sufficient knowledge or hardware for chip design. But he does not need to discuss his system ideas with the design center, because test preparation also lies within his range of responsibility.

Process 5

The customer carries out all steps himself. He works independently in the design center and, according to the requirements set up by the design center, hands over the mask data and the test program to the manufacturer.

This customer utilizes the full scope of the CAD system VENUS. His system ideas remain completely under his control. But he has to be willing to leave a part of his system knowledge to the manufacturer, at least in the form of a test program.

Because the customer himself takes charge also of the chip design, he can achieve optimum results by applying his system knowledge.

Process 6

Contrary to process 5, here the customer generates the test program for his own tester or for the tester of a design center that has been equipped especially for testing support. He allows only the technology to be guaranteed by the manufacturer.

This means that the system knowledge is completely kept by the customer. Only the production data (mask tapes) are handed over the manufacturer.

The design system is built in such a way that for the customers the utilization according to the processes 5 and 6 is made viable. The customer as system designer requires no special knowledge in semiconductor technology; he needs to learn only the application of the user-friendly VENUS software. Similarly, the design system can also be utilized in accordance with processes 1 to 4, depending on the know-how of the user and his capital investment.

6.3 Organisational Preparation of the VENUS Application

Before the actual design activities are started using the documents, equipment, programs and cell libraries of the VENUS design system, one has to take important preparatory measures and decisions. They can be checked with the help of the following check lists.

Preparatory Considerations

The system engineer who has knowledge of the performance range of the devices that have to be developed by means of VENUS has to find answers to the following questions:

Check list 2: Preparatory Considerations

- ☐ Does the system (device) for which the component is meant fulfill the required environmental conditions in terms of temperature, current supply, humidity, chemical air purity, intensity of high-frequency and high-energy radiation and mechanic shock, so that they correspond with the manufacturer's specifications?
- ☐ Is it possible to meet the electric and timing requirements for input level, output level, output driving capability, clock frequency?
- ☐ Does the cell libraries' scope cover the functional requirements?
- ☐ Are RAM, ROM or PLA required on the device?
- ☐ Is the problem suited for a merely digital solution?
- ☐ Is the complexity of the task suitable for a complete integration in one component?
- ☐ Is it possible to carry out a convenient allocation to several components if the complexity is too high (partitioning)?
- ☐ Is the number of required signal and supply pins kept within limits that can be realized on a single chip?

☐ Is there a suitable tester available supported by VENUS (number of pins, frequency)?

☐ Are the functions of each component described completely, and is its behavior specified electrically?

☐ Does a higher level design system expect a description of the component, e.g. for system simulations?

☐ Are there testability requirements for system test and maintenance?

Planning of Resources and Contract Preparation

If these preparatory considerations have been concluded positively, the planning of resources, dates and costs starts. The following list shows some of the questions that should not be neglected:

Check list 3: Planning of Resources and Contract Preparation

☐ Where is access to the VENUS design system possible?

☐ Is it advisable to make the necessary investments for the purchase of CAD workstations?

☐ Should access to the VENUS design system be guaranteed as a service of a design center?

☐ What costs are to be expected for the utilization of the VENUS design system, according to the kind of utilization (customer/manufacturer interface) planned?

☐ When and for which period can the access to the design system/design center be guaranteed?

☐ What time and personnel effort is required for the individual steps (for the customer, for the design center)?

☐ Does the advisory system engineer have the required knowledge, is he available for the provided period of time?

☐ Is it planned, for example for the logic design, to organize a design team? Is a specialized VENUS operator to be trained?

- ☐ Are the delivery promises covered by the semiconductor manufacturer?
- ☐ Are the components meant to be delivered as wafers or in packages?
- ☐ Which package types are possible for the assembly of components (assembly requirements, pins numbers, chip size, stocks at the manufacturer)?
- ☐ Is there a contract with the manufacturer concerning the supply of the necessary packages or the module assembly?
- ☐ Have the cost issues been settled (mask cost, wafer cost, production volume, packaging cost, test cost)?
- ☐ Are the delivered modules meant to be tested or untested?
- ☐ Are the wafer and chip tests agreed upon on the test philosophy of the manufacturer? Which testers are offered by the manufacturer?
- ☐ Which quality standard is guaranteed by the manufacturer in terms of the technology used?
- ☐ For what period of time does the manufacturer guarantee the continuous delivery of components with unchanged specification?
- ☐ Is second sourcing an issue?

Technological Specification as Contractual Basis

As soon as the preparatory considerations, plannings and contract preparations have been concluded, technological specifications can be defined.

It is advisable to create the rough sketch of a technological specification of the design object first, based upon the knowledge of the system environment. This very tentative specification must be the subject of a discussion with the relevant partner of the semiconductor manufacturer as early as possible. The specification must be continuously worked on and corrected; the specification must reach a final state when the mask data are generated because it is part of the contract with the manufacturer and is integrated in key sections of the test program.

Check list 4: Technological Specification

- ☐ device or production number
- ☐ temperature range (environment)

- ☐ other environmental conditions (for example, humidity, mechanical stress, noise at the supply voltage, electrodynamic background radiation)
- ☐ supply voltage range
- ☐ input level
- ☐ output level
- ☐ clock frequency
- ☐ timing of the input and output bit patterns
- ☐ pcb technology
- ☐ number and positions of the supply pads
- ☐ number of signal inputs, signal outputs, bidirectional pads
- ☐ eventually predefined position of signal pads
- ☐ maximum quiescent current
- ☐ maximum dynamic current
- ☐ maximum power dissipation
- ☐ type of package provided, including constructive details such as the size of cavity or the range of angles for the bond wires
- ☐ minimum and maximum die sizes including the relevant bond wire lengths

---

Training

The use of the cell-oriented VENUS design system is so simple that one need not be an expert to develop chips with it. Nevertheless, reading the manuals and documents is not sufficient to turn the engineer who has developed system solutions on pc-boards by means of standard components, into a successful user of this design system for cell components.

Experience proves that in particular two skills can be learned easier by means of short training courses and exercises:

First, this is the *handling of the equipment*, primarily the CAD work station, the graphics terminal, and the alphanumerical terminal for program operation. When using VENUS, some engineers are introduced to computer-aided design methods for the first time.

Second, there are engineers who previously only considered logic design to be the actual content of system design. They must learn to look at the *test aspects* from the very beginning of their considerations. Because of the difficult test conditions of highly integrated circuits, they are confronted with a number of problems which previously had been handled by specialists without system knowledge, after the system development had been finished. Often, these design engineers are introduced to the importance of testing for the first time when using VENUS.

6.4 Technical Preparation for VENUS application

Logic Design

After the preparatory steps listed above have been carried through, the actual design activities can be initiated. The generation of the logic plan is based upon the data sheets of the cell catalogs. The steps of the system design phase – not supported by VENUS – serve as basis.

The descriptions in the cell libraries and the logic model are so exact that it is useless within the logic design to generate and measure the circuit model as a pc board with standard ICs. The logic simulator offers a much more precise verification than would be possible through an examination of a pc board.

The logic plan represents the central documentation of the device developed with VENUS:

- It serves as work basis for the VENUS user.
- It is part of the device specification.
- For the test technology, it helps analyze test problems during the debugging phase of the test program.
- It is the basis for system development (pc board, system) of which the relevant component is a part.
- Finally, it is a part of the system documentation which may reach the system customer.

For these reasons, great care must be used in forming a clear hierarchical structure of the logic plan. The more transparent the functional processes are, the easier it is during logic verification, testability analysis and layout analysis to follow the signal paths and carry out corrections, if necessary.

In addition to the functional parts of the logic plan, i.e. the cells and the signals, parameters for line weighting, affinity data etc., which will be integrated in later chip design, can already be determined. This includes their documentation, too. Moreover, comments and description fields are to be recommended. Those terms already used in the system design should be used here again.

If the networks and cells are named according to the block hierarchy, all further development steps are made easier.

The logic symbols of the data sheets appear only on the lowest hierarchy level as the actual basis for the further design steps. They characterize the cells available in the libraries.

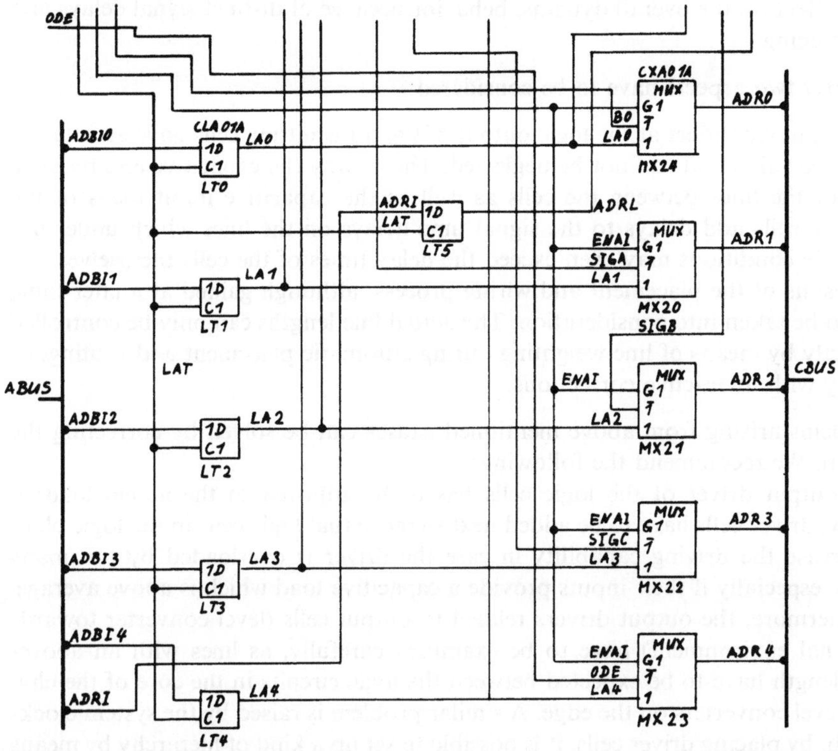

ADUS 1. Sketch of the logic plan

A consequence which should not be underrated results from graphic capture of the logic plan and relevant documentation: the logic symbols on the lowest hierarchy level are arranged in such a way that the design engineer recognizes their functional relationship. The CAD system does not modify this arrangement. Thus the design engineer will always recognize his design.

In the following sections, the explanations of the VENUS procedure steps are illustrated with the ADUS example (see Chap. 1). The test of the example is printed in small size.

Preparation of the Logic Plan

By means of cell symbols as described in the data sheets of the VENUS cell catalog, a sketch of the logic plan is set up. Figure ADUS 1 shows the sketch of a logic plan section (see also Fig. ADUS 4c).

Characteristics of the Conversion of Logic Functions into Integrated Circuits

Before the resulting logic plan is presented to the design system on the graphic terminal, the special time delay conditions of integrated circuits between the cells have to be considered and rated in a first correction step. Contrary to the logic circuits with standard components on pc boards, the connections between the cells – in accordance with the semiconductor technology applied, above all wiring technology – have a negative effect on the overall dynamic behavior because of distinct signal delays and pulse flattening.

Particularly two aspects have to be considered:

- The output amplifiers at the logic output of the transmitting cells have an internal resistance value that can not be neglected. This is why the capacitive and resistive loads of the lines between the cells as well as the capacitive input loads of the receiving cells add delays to the signal time delays on the lines which under unfavorable conditions may even exceed the delay times of the cells themselves.
- The results of the placement and wiring process, although gained at a later time, have to be taken into consideration. The actual line lengths can only be controlled indirectly by means of line weighting during automatic placement and routing, or directly with interactive corrections.

Problems arising from above mentioned causes can be solved by correcting the logic plan. We recommend the following:

The output driver of the logic cells has to be adjusted to the actual load; if necessary, driver cells have to be added next to the actual logic cells in the logic plan; they increase the driving capability in case the driver is overloaded by too many receivers, especially if their inputs provide a capacitive load which is above average.

Furthermore, the output drivers related to output cells (level converter towards the external environment) have to be examined carefully, as lines with an above-average length have to be expected between the logic circuits in the core of the chip and the level converters on the edge. A similar problem is raised by the system clocks for which, by placing driver cells, it is possible to set up a kind of hierarchy by means of tree structures.

Design Rules for Testability

In principle, the design rules for testability must be constantly followed during the logic design. At the latest at this point they must be checked and, if necessary, modifications have to be made. Admittedly, a reduction of the circuit's functions or an increase of complexity must sometimes be accepted. Finally, the module will gain more testability, more functional reliability and more allowance for production variances.

Functional Bit Patterns

The logic plan renders a complete description of the circuit; thus the logic plan would be sufficient for preparing the production data. However, by using the logic plan only without the knowledge of the requirements posed by the system environment the correctness of the component can not be proven.

This is why the system design engineer has also the task of preparing the input bit patterns and appropriate output target bit patterns as well as to prepare and document their temporal relations to each other. The list of the functional requirements must always be as complete as possible; only the design engineer is responsible for checking the completeness of bit patterns which may be later expanded or modified. The subsequent logic verification steps can always verify the design only relative to

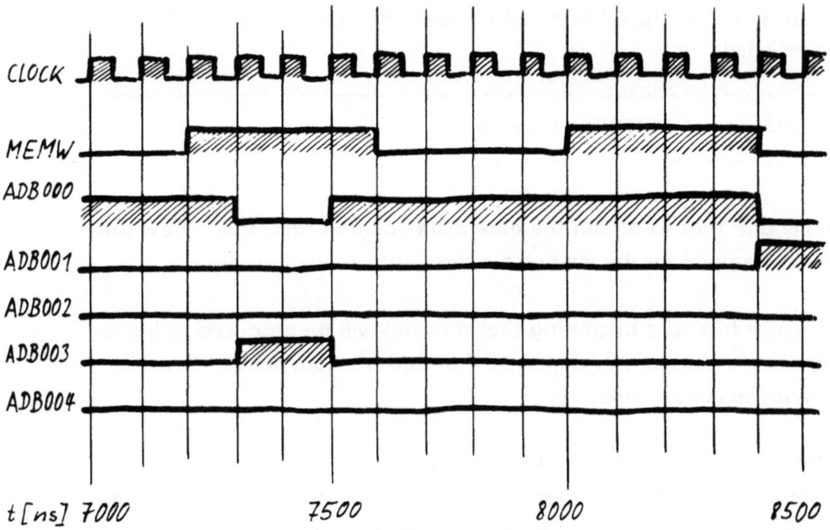

ADUS 2. Sketch of functional bit patterns

this specification. Errors or deficiencies which remain unnoticed are detected only when the completed chip is tested in its later system environment.

Preparation of the Simulation Bit Patterns

Now, one has to prepare functional bit patterns for the logic verification. The bit patterns presented in ADUS 2 correspond with the previously displayed ADUS section. Inputs ADB000 through ADB004 are positioned at signals ADB10 through ADB14 of the bus, going through input drivers; MEMW, together with other input signals, controls the LAT signal (see also ADUS 9, logic simulation).

6.5 Design Steps of Computer-Aided Circuit Design with VENUS

After preparatory consideration and planning has been accomplished and the logic plan has been sketched, the desk activities of the designer have been accomplished. Now, accesss to the computer is necessary (work station, graphic terminal, main frame) on which the VENUS design system is available.

6.5.1 Selection Library and Master

If possible, one must decide which cell library is to serve as basis before beginning the schematic entry. If one chooses the gate-array method the appropriate master has to be determined at the same time.

It is possible, however, to change this decision later: such a change makes appropriate modifying processes necessary in the entry system and data base of the CAD computer. It is obvious that the farther the design has already proceeded when the decision is modified, the higher the effort and the required number of design step repetitions.

For selecting one of the CMOS cell libraries, one may adhere to the steps of the following check list:

Check list 5: Selection of a CMOS Cell Library

- ☐ Usually one tries to obtain the gate-array solution first. This is how one gets first samples for the least cost in the shortest time possible.

- ☐ If it is sure from the beginning that the chip will be produced in higher quantities, a standard or a macrocell design is recommended because it demands less silicon area.

- ☐ If the circuit has more gate functions or more pins than the number provided by the available masters, a gate array solution is out of question. For the standard or macrocell design, the decision concerning which technology to use (one-layer or two-layer metallization) has to follow the dynamic requirements.

- ☐ From middle to upper complexity range, the use of standard or macrocell components with polysilicon/aluminum wiring is advisable whenever the dynamic requirements are not too high. The pertinent process of production is simpler then, i.e. it can be performed cheaper and faster.

- ☐ If the logic structure requires internal tristate pins gate-array solutions are out of question. Cells with appropriate logic outputs can be found only in the standard- and macrocell catalogs.

- ☐ For circuits with very high complexity, gate-array masters are not available. Thus they can be designed only as standard- or macrocell-components.

- ☐ If the logic plan includes macrocells (RAM, ROM, PLA etc.) only a standard- or macrocell design is possible. Gate-array masters with RAM cells etc. are currently not offered.

If there are no clear reasons given for the decision, a preliminary placement of the temporarily chosen master or a preliminary routing with one of the cell libraries can nevertheless be performed after the logic plan input. No effort should be wasted for optimization. Important are the results concerning whether the choice of the master has been right or whether a chip size is obtained which fits the chosen package. If the results are negative it is still possible to return to another master or to another standard cell library without the effort being too high.

Preliminary runs are useful, as generally it is impossible to obtain accurate predictions about the obtainable occupation degrees of masters or chip area for the cell ICs. The wiring space depends too much on the interconnection structure of the logic elements with each other. This applies above all for the heavy use of buses.

Moreover for cell devices a preliminary wiring proves whether the chip is pad-determined. This is the case if the number of signal and supply pads (pad cells) is so high that their entire length exceeds the perimeter of the logic cell core.

6.5.2 System Initialization

The first step of the VENUS procedure is passing administrative data to the CAD system.

The design engineer confers a circuit name, opens the data base for the circuit and sets up passwords for access control. The data resulting during the entire design process is administered under this name by means of service programs.

Initialization of the Data Base

At the beginning of the actual VENUS procedure, a data base is opened for the circuit by means of the function DBNEU.

For this, one has to state (among other things) user identification ($BENKEN.) and the master identification ($MASTERK.) under which the VENUS programs are installed. The input ABIB refers to the master identification library that contains a description of the cell family used. Inputs in the VENUS mask are written in brackets < ... > (see Fig. ADUS 3).

6.5.3 Schematic Entry on a Graphic Terminal of a Work Station

For this step it is assumed that the design engineer has available an almost complete sketch of the logic plan before he applies the design system. So the step "logic design" is mostly realized by means of "paper and pencil". Only the schematic entry is supported by the system.

```
+------------------------------------------------------------------------+
!  VENUS                        MONITOR VERSION XXX                  YYY  !
!                                                                         !
!  USER                          (USER         )                          !
!                                                                         !
!  DATABASE IDENTIFICATION: ($BENKEN.     ) NAME: (ADUSDB    )            !
!  LIBRARY  IDENTIFICATION: ($MASTERKEN.  ) NAME: (ABIB      )            !
!  SHELF    IDENTIFICATION: (             ) NAME: (          )            !
!                                                                         !
!  FUNCTION                      (dbneu        )                          !
!                                                                         !
!  FUNCTION-SPECIFIC INPUTS                                               !
!           (                                                      )      !
!  FUNCTION-SPECIFIC OUTPUTS                                              !
!           OUTPUT ON (PRINTER, FILE, TERMINAL) (P, F, T)        ( )      !
!  FOR OUTPUT ON FILE:  FILE NAME    (                             )      !
!                       VOLUME: (    )        Device:  (           )      !
!                                                                         !
+------------------------------------------------------------------------!
!                                                                         !
!                                                                         !
!                                                                         !
!                                                                         !
+------------------------------------------------------------------------!
```

ADUS 3. Initialization of the data base: DBNEU

The input of logic plans is realized graphically on the CAD work station with the possibility of using all logic symbols of one cell library. In the logic plan, cell and signal names are indicated as well as an optional number of comments. Thereafter, it is converted into a net list and stored in a data base after the necessary tests and plausibility checks have been performed. The net list is the basis for all further design steps.

The output of logic plans is possible on the graphic terminal of the work station or through the plotter. Net lists and other circuit information may also be output from the data base via printer.

During the input of the logic plan on the graphics terminal the logic plan has to be structured conveniently. We recommend the following:

- The functionally non-relevant level converters within the pad cells (signal pads) should be separated from the other parts of the logic; to that end the output pins of the input level converters and the input pins of the output level converters are each combined to signal bundles in such a way as demanded by the functions.
- Furthermore the external connections of the level converters have to be marked by so-called pseudopads. These serve to name the signals which connect the component with the system environment. This naming is absolutely necessary because these names are used for the following development steps of logic verification and testability analysis.

On the highest hierarchy level the functional logic is combined to a block which is stimulated by the input line bundles and which transmits signals to the environment

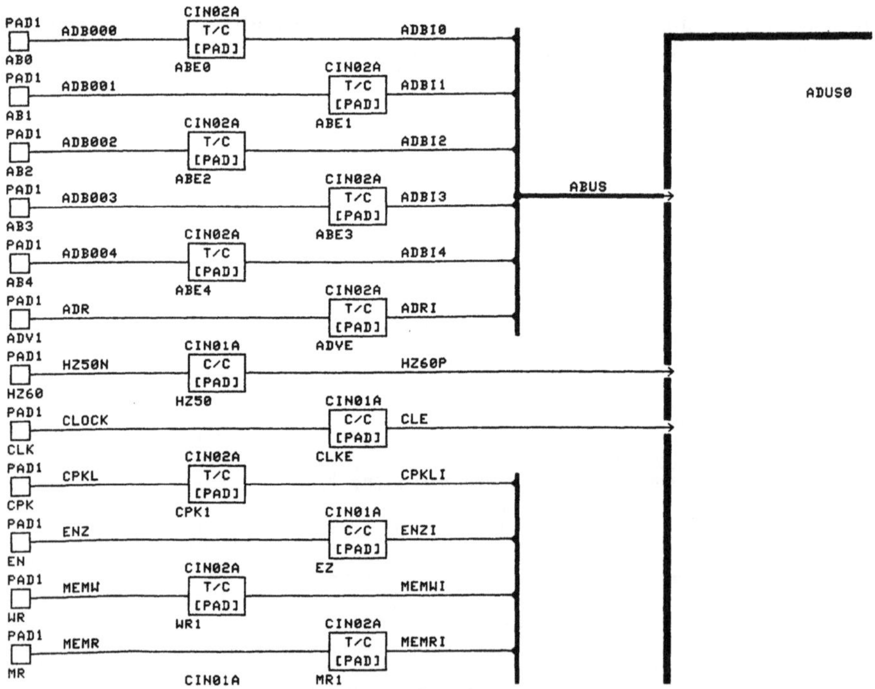

ADUS 4a. Logic plan input: highest hierarchy level

through the output line bundles. This separation of the bundles may, for example, serve for grouping the elements of the next lower hierarchy level in corresponding blocks. Each bundle of external lines feeds exactly one block. Each bundle of lines leading outside originates from one of the blocks or in an additional block. These bundles provide only for a better overview and a clearer graphical logic documentation.

Schematic Entry of Logic Plan

By the logic plan capture on the graphic work station, the circuit is transferred to the design system; here, a hierarchical segmentation of larger circuits is recommendable. On the highest hierarchy level, the connection of ADUS with the chip package is described. If a signal is to be run to the outside through a bond wire one has to insert so-called pseudopads (PAD 2) in addition to the actual input/output drivers (such as CIN02A) in the logic plan. For logic simulation and test data generation, the individual names of the signals between pseudopad and driver are relevant (see ADUS 4a).

The second hierarchy level shows a segmentation of the circuit into various functional units (ADUS 4b).

The resolution of the circuit in single signals and cells is carried out on the lowest hierarchy level. ADUS 4c shows block ADUS 3 from ADUS 4b. Cell CLA01A is a latch, CXA01A a multiplexer and CBC02A a binary counter.

Every designer should structure his representation of the logic plan (i.e. of the connections between the blocks of one hierarchy level or across hierarchy levels) according to certain conventions. The following rules may be applied, for example:

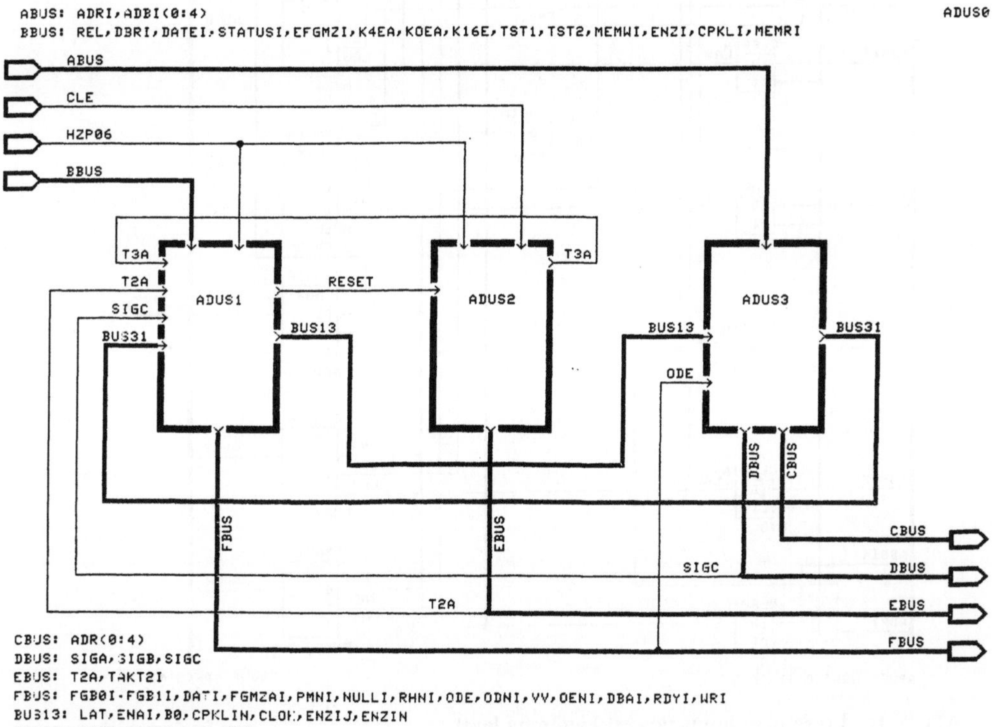

ADUS 4b. Logic plan input: second hierarchy level

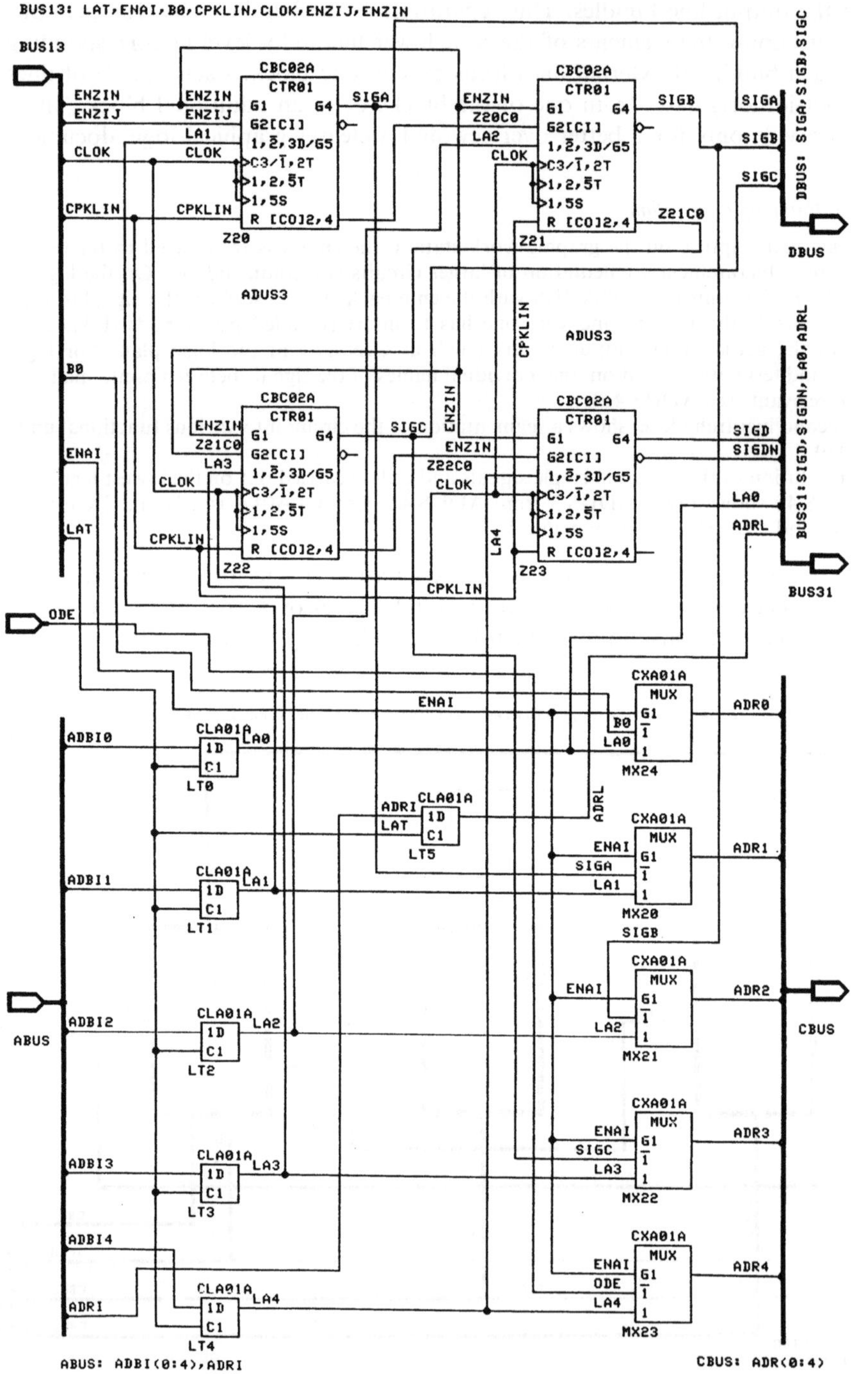

ADUS 4c. Logic plan input: lowest hierarchy level

- Signals and line bundles from other sheets reach blocks only from the top and leave them at bottom.
- Signals and line bundles between blocks on one sheet are run to the blocks from the left side and leave them on the right side.
- Finally, on the lowest hierarchy level the logic symbols of the cells from the cell catalog are displayed. They are stimulated only from the left side by single signals and output single signals to the right side.
- Thus, the main signal flow on the lowest hierarchy level runs from the upper left side to the lower right side.

Generally one is recommended to completely name individual signals and cells in the logic plan and to use comments. It should be possible to relate each sheet of the documentation to a certain development project. Furthermore, it is strictly recommended to provide exact version, date and time data.

The clarity and simplicity of the logic plan as it has been described make it possible to document each modification, correction and extension of the logic immediately after the input. Thus, consistency between documentation and data base can always be achieved.

After the graphic input has been accomplished, the acquired data is made subject to various plausibility checks; for example, warning about open outputs are given or open inputs are reported as errors which have to be removed. Moreover, the system now combines the data of the separately input sheets into an overall logic plan.

6.5.4 Net List Transfer to the Mainframe, Set-up of the Project Data Base

The next design step is the set-up of the project data base in the mainframe. From the overall logic plan in the work station, a net list containing only the functional parts and the parameters for chip design is extracted; after a file-transfer to the mainframe, it is stored in a previously defined data base. Simultaneously, formal checks are again carried out. The data base can be read, statistically evaluated and archived directly.

Logic Plan Capture to the Data Base

On the work station, the net list is extracted from the logic plan. The section shown in ADUS 5 illustrates the stucture of this file:

```
TEB 224                                 CBC02A
LEB                      Z20                     CBC02A
PNB          A1
SSB SIGA
PNB          A2
SSB
PNB          A3
SSB Z20C0'
PNB          E1
SSB ENZIN
PNB          E2
SSB ENZIJ
PNB          E3
SSB LA1
PNB          E4
SSB CLØK
PNB          E5
SSB CPKLIN
```

ADUS 5. Section of net list

```
+-----------------------------------------------------------------------+
!  VENUS                       MONITOR VERSION XXX                YYY  !
!                                                                      !
!  USER                         (USER          )                       !
!                                                                      !
!  DATABASE IDENTIFICATION: ($BENKEN.     )  NAME: (ADUSDB   )         !
!  LIBRARY  IDENTIFICATION: ($MASTERKEN.  )  NAME: (ABIB     )         !
!  SHELF    IDENTIFICATION: (             )  NAME: (         )         !
!                                                                      !
!  FUNCTION                     (busein        )                       !
!                                                                      !
!  FUNCTION-SPECIFIC INPUTS                                            !
!           (                                                  )       !
!  FUNCTION-SPECIFIC OUTPUTS                                           !
!           OUTPUT ON (PRINTER, FILE, TERMINAL) (P, F, T)     ( )      !
!  FOR OUTPUT ON FILE:   FILE NAME     (                       )       !
!                        VOLUME: (      )      Device:   (     )       !
!                                                                      !
+----------------------------------------------------------------------!
!                                                                      !
!                                                                      !
!                                                                      !
!                                                                      !
+----------------------------------------------------------------------!
```

ADUS 6a. Net list transfer: BUSEIN

```
+-----------------------------------------------------------------------+
!  VENUS                       MONITOR VERSION XXX                YYY  !
!                                                                      !
!  USER                         (USER          )                       !
!                                                                      !
!  DATABASE IDENTIFICATION: ($BENKEN.     )  NAME: (ADUSDB   )         !
!  LIBRARY  IDENTIFICATION: ($MASTERKEN.  )  NAME: (ABIB     )         !
!  SHELF    IDENTIFICATION: (             )  NAME: (         )         !
!                                                                      !
!  FUNCTION                     (busein        )                       !
!                                                                      !
!  FUNCTION-SPECIFIC INPUTS                                            !
!           (                                                  )       !
!  FUNCTION-SPECIFIC OUTPUTS                                           !
!           OUTPUT ON (PRINTER, FILE, TERMINAL) (P, F, T)     ( )      !
!  FOR OUTPUT ON FILE:   FILE NAME     (                       )       !
!                        VOLUME: (      )      Device:   (     )       !
!                                                                      !
+----------------------------------------------------------------------!
!  FUNCTION BUSEIN       HAS                                           !
!  BEEN PERFORMED CORRECTLY                                            !
!  PREVIOUS INPUT PARAMETERS:                                          !
!                                                                      !
+----------------------------------------------------------------------!
```

ADUS 6b. Acknowledgment

The counter cell CBC02A is stored with the individual number 224 and its individual name Z20 in the cells TEB and LEB.

In this context, the position of the cells (logic symbols) on the logic plan are irrelevant. In the net list, only the assignment of the signal names to the inputs/outputs of the cells is important, in the first two cells PNB, SSB, for example: signal SIGA is connected to output A1 of the cell. See also Cell Z20 in the ADUS 4c.

By calling the function BUSEIN through the VENUS mask the net list is stored in the data base (ADUS 6a).

```
+-----------------------------------------------------------------------+
!  VENUS                       MONITOR VERSION XXX                 YYY  !
!                                                                       !
!  USER                        (USER        )                           !
!                                                                       !
!  DATABASE IDENTIFICATION: ($BENKEN.     ) NAME: (ADUSDB    )          !
!  LIBRARY  IDENTIFICATION: ($MASTERKEN. ) NAME: (ABIB      )           !
!  SHELF    IDENTIFICATION: (             ) NAME: (          )          !
!                                                                       !
!  FUNCTION                    (einkon      )                           !
!                                                                       !
!  FUNCTION-SPECIFIC INPUTS                                             !
!          (                                                 )          !
!  FUNCTION-SPECIFIC OUTPUTS                                            !
!          OUTPUT ON (PRINTER, FILE, TERMINAL) (P, F, T)     (f)        !
!  FOR OUTPUT ON FILE:  FILE NAME          (adus.prot.einkon  )         !
!                       VOLUME: (     )      Device:  (        )        !
!                                                                       !
+-----------------------------------------------------------------------!
!                                                                       !
!                                                                       !
!                                                                       !
!                                                                       !
+-----------------------------------------------------------------------!
```

ADUS 7a. Formal check of the data base: EINKON

Information about the program run is provided by an acknowledge message in the mask (see ADUS 6b).

The VENUS function EINKON carries out formal checks of the circuit in the data base (see ADUS 7a).

The example of an EINKON protocol in ADUS 7b shows that during the input of the logic plan one signal has been forgotten which resulted in some cell inputs remaining unwired.

6.5.5 Logic Verification

The appropriate type of verification for a cell-oriented design system is the logic simulation. The cells are constructed in such a way that their different combinations, apart from delays due to load capacities, do not affect their behavior. Individual delay times are described with sufficient exactness by the logic models of the cells.

The logic model of a circuit design is generated automatically from the net list and the logic models of the cells, stored in the data base. The circuit designer can determine whether he wants to use only the predetermined delays of the cells or, in additions individual signal delay times which take the load capacities and wire lengths into account (resimulation, see below.). For stimulating the simulation of the logic model, the design engineer has to set up a bit pattern file which must include the input bit patterns, and time-, repetition-, and termination conditions.

For the simulation run one may choose between a cycle- and a time delay-simulation. The cycle simulation basically serves for verifying the logic design, especially with consideration of a later test on the test machine, while the time delay simulation forecasts whether the chosen realization meets the specification requirements, including the line delay times which are calculated according to the chip design. The results of the simulation can be evaluated by means of signal diagrams.

```
          RP = GRID POSITION
          B  = ABBREVIATION FOR COMPONENTS
     BNA  = COMPONENT'S NAME
     ELECTRIC PARAMETERS
          PT = POINT TYPE
          BL = STATIC LOADABILITY
          SE = NUMBER OF SENDERS
          LA = NUMBER OF LOADS
          N  = INTERFACE
     FKZ  = ERROR SIGN
     SKZ  = LOOP SIGN
     M    = TYPE
          REF = REFERENCE
          GBL = OVERALL SHEET
     PLQ  = GRID SQUARE
     RPN  = PIN-REAL
     SNA  = LOGIC ELEMENT NAME
     SPN  = PIN-SYMBOLIC
```

```
    22.07.85      UNIT: ADUS            WGS:  01     DB-NAME: FADUSDB          DB carrier:                         0002+
A     V E N U S                E I N K O N              -BLOCK TEST LOGIC ELEMENT ORIENTED-
 ! TYPE    !    BA-P       !            !          !    !                                                          !F !
 !REF  GBL !  SP !  RP     !     BNA    !   SNA    !PLQX!    ERROR MESSAGES                                        !RL!
 !---------+---------------!------------+----------+----+----------------------------------------------------------+--!
 !AN20     !               !CANO2A      !CANO2A    !    !011xINPUT PIN LOGICALLY NOT CONNECTED                     !  !
 !         !               !            !          !    !                                                          !  !
 !END      !               !CORO2A      !CORO2A    !    !011xINPUT PIN LOGICALLY NOT CONNECTED                     !  !
 !         !               !            !          !    !                                                          !  !
 !ENJ      !               !CDR11A      !CDR11A    !    !011xINPUT PIN LOGICALLY NOT CONNECTED                     !  !
 !         !               !            !          !    !                                                          !  !
 !UND1     !               !CANO2A      !CANO2A    !    !011xINPUT PIN LOGICALLY NOT CONNECTED                     !  !
-----------------------------------------------------------------------------------------------------------------------

    22.07.85      UNIT: ADUS            WGS:  01     DB-NAME: FADUSDB          DB carrier:                         0004+
A    V E N U S                 E I N K O N            -STATISTICS-
 0000 DOCUMENTS       HAVE BEEN CHECKED.            0 ERRORS
!                     AND                           0 WARNINGS HAVE BEEN FOUND.
 0000 COMPONENTS      HAVE BEEN TESTED.             0 ERRORS
!                     AND                           0 WARNINGS HAVE BEEN FOUND.
 0227 LOGIC ELEMENTS  HAVE BEEN TESTED.             4 ERRORS
!                     AND                           0 WARNINGS HAVE BEEN FOUND.
 0218 SIGNALS         HAVE BEEN TESTED.             1 ERRORS
!                     AND                           0 WARNINGS HAVE BEEN FOUND.
```

ADUS 7b. EINKON protocol with warnings

Logic simulations at this stage of the design process allow only a reduced dynamic verification. Only the delay times of the cells are considered as time elements. For them, characteristic SLOW- and FAST-conditions can be chosen. The delays stored in the library are calculated for a output standard load. The actual loads of the transmitting cells according to the net topology (but not the wiring!) and the resulting difference in number and amount of the receiver input loads can already be determined here. Loads due to the actual wiring are at this stage – before chip design – yet unknown.

The simulations have to be carried out as detailed as possible. Any design weakness of the logic has to be discovered here. If the target bit patterns do not appear at the outputs, the interal nodes have to be examined, too, until one discovers which circuit sections do not behave properly. The system developer has to be sure to have specified and realized all requirements, and to have removed any unnecessary function. In particular, redundancies in the logic plan have to be avoided.

Furthermore, it is quite advisable to prove, by means of the SLOW conditions of the library elements that the dynamic specifications can be fulfilled in principle also under unfavorable conditions. Here, too, final clarification can be provided only by the design step layout analysis and by the repetition of the logic verification with real line delays (resimulation).

Logic Simulation

The bit patterns for the logic simulation are stored in the EBIT list. A section of this list is shown in ADUS 8:

```
LIST - EBIT
'DETERMINATION OF CLOCK T BY MEANS OF THE CLOCK-COMMAND'
CLOCK T, ON=0,W=500,P=1000,E=P,OFF=3000000$
CLIN T CLOCK AT 0$
'RESET COUNTERS AND FLIPFLOPS'
CHANGE CPKL,ENZ,TEST1,KOI,K41,MEMR,MEMW,DATE,DBR,ADB004 TO 0 AT 0$
CHANGE ADB000,ADB001,ADB002,ADB003,ADR,HNREL,STATUS TO 0 AT 0 $
CHANGE HZ50N,TEST2 TO 1 AT 0 $
CHANGE ENZ,TEST1 TO 1 AT 500$
'-------------------'
CHANGE ADB000,ADR TO 1 AT 68000 $
CHANGE MEMW TO 1 AT 72000,116000,8000 $
CHANGE MEMW TO 0 AT 76000,116000,8000 $
CHANGE ADB(003-000) TO 8 AT 73000 $
CHANGE ADB(003-000) TO 1 AT 75000 $
CHANGE ADB(003-000) TO 2 AT 84000 $
CHANGE ADB(003-000) TO 4 AT 89000 $
CHANGE ADB(003-000) TO 2 AT 91000 $
```

ADUS 8. Section of stimuli file

By means of the CLOCK command, clock T (clock = Takt) is defined with period P = 1000 simulation steps (a simulation step = 1 ss = 0.1 ns), clock width W = 500 ss and the overall duration of 3.000.000 ss. With CLIN clock T is applied to the signal CLOCK at time $t = 0$. State changes of signals are described with CHANGE; the states of signal bundles such as ADB000, ADB001, ..., may be displayed hexadecimally. CHANGE ADB(003–000) TO 8, for example, results in ADB000 = ADB001 = ADB002 = 0, ADB003 = 1.

The logic verification user interface is mask-controlled (in this example still with the simulator VERDIPUS):

In the time interval which has been determined by a starting point (here 7750 ns) and a time grid (here 25 ns), optional signals can be displayed. The bit patterns in the upper half of Fig. ADUS 9b correspond with those of Fig. ADUS 2. RAM0 ... RAM4 are output signals which are connected to ADR0 ... ADR4 via an output driver. The values of the signals are, for example, represented by the following symbols:

- – logic 1,
- · logic 0,
- × undefined,
- : transition which can not be displayed in the current grid.

```
V E R D I P U S      FBG-NR,VERS /$ADUS/-///.T-EST/-//-/.04,001/ NEXT MASK/    /
SIMULATION MASK      BIBLIO.VERS /CMOS,001/   PROCESSOR /FOR/    BATCH (Y/N) /N/

PREPARATION MODEL          LOBE / /    ENS / /    DB / /
- - - - -                TIME DELAY(LS,LM,-S,-M,+S,+M) /    /  LINE DELAYS     / /
            (DB)         UNIT NAME /                   /  WGS /   / VERSION     /   /

            BIT PATTERN  BITE / /    EBIT / /    STATIC      / /  EZ-MIN /         /

            FKAH                / /  SHORT CIRCUIT ERROR     / /

SIMULATION  LOGIC         /*/         ERROR                  / /
- - - - -   RESTART       / /         STATE GUARENTEE(MA)  / /
            FACTOR OF DELAY MODIFICATION /      %   HAZARD PROPAGATION /· /

EVALUATION  FORMAT   !   CYCLES          / /        SIGNAL01          / /
- - - - -   DA / /   !   MAGNIFIER       / /        HAZARD            / /
            SA / /   !   CYCLE LENGTH    / /        ERROR STATISTICS  / /
                     !   CLOCK PROTOCOL / /         ERROR CATALOG     / /

SPECIAL FUNCTIONS
- - - - - - - - -
            SORTSINA  / /            ERGVOR  / /
ACK:
```

ADUS 9a. Logic simulation: VERDIPUS mask

```
+--------------------------------------------------------------------------------+
!P R O S I M  VERS XXXX       TIME DIAGRAM                      NEXT MASK /SIW/!
!PROJECT /ADUS    ,VERS: 001/                     TEST PROGRAM /TEST    ,VERS: 04/!
!TIME /  7750 NS/          GRID / 25 NS/             NEXT PAGE /   /       COPY / /!
!                                                                                 !
!CLOCK           --.--.--.--.--.--.--.--.--.--.--.--.--.--.--.--.--.--.--.--.--..!
!                                                                                 !
!MEMW            ........----------------................----------------........!
!ADB000          ------------........------------------------------------........!
!ADB001          ........................................................--------!
!ADB002          ................................................................!
!ADB003          ............--------............................................!
!ADB004          ................................................................!
!                !         !         !         !         !         !         !   !
!              7000      7250      7500      7750      8000      8250      8500  !
!                                                                                 !
!RAM0            xxxxxxxxx:-----------------------------------------------:......!
!RAM1            xxxxxxxx:................................................:------!
!RAM2            xxxxxxxx:.......................................................!
!RAM3            xxxxxxxx:.......................................................!
!RAM4            xxxxxxxx:.......................................................!
!                                                                                 !
!                                                                                 !
!                                                                                 !
+--------------------------------------------------------------------------------!
```

ADUS 9b. Logic simulation: PROSIM mask

Simulation results which have been obtained with regard to line delays and maximum cell delay times are shown in ADUS 13.

6.5.6 Testability Analysis

After the conclusion of the logic development, testability has to be ensured. After production the chips are functionally pre-tested on the wafer by means of probe cards. Moreover, after the assembly of the component, an intensive test under the specified dynamic conditions must be possible. For this only those relatively few supply and signal pads are available which can be reached by means of the probe card during the wafer test or through the package pins during the component test. Access to the internal parts of the circuit is not possible during the test.

For testing, each circuit node has to be toggled at least once within the shortest period of time possible, in order to allow outside observation of the maybe faulty behavior.

For the purpose of testability analysis, first a program checks for compliance with the test design rules. The rule violations discovered indicate clear design risks. It is strongly recommended to correct the logic accordingly.

Finally, a fault simulator indicates for each node whether errors at these nodes are detected by means of the previously supplied bit patterns – if used as test patterns. If the degree of error detection is not sufficient this may direct the attention of the developer to redundancies in the logic or to input bit patterns which have so far not been considered. In this case, further appropriate bit patterns must first be provided by the developer.

Finally, a test pattern generator may be applied, too. It generates from the circuit structure related input and output bit patterns for nodes which before could not be controlled or observed. These bit patterns make, however, not always sense from the viewpoint of functionality. The degree of error detection is increased. Thus, it is ensured that the logic is covered sufficiently by the still to be generated test program, and this not only functionally but also in terms of the required fault coverage.

Testability Analysis

The fault simulator calculates the error detection degree that can be obtained with the given bit patterns. For the ADUS example, more than 5000 test patterns are gained from the EBIT list of the logic simulation. The degree of error detection is displayed graphically by the testability analysis program (ADUS 10a).

Additionally a test pattern generator can be applied in order to generate further stimuli (ADUS 10b).

Storing elements in the circuit make it necessary that, for setting a circuit node to a required condition, generally a sequence of several stimuli (SET) is applied. With the test patterns supplemented by the stimuli generator, an improvement in the degree of error detection can be obtained (ADUS 10c).

6.5.7 Chip Design

For cell-oriented design, the chip design (also: chip construction) consists mainly of the placement of the cells – including the padcells on the chip frame – and wiring them. Placement and routing are carried out automatically. Modifications using the color graphics terminal are permissible. They may affect the placement of the cells as well as the wiring, for example, for time-critical signals.

```
(IN)     EXEC $MASTERK.FEHLERSIMULATOR
(OUT)
(OUT)
(OUT)    START AT 13:21:46 DD/MM/YY
(OUT)    FAULT SIMULATOR
(OUT)    VER. XXX
(OUT)    ENTER OPTIONS
(IN)     MT 5 END
(OUT)    ***DIAG--THERE IS NO RUN HISTORY
(OUT)    THIS RUN WILL BE STARTED ON PATTERN 1
(OUT)    NUMBER OF COLLAPSED FAULTS          592
(OUT)    PREPROCESSING TIME IN SECONDS         0
(OUT)          1.   19=*********
(OUT)          2.   20=**********
(OUT)          3.   20=**********
(OUT)          4.   21=**********
(OUT)          6.   21=**********
(OUT)         10.   32=****************
                .
                .
                .
(OUT)       5014.   84=******************************************
(OUT)       5022.   85=******************************************
(OUT)       5039.   85=******************************************
(OUT)       5047.   85=******************************************
(OUT)       5527.   86=*******************************************
(OUT)    TERMINATED ON PATTERN             5571
(OUT)    LAST DETECTION ON PATTERN         5527
(OUT)    DETECTED COLLAPSED FAULTS          510
(OUT)    NUMBER OF FAULT SETS GENERATED       0
(OUT)    TOTAL NUMBER OF FAULTS             841
(OUT)    TOTAL NUMBER OF DETECTS            733
(OUT)    PER CENT DETECT=                    87
(OUT)    RUN TIME IN SECONDS                301
```

ADUS 10a. Testability analysis: fault simulator

The 'chip design' development step brings the VENUS-user into contact with the characteristic boundary conditions of integrated circuits for the first time. Here again, he does not need to concern himself with semiconductor physics or semiconductor technology.

From the viewpoint of chip design, cells are formations of varying extensions with precisely determined positions for signal inputs and signal outputs. The logic connections of the symbols in the logic plan become wirings, with certain lengths which can cross each other. Due to crossing, the wires have to change to another level of technology (Poly-Si and Al or Al1 and Al2) and thus have contacts at the transitions (from Poly-Si to Al or from Al1 to Al2).

Furthermore, geometrically layed-out wires possess resistive and capacitive loads. Together with the internal resistances of the outputs and the input capacities of the inputs, they determine the dynamic behaviour between the cells on the chip being constructed.

At this stage of chip development the VENUS-user for the first time must think of the power supply of the cells which of course are active electric circuits. Although the current supply of all elements is guaranteed by the CAD system, it may be necessary to provide the output drivers on the chip boundary (level converters,

```
(IN)     EXEC $MASTERK.STIMULIGENERATOR
(OUT)
(OUT)
(OUT)    START AT 14:24:49 TT/MM/YY
(OUT)    AUTOMATIC STIMULUS GENERATOR
(OUT)    VER. XXX
(OUT)    ENTER OPTIONS
(IN)     ST 30 MT 20 CYCO MPTH 1 END
(OUT)
(OUT)
(OUT)                                    MODEL FILE NAME                          ADUS
(OUT)                                    NUMBER OF NODES                          1384
(OUT)                                    LENGTH OF MODE TABLE                     2214
(OUT)                                    NUMBER OF INPUT NODES                      22
(OUT)                                    NUMBER OF OUTPUT NODES                     24
(OUT)                                    MAXIMUM NUMBER OF PATTERNS PER SET       1000
(OUT)                                    SET TIME ALLOTMENT IN SECONDS              30
(OUT)                                    RUN TIME ALLOTMENT IN SECONDS            1200
(OUT)                                    MAXIMUM PATTERN/DETECTION RATIO         10000
(OUT)                                    DETECTION GOAL                            100
(OUT)                                    NUMBER OF POSSIBLE FAILURES               107
(OUT)                                    PROCESSING WILL BEGIN ON THE FIRST OUTPUT PIN
(OUT)                                    PREPROCESS TIME=                         1.35
(OUT)
(OUT)
(OUT)
(OUT)
(OUT)    SET   OUTPUT NODE  STATE    PATTERNS  TOTAL     DETECTS   PERCENT      TIME
(OUT)                                 IN SET  PATTERNS
(OUT)      1 SIA            (0)          9      10         1        0.93       1.00
```

ADUS 10b. Testability analysis: stimuli generator

```
(IN)     EXEC $MASTERK.FEHLERSIMULATOR
(OUT)
(OUT)
(OUT)    START AT 07:57:01 DD/MM/YY
(OUT)    FAULT SIMULATOR
(OUT)    VER. XXX
(OUT)    ENTER OPTIONS
(IN)     MT 5 END
(OUT)    NUMBER OF COLLAPSED FAULTS          592
(OUT)    PREPROCESSING TIME IN SECONDS         0
(OUT)    ...CORE EXPANDED TO 601300
(OUT)       5687.  86=*******************************************
(OUT)       5871.  86=*******************************************
(OUT)       6013.  87=*******************************************
(OUT)       6024.  87=*******************************************
(OUT)       6052.  87=*******************************************
(OUT)       6075.  88=*******************************************
(OUT)       6085.  88=*******************************************
(OUT)       6096.  88=*******************************************
(OUT)       6120.  89=*******************************************
(OUT)       6221.  89=********************************************
(OUT)    TERMINATED ON PATTERN              6250
(OUT)    LAST DETECTIONN ON PATTERN         6221
(OUT)    DETECTION COLLAPSED FAULTS          531
(OUT)    NUMBER OF FAULT SETS GENERATED        0
(OUT)    TOTAL NUMBER OF FAULTS              841
(OUT)    TOTAL NUMBER OF DETECTS             765
(OUT)    PER CENT DETECT=                     90
(OUT)    RUN TIME IN SECONDS                  21
```

ADUS 10c. Testability analysis: fault simulator

padcells) with additional supply pads due to their considerable current consumption. For the wiring of the signal connections, one must also consider the fact that the supply lines are principally routed at a fixed metal level (Al1 or Al2). Therefore the signal crossovers have to use the other available level.

The chip design development step specifically includes:

- Automatic conversion of the net list into an input file;
- Automatic chip construction runs which use *fixed constraints*, the net list and the *user inputs* as parameters;
- In the case of unrouted connections: the completion of the program result – realized preferentially by the developer on the graphics terminal;
- Qualitative improvements of the complete construction results by interactive modifications on the graphics terminal.

The cells themselves always remain unchanged, only their arrangement and the geometric realization of their connections as determined by the net list are affected.

From the verified net list stored in the data base (layout parameters such as pad arrangement, cell clustering, line weighting, choice of gate array master, taken into account) placement and wiring suggestions are delivered by the system – controlled by additional parameters like number of standard cell rows.

Mainly for complex circuits, sometimes missing wires have to be supplemented by interactive, though system-controlled, corrections. These modifications are compared immediately with the data of the net list in the data base, and automatically checked against the geometric design rules. Through further interventions the developer can, beyond the effects of the aforementioned assertions, try to optimize the entire chip area, single nets or cell arrangements. Moreover it is, for example, possible in the case of the Poly-Si/Al wiring of the standard-cell chips to run the clock lines preferentially in Al; the current supply lines must, however, still be crossed in Poly-Si.

The result of the chip construction is a file which, after having been merged with the cell layouts, generates the production data.

Chip Construction

The placement and routing of the cells is done automatically controlled by user supplied file. In the example, the number of cell rows has been set to 7 (ADUS 11 a).

Obviously, the area of the chip is determined by the padcells. A further possibility for intervening with the routing process through the control file is the formation of clusters; here, certain cells can be placed next to each other in one row according to a predetermined order (ADUS 11 b):

The effect of the cluster formation can be seen in the following two figures (ADUS 11 c, d): the first one shows a section of the ADUS circuit with four cluster cells, without cluster predetermination (ADUS 11 c).

In the second figure, all six cells of the cluster are in one row (ADUS 11 c).

The colors have the following meaning: blue is for aluminum, red is for polysilicon, and light blue is for the polysilicon – aluminum contact holes.

6.5.8 Layout Analysis and Resimulation

For a standard design system, the layout analysis is limited to the delay time analysis, since the design system guarantees a layout which follows the geometric design rules. Only the interconnect lines between the cells are analyzed. The cells themselves have been tested, are verified and assumed to be correct.

For this purpose, an analysis program extracts the geometric data from the actual wires. By means of the known technological data, RC tree structures are then calculated. The capacitive loads of the respective cell inputs are added. Then the time behavior of these tree-structures is calculated with approximation formulas consideration of the cell output characteristics (inner resistance with different values for falling and rising edges). For each output–input combination, a time delay for the relevant logic signal results, again by approximation. These real time delays are on one hand available as readable list, on the other hand they are automatically delivered to the model preparation for the logic simulator.

These time delay lists give the designer immediate access to information about those time delays that may be too long or have assymetries, or clock sources which may be not properly timed. Then, the designer can interactively improve the chip design or start program runs with changed parameters – for example, parameters for line weighting or prepositioning of cells.

Time Delay Analysis

The CALDEL program calculates the delay time of each connection, with reference to the capacitive loads of the cell inputs (ADUS 12a).

CALDEL may be applied before wiring (CON state: only the load due to the cell inputs is considered) or afterwards (PLV or WEG state: also considers line resistances and capacities). A section of the CALDEL protocol with characteristic, maximum and minimum delay times is shown in the ADUS 12b.

By allocating line weights in the control file for the chip design the user may obtain shorter delay times for selected signals. Generally one has, however, to accept an increase of the other time delays and of the chip area (ADUS 12c, d).

Due to the developer's system knowledge regarding critical paths such a time delay analysis is most of the time especially valuable and therefore highly recommended. Its results should be passed to the logic/timing simulator for a resimulation.

Resimulation with Line Delays

The logic simulation considers, if requested, the calculated line delay in addition to the cell delays contained in the libraries. For comparison purposes, the results of the simulation are shown:

- without line delays, for characteristic cell delays TYP (ADUS 13a),
- with line delays, for maximum cell delays SLOW (ADUS 13b),
- with line delays, for characteristic cell delays TYP (ADUS 13c),
- with line weighting, for characteristic cell delays TYP (ADUS 13d).

For the time grid, a higher degree of precision than in ADUS 8 has been chosen in order to make the small time differences more distinctive. The deviation between cases a) and b) is much higher than that between a), and c) or d). The case of maximum cell time delays thus represents the worst case.

In the following, two other verifications are recommended. Because the line delay times, though not affected by the conditions that determine the slow- and fast-values, are also subject to production variations, a SLOW-logic simulation has to be carried out with the maximum possible line delays, and a FAST-logic simulation has to be performed with the minimum possible line delays. With the plausibility check, the designer then has to test how the dynamic behavior of the chip is affected by the electric behavior of the chosen chip package and the necessary bonding wires.

```
(IN)     EXEC $MASTERK.PH.CALDEL
(OUT)    % P500 CALDEL/XXX/YY-MM-DD LOADED
(OUT)    PROGRAMM CALDEL   START (VERSION   A.BB.CC   =DD/MM/YY  HH:MM:SS=) DD/MM/YY HH:MM:SS
(OUT)    CIRCUIT NAME  ?
(IN)     adus
(OUT)    VERSION OF THE INPUT FILE "ADUS.PAR." ? (NNNN/"NONE"/?)
(IN)     1
(OUT)    WHICH STATE TO CONTINUE WITH?  (C(ON)/P(LV)/W(EG))
(IN)     p
(OUT)    VERSION OF THE INPUT FILE "ADUS.PLV." ? (NNNN/"NONE"/?)
(IN)     1
(OUT)    PROGRAM CALDEL       END
```

ADUS 12a. Time delay analysis: CALDEL call

DELAY IN PIKO SECONDS					TYP		SLOW		FAST	
NET NAME	A CELL	APIN	E CELL	EPIN	DELAY RISE	DELAY FALL	DELAY RISE	DELAY FALL	DELAY RISE	DELAY FALL
LAT	DLAT	A1	LT0	E2	3719	3485	6730	6021	2718	2613
LAT	DLAT	A1	LT1	E2	3396	3158	6282	5564	2448	2341
LAT	DLAT	A1	LT2	E2	3720	3486	6737	6023	2719	2614
LAT	DLAT	A1	LT3	E2	3375	3140	6251	5543	2430	2325
LAT	DLAT	A1	LT4	E2	3720	3486	6731	6023	2719	2614
LAT	DLAT	A1	LT5	E2	4285	4045	7551	6822	3175	3069
LA0	LT0	A1	AN44	E2	483	390	1474	1197	213	169
LA0	LT0	A1	MX24	E3	645	552	1698	1420	350	307
LA1	LT1	A1	MX20	E3	1343	1206	3068	2654	839	776
LA1	LT1	A1	Z20	E3	1259	1122	2948	2535	770	708
LA2	LT2	A1	MX21	E3	1714	1547	3871	3362	1077	1002
LA2	LT2	A1	Z21	E3	1762	1594	3940	3431	1117	1041
LA3	LT3	A1	MX22	E3	1952	1775	4266	3735	1257	1178
LA3	LT3	A1	Z22	E3	1817	1643	4071	3544	1149	1070
LA4	LT4	A1	MX23	E3	1944	1772	4231	3711	1255	1179
LA4	LT4	A1	Z23	E3	1978	1806	4278	3756	1283	1206

ADUS 12b. Time delay analysis: CALDEL protocol

```
%--------------------------------%
%S I G N A L W E I G H T%
%--------------------------------%
/SG
LA0           9
LA1           9
LA2           9
LA3           9
LA4           9
LAT           9
```

ADUS 12c. Time delay analysis: line weight command

```
=====================================================
```

DELAY IN PIKO SECONDS

NET NAME	A CELL	APIN	E CELL	EPIN	TYP DELAY RISE	TYP DELAY FALL	SLOW DELAY RISE	SLOW DELAY FALL	FAST DELAY RISE	FAST DELAY FALL
LAT	DLAT	A1	LT0	E2	1106	944	2691	2138	674	606
LAT	DLAT	A1	LT1	E2	2430	2240	4594	4025	1734	1648
LAT	DLAT	A1	LT2	E2	1001	839	2553	2000	588	521
LAT	DLAT	A1	LT3	E2	2415	2227	4572	4007	1723	1637
LAT	DLAT	A1	LT4	E2	1128	963	2722	2172	689	624
LAT	DLAT	A1	LT5	E2	2416	2228	4573	4008	1724	1638
LA0	LT0	A1	AN44	E2	722	620	1894	1585	394	347
LA0	LT0	A1	MX24	E3	775	672	1968	1659	438	391
LA1	LT1	A1	MX20	E3	787	678	2045	1717	433	384
LA1	LT1	A1	Z20	E3	787	679	2046	1718	434	384
LA2	LT2	A1	MX21	E3	746	591	2361	1910	308	244
LA2	LT2	A1	Z21	E3	1274	1122	3100	2646	747	677
LA3	LT3	A1	MX22	E3	542	445	1602	1309	254	210
LA3	LT3	A1	Z22	E3	543	445	1602	1309	254	210
LA4	LT4	A1	MX23	E3	639	541	1740	1445	333	287
LA4	LT4	A1	Z23	E3	555	457	1624	1329	263	218

ADUS 12d. Time delay analysis: CALDEL protocol with line delays

```
+-------------------------------------------------------------------------------+
!P R O S I M  VERS XXXX        TIME DIAGRAM                      NEXT MASK /SIW/!
!PROJECT /ADUS     ,VERS: 001/                   TEST PROGRAM /TEST     ,VERS: 04/!
!TIME /  7240 NS/          GRID /   1 NS/           NEXT PAGE /   /       COPY / /!
!                                                                               !
!CLOCK           ---------------------------------------........................!
!                                                                               !
!MEMW            ---------------------------------------------------------------!
!ADB000          ---------------------------------------------------------------!
!ADB001          ...............................................................!
!ADB002          ...............................................................!
!ADB003          ...............................................................!
!ADB004          ...............................................................!
!                !         !         !         !         !         !         !  !
!             7210      7220      7230      7240      7250      7260      7270  !
!                                                                               !
!RAM0            xxxxxxxxxxxxxxxx-----------------------------------------------!
!RAM1            xxxxxxxxxxxxx:.................................................!
!RAM2            xxxxxxxxxxxxx:.................................................!
!RAM3            xxxxxxxxxxxxx:.................................................!
!RAM4            xxxxxxxxxxxxx:.................................................!
!                                                                               !
!                                                                               !
!                                                                               !
+-------------------------------------------------------------------------------!
```

ADUS 13a. Resimulation: without line delays, TYP (PROSIM mask)

```
+-------------------------------------------------------------------------------+
!P R O S I M  VERS XXXX        TIME DIAGRAM                      NEXT MASK /SIW/!
!PROJECT /ADUS     ,VERS: 002/                   TEST PROGRAM /TEST     ,VERS: 02/!
!TIME /  7240 NS/          GRID /   1 NS/           NEXT PAGE /   /       COPY / /!
!                                                                               !
!CLOCK           ---------------------------------------........................!
!                                                                               !
!MEMW            ---------------------------------------------------------------!
!ADB000          ---------------------------------------------------------------!
!ADB001          ...............................................................!
!ADB002          ...............................................................!
!ADB003          ...............................................................!
!ADB004          ...............................................................!
!                !         !         !         !         !         !         !  !
!             7210      7220      7230      7240      7250      7260      7270  !
!                                                                               !
!RAM0            xxxxxxxxxxxxxxxxxxxxxxxxxxxxxxxxxxxxxxxxxxxxxxxxxxxxxx:........!
!RAM1            xxxxxxxxxxxxxxxxxxxxxxxxxxxxxxxxxxxxxxxxxxxxxxxx...............!
!RAM2            xxxxxxxxxxxxxxxxxxxxxxxxxxxxxxxxxxxxxxxxxxxxxxxx...............!
!RAM3            xxxxxxxxxxxxxxxxxxxxxxxxxxxxxxxxxxxxxxxxxxxxxxxx...............!
!RAM4            xxxxxxxxxxxxxxxxxxxxxxxxxxxxxxxxxxxxxxxxxxxxxxxx...............!
!                                                                               !
!                                                                               !
!                                                                               !
+-------------------------------------------------------------------------------!
```

ADUS 13b. Resimulation: with line delays, SLOW (PROSIM mask)

```
+-------------------------------------------------------------------------------+
!P R O S I M  VERS XXXX        TIME DIAGRAM                     NEXT MASK /SIW/!
!PROJECT /ADUS    ,VERS: 051/                      TEST PROGRAM /TEST    ,VERS: 06/!
!TIME /  7240 NS/          GRID /  1 NS/              NEXT PAGE /   /         COPY / /!
!                                                                               !
!CLOCK          ----------------------------------------........................!
!                                                                               !
!MEMW           ----------------------------------------------------------------!
!ADB000         ----------------------------------------------------------------!
!ADB001         ................................................................!
!ADB002         ................................................................!
!ADB003         ................................................................!
!ADB004         ................................................................!
!               !         !         !         !         !         !         !   !
!             7210      7220      7230      7240      7250      7260      7270  !
!                                                                               !
!RAM0           xxxxxxxxxxxxxxxxxxxxxxx:----------------------------------------!
!RAM1           xxxxxxxxxxxxxxxxxxxxxx:.........................................!
!RAM2           xxxxxxxxxxxxxxxxxxxxx:..........................................!
!RAM3           xxxxxxxxxxxxxxxxxxxxxxx:........................................!
!RAM4           xxxxxxxxxxxxxxxxxxxxxx:.........................................!
!                                                                               !
!                                                                               !
!                                                                               !
+-------------------------------------------------------------------------------!
```

ADUS 13c. Resimulation: with line delays, TYP (PROSIM mask)

```
+-------------------------------------------------------------------------------+
!P R O S I M  VERS XXXXX       TIME DIAGRAM                     NEXT MASK /SIW/!
!PROJECT /ADUS     ,VERS: 500/                     TEST PROGRAM /TEST    ,VERS: 06/!
!TIME /  7240 NS/          GRID /  1 NS/              NEXT PAGE /   /         COPY / /!
!                                                                               !
!CLOCK          ----------------------------------------........................!
!                                                                               !
!MEMW           ----------------------------------------------------------------!
!ADB000         ----------------------------------------------------------------!
!ADB001         ................................................................!
!ADB002         ................................................................!
!ADB003         ................................................................!
!ADB004         ................................................................!
!               !         !         !         !         !         !         !   !
!             7210      7220      7230      7240      7250      7260      7270  !
!                                                                               !
!RAM0           xxxxxxxxxxxxxxxxxxxxxx------------------------------------------!
!RAM1           xxxxxxxxxxx:....................................................!
!RAM2           xxxxxxxxxx:.....................................................!
!RAM3           xxxxxxxxxxx:....................................................!
!RAM4           xxxxxxxxxx:.....................................................!
!                                                                               !
!                                                                               !
!                                                                               !
+-------------------------------------------------------------------------------!
```

ADUS 13d. Resimulation: with line delays, TYP, with line weighting (PROSIM mask)

With all these verification steps, a reliable test of the previously determined dynamic specifications is possible with respect to the chip's behavior. The signals in the chip core, however, are still treated in the manner of a logic simulator: actual dynamic effects such as smoothing of the edges, overshoots or voltage breakdown due to excessive dynamic current consumption are not considered. Since the circuits are usually quite complex, this would not be a realistic type of analysis. The experience gained so far proves that a more precise simulation on circuit level is not necessary. For this reason, with cell concepts, a certain safety margin concerning the maximum of technologically feasible has to be maintained. This applies to the cells upon which VENUS is based, as well as to the circuits which are constructed with VENUS.

6.5.9 Production Data Generation

Before the next step of the design process, the generation of production data, is initiated, the following preparatory measures have to be taken according to check list 6:

Check list 6: Preparation of Production

- ☐ It has to be decided whether only a check plot with or without cell layout representation needs to be generated from the existing design data. Are single levels or enlarged sections required?

- ☐ Is the generation of mask data desired by the semiconductor-manufacturer's mask center or is the data only needed for test runs.

- ☐ The arrangements with the semiconductor-manufacturer have to be reassured. A project number related to a special master has to be assigned to gate-array chips; a production number has to be assigned to the standard cell chips. During the mask data generation they are made known to the system by the user, and later they appear on the chip - in the highest aluminum level.

- ☐ Before production begins, the final chip specification must be decided upon in cooperation with the manufacturer. This specification makes it clear that we deal with a development result of the VENUS standard design procedure, and not with a manual design. For VENUS chips, the manufacturer does production without a number of otherwise compulsory checks. The reason for this simplification is the quality standard, reached through many internal quality-assuring procedures, as well as the cell libraries (and possibly the masters) which have been previously released by the manufacturer.

6-1

- ☐ Furthermore now, at the latest, the decision on a specific package is due. The exact chip size is known. The arrangement and the positions of the power supply- and signal- pads can be gathered from a so-called "bond plan" which, in connection with the mask data, is generated automatically, if requested.

- ☐ Finally, arrangements have to be made with the manufacturer concerning the exchange of documents (forms, mask plots), the test program (electric parameters, test patterns, test device) and the construction of the component-specific load board for the chosen test machine.

The actual generation of the requested plots and mask data within the design system is a routine process. The user must only input the product or project number.

During the generation of the mask data, a frame structure is automatically added to the chip data (overall layout) developed by the user. This includes production-specific alignment structures, scribing frames, measurement squares etc. Furthermore, a technology test structure, which has been supplied by the manufacturer, is merged into the scribing frame – i.e. outside of the VENUS-user's range of responsibility. It provides the manufacturer with a test method for guaranteed technology parameters for each chip of the wafer. These structures will get lost when the wafer is divided into chips before assembly.

Without the user noticing it, during the mask tape generation all geometric data – specified for the single masks – are adjusted to the relevant conditions of production, i.e. they are expanded or shrunk at the edges of the figure.

In addition to the production data (mask tape) the user is recommended to generate control plots of the overall layout (all levels combined) as well as control plots of the individual mask levels. Through the latter, he is supported in the release of the mask plots, as demanded by the mask center.

Production Data Generation

The program PLOTMASK makes possible the generation of plotter tapes ("PLOT CALCOMP 925 FORMAT") and for the mask generation ("MEBES", "PATTERN-GENERATOR") (ADUS 14a). ADUS 14b shows a CALCOMP-PLOT of the aluminum level of ADUS.

In ADUS 14c, the aluminum (blue) and polysilicon (red) levels, and the polysilicon/aluminum contact holes (light blue) are shown. The chip section of ADUS 14c is the same as that of ADUS 11c.

6.5.10 Test Data Generation

In addition to the mask tape which is based on a certain process technology according to library choice, the manufacturer expects an executable test program, a specific load board and a probe card for the wafer test. The compatibility with the previously determined test machine must be guaranteed.

As the entire system knowledge about the developed chip lies in the hands of the VENUS user, i.e. the design engineer, he is most fit for generating the entire test program.

The VENUS design system offers the user the possibility of generating the entire test program on his own. For this purpose, he is supplied with those (prepared) tests that are expected by the manufacturer. The actual values to be entered are a part of the arrangements with the manufacturer. Only the design engineer is able to see to it that, if possible, each node is checked for unavoidable stochastic production failures.

The tools which provide test patterns (fault simulator, bit pattern generator) which were mentioned when verification and testability analysis were discussed are applied again, if necessary.

The functional bit patterns are a part of the test program. However, they must be changed into cyclic bit patterns because of the purely static test on the test machine.

Finally, a last fault simulation has to document the obtained degree of error detection. The latter must be a part of the arrangements with the manufacturer.

The VENUS design system automatically combines all prepared tests, the library-specific technology values, and the test patterns into an executable test program.

Test Data Generation

The program PROGENIC generates a test program for the test object developed with VENUS. The user fills the fields in parantheses (...) (see ADUS 15a).

The input files SCHA and REFE contain a chip description which is specific for the test system; the BITA list holds the test patterns; the file PINZUORD assigns the signal names of the inputs/outputs to the pin names of the test object and ICSPEC specifies electric parameters which are necessary for the test. In the following, two sections of the test program are explained: contact test and function test:

- Contact test (ADUS 15b)
  In this first test, all pads are tested for open and short-circuit errors. If the result of the contact test is negative it is normally useless to continue testing.

```
+----------------------------------------------------------------------------+
!****************************************************************************!
!*                            PROGENIC CMOS                                 *!
!****************************************************************************!
!*                                                                          *!
!*  COMPONENT NAME:  (ADUS  )                                               *!
!*                                                                          *!
!*  INPUT FILES:                                                            *!
!*                                                                          *!
!*   SCHA               :  (ADUS.SCHA                             )         *!
!*   BITA               :  (ADUS.BITA                             )         *!
!*   PIN ALLOCATION     :  (ADUS.PIN ALLOCATION                   )         *!
!*   IC SPECIFICATION   :  (CMOS.IC SPECIFICATION                 )         *!
!*   IF DYN. SUPPLY-CURR.-MEASUREMENT REQUIRED, ENTER NAME  OF THE FILE WITH *!
!*   DYN. BIT PATTERNS :  (                                       )         *!
!*   IF SHMOOPLOT REQUIRED, ENTER NAME OF FILE WITH BIT PATTERNS FOR        *!
!*   SHMOOPLOT          :  (                                      )         *!
!*   REFE               :  (ADUS.REFE                             )         *!
!*  VERSION NUMBER OF THE TEST PROGRAM (PERMISSIBLE ENTRIES 1-Z)  (1)       *!
!*                                                                          *!
!*                                                                          *!
!*  DO YOU WANT FURTHER PROGRAMS TO BE GENERATED (Y/N)  ?  (N)              *!
!*                                                                          *!
!****************************************************************************!
!                                                                            !
x----------------------------------------------------------------------------x
```

ADUS 15a. Test data generation: PROGENIC mask

```
*                                                                            ;
SET DA* 11111111111111111111 11111111111111111111 11111110000000000000
        00000000000000000000 00000000000000000000 00000000000000000000 ;
SET MA*  (120:0);                        REM DON'T CARE PINS             ;
SET R*   (120:0);                        REM ALL R-RELASIS OPEN          ;
SET S*   (120:0);                        REM CLEAR ALT.REF.VOLTAGE       ;
SET CRO* (120:1);                        REM ALL COMP.RELAIS OPEN        ;
                                                                       REM
   TEST SEQUENCE (PMU-MEASUREMENT):
   = = = = = = = = = = = = = = = =
***   MEASURE ON ALL PINS WITHOUT  VDD & VSS  !   ***                    ;

ENABLE TRIPI3 GT IDYNMAX,RNG3;
FORCE E0 VI, RNG2;                       REM INPUT VOLTAGE               ;

SET PERIOD DICYC,RNG0;                   REM SET DUMMY-TEST CYCLE        ;
ENABLE DA,MA; SET F 0; ENABLE TEST;      REM
*                                                                        ;
SET DELAY DCTIME, DC;                    REM OVERWRITE HARDWARE DELAY    ;
FORCE CURRENT 0, RNG2; SET PMU SENSE, RNG2;
ENABLE DCT0 LT VLOLIMIT; ENABLE DCT1 GT VUPLIMIT;                      REM
*                                                                        ;
CPMU PIN 025; CHN = 025;
FORCE CURRENT IFORCE, RNG2; MEASURE VALUE; FORCE CURRENT 0, RNG2;
CPMU PIN 026; CHN = 026;
FORCE CURRENT IFORCE, RNG2; MEASURE VALUE; FORCE CURRENT 0, RNG2;
CPMU PIN 027; CHN = 027;
FORCE CURRENT IFORCE, RNG2; MEASURE VALUE; FORCE CURRENT 0, RNG2;
CPMU PIN 028; CHN = 028;
FORCE CURRENT IFORCE, RNG2; MEASURE VALUE; FORCE CURRENT 0, RNG2;
```

ADUS 15b. Test data generation: contact test

```
INF  LM-ADR SET MODULE 4096, SPM; SET MPIN 120; SET REL '    08/02/85';        REM
            *
              L M I = F I L E  *** 01 ***   F O R   F U N C T I O N  T E S T S
              ==============================================================
            *
            *********************               00ADUS          *********************
            *
            ** LSI-
            ** PIN              11111111112 222 2222233333333334 4444444
            ** ***    12345678901234567890 123 5678901234567890 1234567
            *
            ** LSI-
            ** ***
            ** PIN
            *                                                                        ;
            ENABLE MA;                          REM  MA-REGISTER IS USED FOR LM-LOAD;
            LSET MA (120:0);                    REM  MA-REGISTER IS RESET            ;
   1 000001 LSET DA* 00000011111111010000 00001111111111001000 00100100000000000000
                     00000000000000000000 00000000000000000000 00000000000000000000;
   1 000002 LSET MA* 11111100000000101111 11100000000000110111 11011010000000000000
                     00000000000000000000 00000000000000000000 00000000000000000000;
   1 000003 SET F    00001010101001000000 01100000000001000011 00000000000000000000
                     00000000000000000000 00000000000000000000 00000000000000000000;
   2 000004 SET F    00001010111000000000 01100000000000000011 00000000000000000000
                     00000000000000000000 00000000000000000000 00000000000000000000;
   3 000005 SET F    00001010111000010000 01100000000000000011 00000000000000000000
                     00000000000000000000 00000000000000000000 00000000000000000000;
   4 000006 SET F    00001010111001010000 01100000000010000011 00000000000000000000
                     00000000000000000000 00000000000000000000 00000000000000000000;
   5 000007 SET F    00001010111000010000 01100000000010000011 00000000000000000000
                     00000000000000000000 00000000000000000000 00000000000000000000;
```

ADUS 15c. Test data generation: LMI file

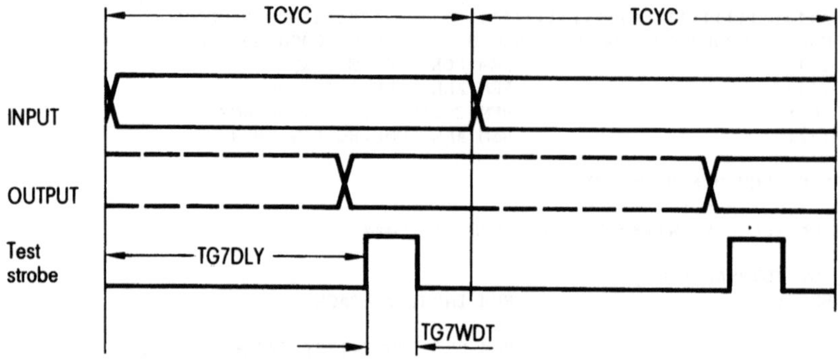

ADUS 15d. Test data generation: cycle simulation: time diagram

```
                                REM  CONN. INPUT PINS (I+B)               ;
SET DA* 00000011111111010000 00001111111111001000 00100100000000000000
        00000000000000000000 00000000000000000000 00000000000000000000 ;
SET MA*  (120:0);               REM DON'T CARE PINS                       ;
                                REM  SET RELAIS  (VDD & VSS)              ;
SET R*  00000000000000000000 00010000000000000000 00000001000000000000
        00000000000000000000 00000000000000000000 00000000000000000000 ;
                                REM  SET ALT.REF. INPUT VOLTAGE           ;
SET S*   (120:0);                         REM CLEAR ALT.REF.VOLTAGE       ;
                                REM  POWER-PIN(VDD&VSS) COMP.RELAIS OPEN;
SET CR0*00000000000000000000 00010000000000000000 00000001000000000000
        00000000000000000000 00000000000000000000 00000000000000000000 ;
                                                                       REM
*
   TEST SEQUENCE:
   ==============
*                                                                         ;

ENABLE TRIPI3 GT IDYNMAX,RNG3;
FORCE VF1 VSS ,RNG3; FORCE VF3 VDD , RNG3;
FORCE E1  VIH   ,RNG2; FORCE E0  VIL,RNG2;

SET   S0  VOLMAX  ,RNG2; SET   S1  VOHMIN  ,RNG2;
IFC S7 THEN BEGIN
SET TG7 DELAY TG7DLY,RNG0; SET TG7 WIDTH TG7WDT TG7WDT,RNG0;
ENDC;
IFC S8 THEN BEGIN
SET BTG7 DELAY TG7DLY,RNG0; SET BTG7 WIDTH TG7WDT,RNG0;
ENDC;
IFC S20 THEN BEGIN
SET TG7 DELAY TG7DLY,RNG0; SET TG7 WIDTH TG7WDT,RNG0;
ENDC;
SET PERIOD  TCYC,RNG0;                  REM SET DUMMY-TEST CYCLE & TCYC   ;
ENABLE DA,MA; SET F 0; ENABLE TEST;


AT 0;
EXEC LMLOAD (001, 2,PAMAX01,FILEID  );
SET START 0; SET MAJOR 1,PAMAX01;
ENABLE TEST;
```

ADUS 15e. Test data generation: Function test

With the command SET DA *, the pins which have to be considered as inputs in this test, are stated. After setting various test parameters, first pin 25 is selected and the current IFORCE is switched on ("FORCE CURRENT IFORCE; RNG2"; the value of IFORCE is defined elsewhere in the program, RNG2 defines a measurement range). The voltage drop is measured ("MEASURE VALUE"), and the current is set to 0 before the beginning of the next pin's test. A drop below the lower voltage limit VLOLIMIT is interpreted as a short-circuit error; exceeding the upper voltage limit VUPLIMIT is interpreted as a break.

- Function test (ADUS 15c, d, e)

The function test is based upon the test patterns of the BITA file which are entered in the LMI file by PROGENIC. During time TCYC, the input bit pattern is set to the pin predetermined by the SET DA * command; after time TG7 DLY, during interval TG7 WDT, the states of the pins selected with the SET MA * command are measured and compared with the data from the LMI file for these pins. The command blocks IFC ... ENDC of the program assign, depending on the test machine chosen, the times TG7 DLY and TG7 WDT. With LMLOAD the LMI file is read; the next line defines the first ("0") and the last ("PAMAX01") bit pattern to be used in this test, and ENABLE TEST initiates the actual test.

6.6 Sample Fabrication

A characteristic of chips which are developed by means of a standard design procedure is their low production volume. Either only a few samples are needed; or a first sample series with low volume can be expected within a very short period of time.

Shared-Silicon, First Sample

In order to allow a cost-effective approach to chips in low volume, the manufactorer offers a flexible process. The masks of one mask set do not merely contain the production data (mask tapes) of one project.

- For gate-array chips, the specific personalization levels (Al1 and Al2 level as well as contact level) of various users are projected to various sections of the master-wafer.
- For standard cell chips, one or several stripes of various widths on the wafer are occupied by one project each (Fig. 6.2).

Thus, in addition to the mask cost, the cost of a wafer run is also shared by several users. After separation of the wafer sections, each user receives those chips for assembly that are allocated to him. At the same time, this guarantees that the chips of any user are available exclusively for that very user. The masks and the whole wafers do not leave the manufacturer's sphere of responsibility.

Mass Production

If production, test and sample supply meet the requirements of the system designer, one has to consider whether, and if so how many, additional chips are required, unless a functional redesign is necessary (check list 7)

If a decision is made in favor of an optimized transistor-oriented redesign one has to think of the redesign effort which will then be quite high. The development risk (no

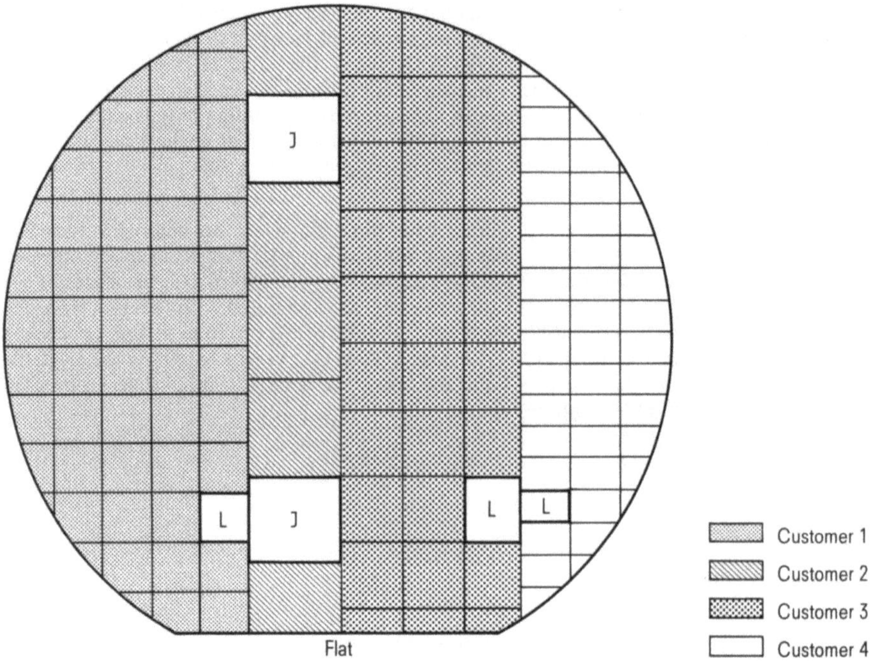

Figure 6.2 Shared-silicon principle: Example with four different circuits (J: adjustment fields, L: fields for chip naming)

Check list 7: Mass Production

- ☐ Without additional development and mask cost, but with wafers that are still shared by several projects, the same masks can still be used.
- ☐ Without new development cost, but with additional mask cost which then are paid by one project only, the wafers can be utilized by one user completely.
- ☐ For medium-high production volume one may consider if some gate-array designs can be turned into standard cell designs which would result in lower unit cost in the long run.
- ☐ Finally, for volumes that are expected to be very high in the long run one must try to lower the unit cost by an optimized redesign on the transistor level (full-custom).

---

consistent logic plan documentation, no data base, no proven quality standard, unforeseeable electrical effects, no guarantee of testability, no automatically generated test program) is higher than for the described VENUS standard design procedure. The design productivity for a full-custom design is characteristically 300 gate functions per designer and year. For designs which are based on standard design procedures it is about ten to twenty times higher! Since transistor circuits are developed from scratch within a full-custom design, not only a check on stochastic production errors is necessary but also a functional verification after each redesign – and redesigns usually are necessary more than once! If problems are complex this may lead to a high effort in terms of labor and experiments. Experts may be necessary who master special procedures (such as the electron beam prober/potential contrast method). In each project schedule for full-custom designs one has to provide for redesign and functional verification times!

Manufactured Chip

ADUS 16 shows a manufactured chip, glued in the package, with the bond wires establishing the connection between the pads and the package:

6.7 Test

Any integrated circuit may show stochastic production errors. During application such errors are discovered only with great difficulty. Since they can only be repaired by exchanging the entire chip, each chip has to be checked at all its internal nodes after production. For this, the test on the test machine is used.

Chips which have been developed by using the VENUS design system guarantee the correctness of those functions which have been excercised during logic simulation. Only after a finished chip has been integrated in its system environment (pcb, device), may one discover that certain functional requirements have been missed during logic design. In order to avoid such a problem, labôratory measurements should be carried out as early as possible.

The measurements and tests referred to are not meant for discovering design errors or design weaknesses. These have to be found during logic simulation.

Measurements of First Chips and Components

As soon as the designer has first samples of the manufactured semiconductor circuit available, he can check individual chips for their current consumption, voltage levels, and possibly individual functions.

The first packaged chips are usually integrated quite early into a hardware environment which imitates the system in its operating conditions quite well. Here the designer, with his knowledge of the critical paths and the pertinent selected bit patterns, can first check the specified dynamic characteristics of the chip. Tests of this kind are not included in the static production test on the test machine.

Furthermore, the system should be reproduced as completely as possible, with a complete environment (temperature, supply voltages) and input bit patterns, so that the designer can be pointed to errors or missing functions in his logic design at this time.

Activating the Test Program

The VENUS design system provides three types of test programs:

- wafer test,
- chip test,
- random test for quality assurance.

The user of the VENUS design system should run the test programs for the first two cases. The third one is run by the manufacturer. The user is advised to have previously planned and developed the test assemblies specifically for his chip, together with the operators of the test machine.

The wafer test serves for selecting the chips in advance when they are still on the silicon wafer, immediately after production. Here, the chips are contacted via probe cards. Gross production failures can be recognized. After all chips on one wafer have been separated, only those chips are packaged that are expected to meet the electric specification. Thus bonding and packaging costs are saved. The functionality of test assembly, probe card, test program, test machine operating systen and tester-configuration may be tested already in advance without the wafers. The same applies to the component test with regard to the required test assembly. Figure 6.3 shows such a test assembly. The component test includes the static test of if possible all internal nodes by means of test patterns supplied during test data generation, and the test of the electric specification of all pad cells.

Literature – Chapter 6

6.1 SEMICUSTOM. Zellenorientierter Bausteinentwurf. Standardzellen und Gate-Arrays. Handbuch 1: Schaltungsentwicklung, Ausg. April 1985. Siemens, München.
6.2 VENUS. Arbeitsunterlagen für den Benutzer. Siemens, München.

7 Outlook

IC design systems are embedded in environments which present different demands, and which in addition are subject to continuous change. The CAD system VENUS which is the subject of this book supports the design of integrated circuits from the logic level down to the mask level and the generation of the production and test data. The physical design is almost completely automated, whereas the logic design is still carried out interactively. Higher levels of design include complexities which are still not subject to IC design, at least not for standard IC design systems. However, there is no doubt that before long, higher levels such as the module level will also be a part of the IC design. Modules, which today are still realized on pc-boards, will then be complex parts (macrocells) of the chip. This expansion of the IC design has been accompanied by an increase in the automation of the design process by means of intelligent analysis and synthesis programs: thus the *functional scope* of IC design system is enlarged. This expansion is occurring in three areas:

- First, design methods with higher degrees of freedom with respect to the application scheme are provided. They will allow a better utilization of technological potential.
- Second, more support functions are also offered for the design of the higher levels up to the system and architecture design.
- With increasing chip complexity, programmable components for the implementation of chip functions are used more and more. Programming and testing these components requires resources such as those found today in microcomputer development systems.

Trends in expanding the functional scope are described in Sect. 7.1.

In spite of this or similar expansion of the performance profile, standard design systems are not meant to be a tool only for specialists; they have to be kept directed towards *wide-scope application*. Easy learning, reliable design processes and high quality of the results are partial goals. In this book, special attention is given to wide-scope application. Relevant trends and problem areas are discussed in Sect. 7.2.

Integrated circuits which are realized on silicon chips have to be housed in packages, placed and wired on pc-boards, and finally joined to larger system units. Although ICs at the present technological level contain more transistors than whole cabinets full of electronics in the beginning of the seventies, this did not result in all computers being reduced to chip size: with the number of ICs per pc-board increasing and the number of pc-boards per system decreasing, the system complexity has been raised as much as IC complexity. As an example Fig. 7.1 shows the development of some CPUs since 1971. For comparison, a chip, the peripheral processor PP4, which was realized in the Siemens research laboratories in 1983, is drawn in. Although it is not equivalent to the functional scope of a CPU, it clearly demonstrates progress. It contains about 300 000 transistors, i.e. about five times as much as the CPU 4004 in

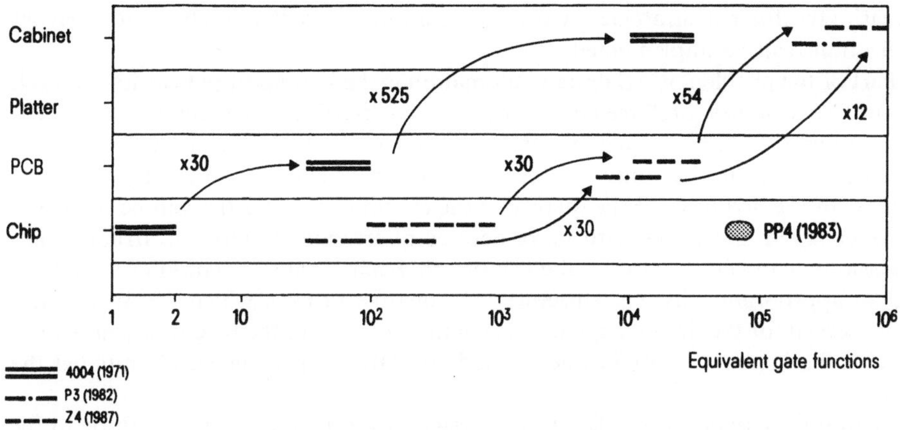

Fig. 7.1. Trends in computer systems development

1971. Also in the future, the "world" of the electronic design will not end at the boundaries of chips or components. The CAD system for IC design thus has to be embedded in the environments of relevant design processes and CAE systems for design, production and test of pc-boards, platters, cabinets etc. The trend of embedding IC design systems *in CAD systems for the system design* is described in Sect. 7.3.

Finally, an IC design system has to include the ability of supporting production process variants or even physically varying production technologies. It has to be possible to adjust the CAD system in such a way that neither its architecture nor its user interface has to be changed greatly. For standard design systems this ability is guaranteed through adapted cell libraries and CAD functions for chip construction (placement, wiring, verification). In this context, the aim should be to jointly utilize existing data structures and as many function programs as possible. The development of the cells themselves as well as quality control, however, have to be managed in close cooperation with the development and maturization of corresponding technological production processes. The *embedding* of IC design systems *in the technology development* is described in Sect. 7.4.

7.1 Functional Scope

In many fields of technology standardization makes it possible to reduce the variety of types and to increase production volumes. This reduces the cost of production, storage, distribution and maintenance. For IC design, these aspects are of minor importance. Apart from the special position of the gate arrays, the production cost of an integrated component is not affected whether it consists of standardized or individual elements. In this case, restrictions of freedom do not lower the IC-production cost but rather simplify the design tools. At first, a drastic simplification of the design process was the prerequisite for initiating a design automation process. Since then, one has gained some years of experience with design systems. Now, increasingly efficient strategies which in addition are better adjusted to relevant applications can be implemented.

Part of the problem in IC design automation is the automatic generation of a cell layout. The structures preferred up to now are characterized by a regular arrangement of the transistors, for example matrix-like or in rows. Because regularity by itself is not a quality feature but only creates the design more simply and clearly, the compactness of the layouts and the quality of the circuit characteristics can be increased by clever layout strategies with more degrees of freedom in term of arrangement, dimensioning and connection of the transistors. Such a transformation of a function into an appropriate physical structure is a creative process which up to now has been reserved for the design engineer. The improvements in the field of layout strategies and the increase of applicable methods of "artificial intelligence" initiated the change.

With the completion of a layout the design of a cell is not finished. Prior to the installation of the cell, its characteristics have to be analyzed and made available, i.e. from the geometric layout data and the characteristics of the intended production process, the characteristics of the individual circuit elements have to be ascertained. With this data, and automatic network analysis can then be carried out which may, for example, provide information on gate delays and power dissipation.

The reliability of such precalculations has to meet rather high demands, because incorrect characterization of a cell may cause negative effects on the operability of the entire chip. Up to now, it has been possible to produce and measure cells from cell libraries on test chips before they were released. Such a procedure is no longer practical for flexible cells which can be parameterized. The high number of possible cell variants only allows the realization on silicon and the measuring of the cell characteristics at random. The increasing accuracy of circuit models and the decreasing price of computer time make the simulations increasingly reliable. It can be expected before long that one will be able to do without a preproduction in silicon.

The data gained from network analysis is too manifold and too detailed for higher design levels. The problem of abstraction and combination of cell characteristics, resulting in some characteristic values, has to be solved for each circuit type individually. In order to obtain a computer's support or even automation, basic research work has still to be carried out.

From the trends described, one may deduce that rigid cells are replaced by flexible cells, i.e. electric and geometric behavior is determined only when the cell is applied. For complex cells, major structural characteristics are influenced by the demands raised by the user, such as type and number of functions to be realized, their arrangements and their connections with each other and with neighbour cells.

The better the physical design can be carried out by the design systems, the more the design engineer can concentrate upon the higher design levels. The shift of the circuit description from gate level to register-transfer level is currently realized. Instead of single signals, signal bundles are considered, and the boolean functions are supplemented by arithmetic and shift operations. On the next higher levels the type of the elements and the description of their mutual relations are still highly determined by the corresponding fields of application. Component specifications can be found described as instruction sets, algorithms, graphs (for filters, for example) or statistic values (data rates, degree of utilization, availability).

The design tools for the higher design levels will support a modular and hierarchical design style. The question as to what this support will be like is still unanswered.

With the bottom-up style, single modules are defined first, before they are combined to more complex modules. As the structuring is primarily carried out by the design engineer, the design system must not only provide assistance for the design, but also fulfill tasks of *analysis, abstraction* and *verification*. Design systems with a bottom-up strategy are often called chip assembler.

The top-down style provides a more demanding approach for design automation. A functional description of the chip is primarily automatically refined step by step by the design system. In this context, the design system has to solve difficult *synthesis problems*. In exchange for this, analysis and verification are simplified or avoided (*correctness by construction*). Design systems of this kind are called chip compilers or silicon compilers; we might add that due to the inflation of terms, even design systems which allow a design only on register-transfer level have also been given this name.

The two extremes of design style which have been described here do not correspond with the usual design process. Here one repeatedly switches from top-down to bottom-up design and vice versa, to obtain favorable results. The design in this "yoyo" style includes the problems mentioned above and, in addition, problems of consistency due to the possibility of intervention on various hierarchical design levels. As this design style includes the other two styles (according to the kind of application and constraints, either the bottom-up or the top-down component can be emphasized), it should be the long-term goal for universal design systems.

The realization of complex functions can be simplified through the utilization of programmable components. A problem-independent hardware structure is supplemented by problem-specific digital data. The "personalization" of a ROM or a PLA is a lower-level example of this process. In the future, however, even microprocessors and their peripheral components will also be integrated in the chip. Thus, we obtain custom computer systems on a chip, which have to be programmed and tested.

This also results in the requirement that development support be offered, corresponding with the μP-development systems of the seventies; this shows the need for another user technology, the "integrated microcomputer development systems" of the 90s (Fig. 7.2). The IμC-development must not only include tools which support the

Technical medium			System support for application technology	Date of origin
Digital circuit	Processor		SW programming system	Since 1963
	Integrated circuit	Micro-processor	μP development system	Since 1973
		Specific circuit	IC design system	Since 1983
		Custom integrated microcomputer	IμC development system	?

Fig. 7.2. Future development of the user technologies for digital electronics

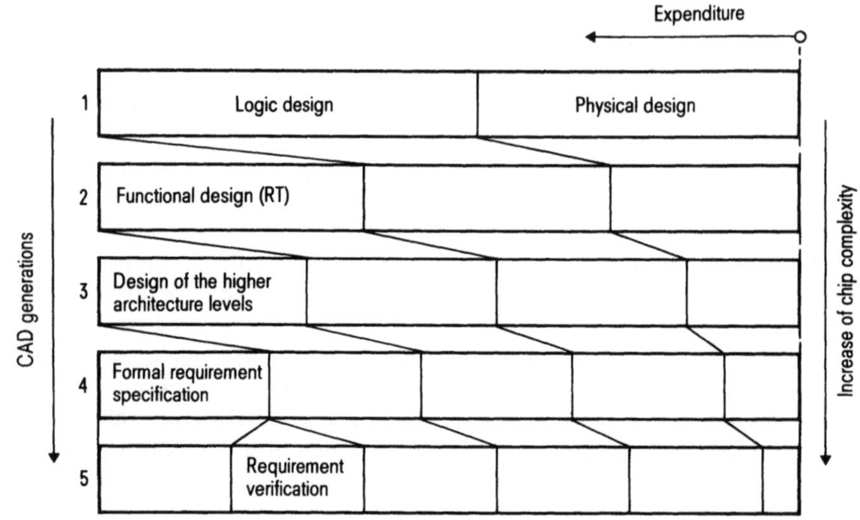

Fig. 7.3. Generations of IC design systems

chip design, but also a SW programming system: programming languages, compilers, specification, test and documentation tools and tools for the hardware/software integration and the system test.

The development trend of functional expansion is illustrated by the generation sequence of the CAD systems (Fig. 7.3). The obvious target of lowering the effort that is to be invested for the physical design has been reached to a great extent with the predevelopment of the cells and with efficient algorithms for placement and routing. Thus the tools for automating the logic design and for supporting the functional design on register-transfer level become more and more important (second CAD generation). Similarly, from the third CAD generation onwards the design (construction and verification) of the higher architecture levels is backed by the computer. The effort to be invested on part of the design engineer for RT design, logic design and physical design is lowered due to increasing automation. Finally the requirement specifications are formalized to a great extent which also makes it a candidate for verification.

With complete computer support for all design steps, the entire effort invested in the chip design is reduced sharply while chip complexity remains the same. The sweeping development of integration technology during the last few years indicates anything but a stagnation in chip complexity. Rather, we have to assume that the increase in design productivity with the same design effort will lead to standard mastering of complex systems on one chip.

7.2 Aspects of Wide-Scope Application

The central goal of design process standardization by means of computer support is to provide IC design access to the non-specialist, too, and thus make it possible for the latter to concentrate on his application. The relevant procedure remains the same for many fields of technology: characteristic partial problems are isolated from the application field; for these problems standard solutions are developed. By combining these standard solutions, various solutions for the user problems can be found.

It is the nature of standard solutions to represent an optimal solution, perhaps not technologically, but rather economically. Because the development engineer is often technology-oriented, he has certain reservations against standard design systems. In order to overcome this attitude, a design system has to offer not only a considerable advantage in terms of time, but also and above all a good user interface. These are the (partly contradictory) requirements for future development systems. In order to increase acceptance, the design system should hold to the usual working style of the development engineer. For a cell-oriented design system, for example, this is true. The functional scope and the operation of the cells is similar to those of the logic series common for SSI and MSI chips. The graphic-oriented input of the logic plan complies quite well with the engineer's way of thinking. These analogies however should not be exaggerated. A new technology often brings about new circuit and design techniques. For VLSI technology these are above all:

- the use of regular or programmable structures,
- circuit development on more abstract levels,
- complex simulations and increased consideration of the circuit's testability.

The design engineer has to adjust his work style to these changes.

A key requirement is that users with quite different qualifications are meant to operate the design system. In this context, one has to think not only of various persons, but also of the process of change to which the user is subject while dealing with the design system. In the beginning the design engineer wishes to start operating the system with as few commands but as much assistance as possible. He wants decisions which he can not yet take due to his lack of knowledge and experience to be taken by the system. During the process of handling his problems and dealing with the design system he gains experience and wants to increase his participation in the decision-making process. When he is finally familiar with the design system, he wishes to have abbreviated command sequences and special solutions tailored to his needs.

Such a wide operation spectrum can be realized if the majority of the control parameters are first set by the system by default, and only a few parameters are offered to the user as a menu. When the user has utilized this free space, he may ask the system which further parameters exist and what the relevant effects are like. Such a design system must therefore be able to explain its implemented strategies. In addition to design functions it also assumes teaching functions.

The effort to be invested for these features at first seems to be quite high. In principle, this process is nothing but embedding the manual in the program system. The advantage for the design engineer is that he needs to learn only what is relevant for his problems. For the manufacturer, there is the additional aspect of keeping the manual permanently up-to-date is much easier.

Ease of operation is further increased by storing operation sequences in command files. The experienced user is thus able to create his own user interface to a certain degree.

A problem affecting the functional scope as well as the operation of the design system is the control of the design process by the user. The more complex a chip is, the more abstract it is described and the more details of realization are hidden from the user. The user will be satisfied with the majority of decisions taken by the CAD system, for the implemented know-how often exceeds his own know-how, or technically possible improvements mean nothing when compared with the development effort which is then necessary for special solutions. But sometimes a subcircuit brings up a bottle-neck situation and impairs, or even jeopardizes, the functionality of the entire chip. The characteristics of this subcircuit, however, can not be adjusted on the upper design levels, because, among other reasons, there are no adjustment parameters provided for this subcircuit on these levels, or even because this subcircuit is not visible. One could find a solution if one ignores the hierarchical process and allows "lateral" interventions. This, however, would result in considerable consistency problems. As long as these problems have not been solved in general, it is advisable not to release the "lateral" interventions. Otherwise the design security could no longer be guaranteed. Another possibility is to enable the design system to recognize and anticipate bottle-neck situations. A good basis for this may be expert systems and learning systems.

7.3 Embedding in the System Design

In a large system, which generally is a computer, a chip is just a component. A computer consists of several processing units and modules which are in turn composed of various submodules, for example, pc-boards. Each of these pc-boards contains various integrated circuits which are connected by means of tracks and via holes in multiple layers, thus making up a logic unit. Thus, when handling the system design, it is necessary to master and support the design of other physical carriers. In contrast to chip design, the design of large systems brings about additional problems, primarily due to the simultaneous mastering of different circuit technologies and due to the still higher degree of complexity.

Figure 7.4 shows the design steps of a system design process from the abstract design level, the system architecture, to production. From this figure it becomes clear that the chip design is only a part of the entire design process [7.1, 7.2].

The design of each level can be divided in substeps:

- *Decomposition path:*
  - design,
  - verification,
  - partitioning.
- *Synthesis path:*
  - integration,
  - verification.

Although the design and verification steps can also be found on the various levels of the chip design, partitioning plays only a minor role there: for chip design, the elements of the overall circuit are also grouped in subcircuits, according to function,

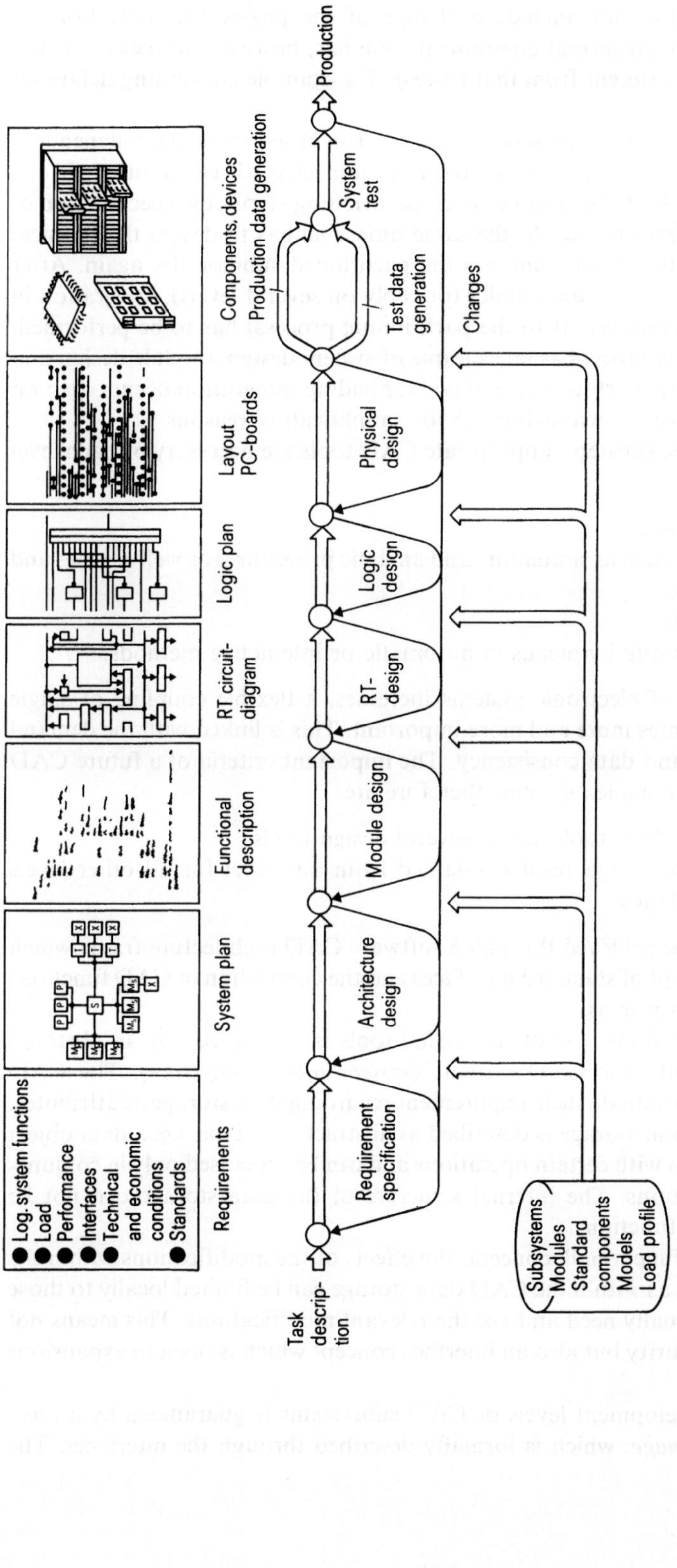

Fig. 7.4. System design process

but this partitioning does not include a change of the physical carrier. For the partitioning of a circuit on several components one has, however, to reckon with a behavior considerably different from that *on-chip*, for example concerning delays on wires.

The design and partitioning steps do not merely run in sequence and independent from each other, but rather, they are inseparably connected with each other. If one follows a top-down method, the result of each partitioning step is the specification of the subcircuits of the level below. At the same time, one has to design the physical carrier and/or the wiring which combines the partitioned submodules again. After successful construction of the submodules (possibly on several levels), integration in the overall system (as counterpart to the partitioning process) has to be performed. Figure 7.5 illustrates this process as an example of system design, module design and chip design. Increasingly, partitioning and corresponding integration occurs even on the chip level; this is not shown in Fig. 7.5 for simplification reasons.

For each of the three substeps, appropriate CAD tools are necessary on each level [7.3], such as:

- interactive design tools,
- verification methods such as simulators and analytic procedures as well as rules and plausibility checkers,
- test preparation tools,
- assistance in partitioning by means of automatic or interactive methods.

As the complexity of electronic systems increases, a flexible coupling of single CAD subsystems becomes more and more important. This is linked with the demand for system continuity and data consistency. The important criteria of a future CAD system for developing complex systems therefore are:

- the use of the same CAD tools across several design levels,
- the utilization of the design results obtained from one level for all other levels without data redundancy.

These targets can be achieved through a software CAD architecture frame which is based upon the concept of standard interfaces and the connection of CAD functions as well as of CAD subsystems.

The exchange or multiple use of functional tools is guaranteed by an abstract, functional interface with the CAD systems (*system-internal interface*). The CAD functions are able to formulate their requirement regarding data storage as attributes. At this interface, the data storage is described as abstract data type, i.e. a data object is defined in connection with certain operations and can be processed only in conjunction with these operations. The internal structure of the data storage can not be accessed by the CAD functions.

By means of this "functional" concept, the effects of the modifications within the CAD functions as well as within the CAD data storage can be limited locally to those components which actually need and use the relevant modifications. This means not only an increase of security but also an interface concept which is open to expansions and user-friendly.

The change of development levels or CAD subsystems is guaranteed by a *functional description language*, which is formally described through the interfaces. The

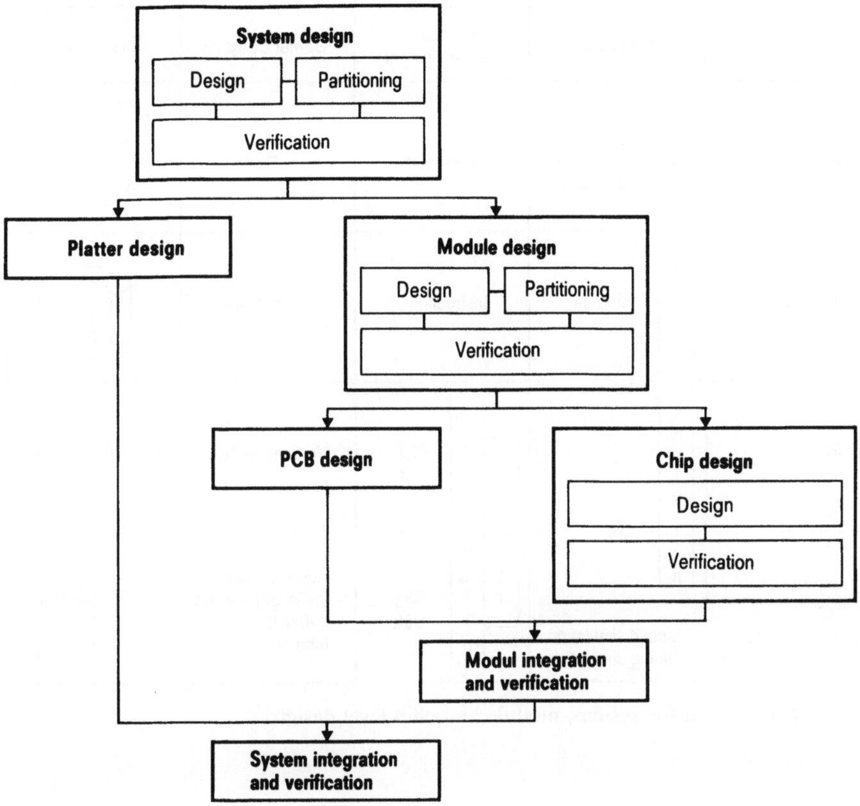

Fig. 7.5. Partitioning and integration

language concept is based upon the principles of modern and higher-level programming languages, and allows a flexible data exchange between CAD systems. In principle, such an interface consists of a *declaration section*, describing the format of the data that is to be transferred, and a *data section*, containing the data in a specific form. The conversion into the target system can be performed by means of *individual adapter modules*. Realization is supported by compiler technologies.

This *external standard interface* makes it possible to exchange models and data between the CAD systems. Thus it is possible, for example, for the module- or system-level to access models which have been designed on the chip level. An interface concept of this kind also allows the configuration of CAD systems in accordance with their function and performance requirements.

Figure 7.6 shows a CAD formation which consists of three autonomous CAD procedures for the system level (HERMES), the module level (PRIMUS) and the chip level (VENUS). To each of these systems one allocates CAD functions which can also be applied in other systems (simulators, for example [7.4]). Results of the individual design levels (CAD subsystems) are also utilized in other levels.

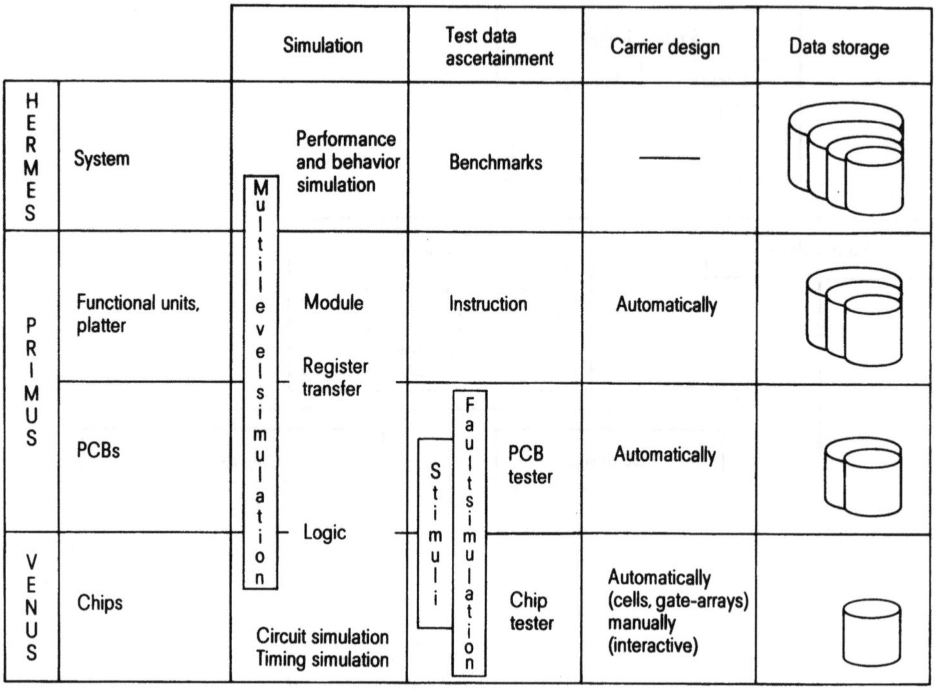

Fig. 7.6. CAD formation for system, module and chip level design

7.4 Embedding in the Technology Development

The semiconductor technology has a considerable effect on IC design systems due to the rapid progress in the development of processing technology. We distinguish two different influences on the CAD system:

- Indirect effects of lower, more physical design levels on the higher levels.
- Direct dependence of cell development on technology development.

Influence of Physics on the Higher Design Levels

The design process is a step by step process towards a physical realization. Physical constraints are noticeable already during the design of higher levels. Here, the design engineer's experience becomes useful; he has to estimate all these influential factors without having designed all or even one of the lower levels. Take *floorplanning* for example: With the number of gate functions and the complexity of the wiring, one may extrapolate the space required for a certain block. The same applies with regard to estimated line delays as early as logic design. "Estimators", which are able to advise the design engineer due to their implemented design know-how, are suitable because of methods of "artificial intelligence". The IC design process demands compromises between often contradicting design results. Apart from a purely algorithmic procedure (for placement and wiring, for example), heuristic methods play an ever increasing role. Expert systems supporting various phases of the design process as design assistants exist only in the research stage [7.5–7.8]. The objective is the integration of algorithmic and heuristic methods.

The complexity of the design process is partly based upon the fact that the basic estimation values can only be checked and corrected by means of the actual design results. This leads to iteration loops, in many cases across several levels. This cyclic design verification can be reduced by means of simultaneous simulations on various design levels (mixed-mode simulation). Thus, one may first carry out a rapid simulation on a clearly and distinctively structured design level, and second perform a refined and more precise simulation ("*zooming in*") for critical sections.

It is possible to extend this partially refined simulation to the lower levels and thus consider, for example, the influence of physical parameters. This makes it possible to recognize and remove problems which otherwise would only be discovered later. Moreover, by applying mixed-mode simulation, the data quantities and the computing times are reduced; otherwise it would hardly be possible for future VLSI circuits to manage those data quantities and computing times on lower levels.

Mixed-mode simulators are nowadays realized for logic and circuit levels as timing simulators [7.9] (DIANA, MOTIS, SPLICE). In this case, however, problems occur during the signal transformation between logic and electric signals and during event control. In order to decide which part of the circuit to simulate on which level, one may again access expert systems [7.10].

Mixed-mode simulation has already been successfully utilized in the circuit and device levels (MEDUSA [7.11]). The rudiments of a mixed-mode simulation in more than two levels (from system to circuit level) are implemented in the CAD system CASCADE [7.12].

Coupling of Process and Cell Development

The development of physical models for electronic components is a time-consuming process. It is repeated for each new process development or for modifications of existing processes. Because the parameters of the elementary components form the basis for the cell development and subsequent chip development, a mainly sequential procedure results (Fig. 7.7 a); the process development precedes the determination of the process parameters. Device modeling is joined with a final parameter determination. The subsequent cell development already overlaps the development of the first chips.

If it were possible now to anticipate the results of the process and device development and the determination of the corresponding parameters by applying rapid simulation processes, cell and chip development could start considerably earlier, although only with tentative parameters. As soon as the actual process and device development has been accomplished, the cell and chip development, which continued in the meantime, has to be adjusted to actual conditions (Fig. 7.7 b). This is possible only if there is data continuity for the CAD procedures for process and device development and simulation.

In addition to this trend towards integration of CAD process sections, there is a trend towards simplification of component modeling. Although physical models are

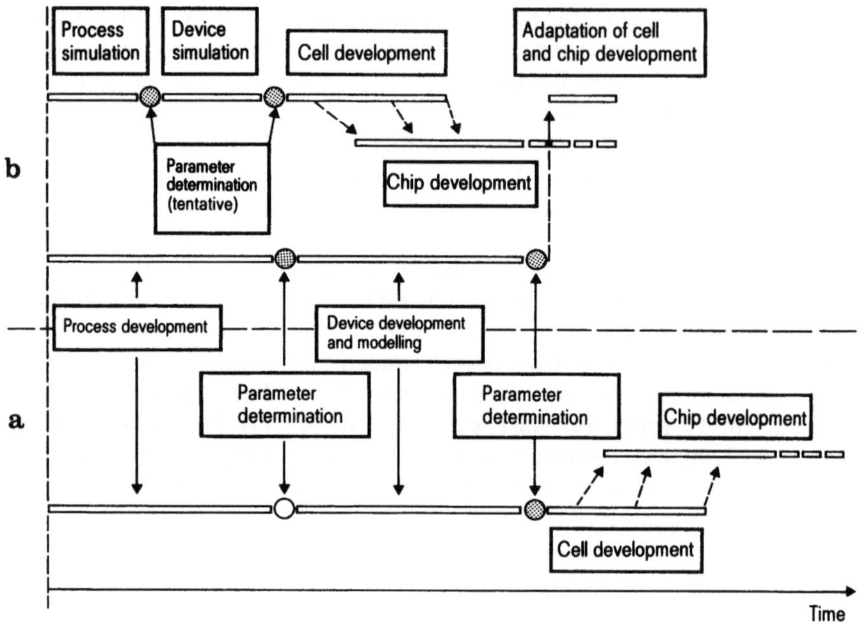

Fig. 7.7. Impacts of coupling of process and cell development

desirable for a deeper understanding of the phenomena, simpler table oriented models are in many cases sufficient for a fast estimation of behavior. The relevant data is made available immediately for development and simulation without a complicated adjustment process [7.13].

As state of the art for the process and device simulation, there are simulators applied for one and two (spatial) dimensions [7.14, 7.16]. Because additional physical effects have to be taken into account for increasingly refined structures, a three-dimensional structure becomes necessary. A prototype for a 3D device simulator is introdcued in [7.17]. Furthermore, CAD systems are being applied or developed, combining process, device, and circuit simulation [7.18, 7.19].

The integration of the CAD procedures from process to chip development includes another advantage besides the shortening of the entire development time. As always, the effect of process modifications and scattering (which occurs even for stable processes) on the behavior of already completely developed integrated circuits can be ascertained quickly and precisely. This indicates the possibility of closing the control loop from process data survey to functional behavior of the chip back to process control.

Literature – Chapter 7

7.1 Gonauser, M.; Kober, R.; Wenderoth, W.: A Methodology for Design of Digital Systems and Requirement for a Computer Aided System Design Environment. Proc. IFIP WG 10.1 Working Conf. Methodology for Computer System Design, Lille, France, 1983.

7.2 Siewiorek, D. P.; Guise, D.; Birmingham, W. P.: Proposal of Research on DEMETER: Design Methodology and Environment. Carnegie-Mellon-University, Jan. 22, 1983.

7.3 Gonauser, M.; Sauer, A.: Needs for High Level Design Tools. Proc. IEEE Int. Conf. Computer Design, 1983.

7.4 Egger, F.; Frantz, D.; Gonauser, M.: SMILE-A Multilevel Simulation System. Proc. IEEE Int. Conf. Computer Design (ICCD) 1984, pp. 188–193.

7.5 Fujita, T.; Goto, S. W.: Knowledge – Base and Algorithm for VLSI Design. Proc. ISCAS 85, pp. 877ff.

7.6 Lob, C.; Spickelmier, R.; Newton, A. R.: Circuit Verification Using Rule-Based Expert System Critic. Proc. ISCAS 85, pp. 881ff.

7.7 Kowalski, T. J.; Geiger, D. J.; Wolf, W.; Fichtner, W.: The VLSI Design Automation Assistant: A Birth in Industry. Proc. ISCAS 85, pp. 889ff.

7.8 Bushnell, M. L.; Director, S. W.: ULYSSES: An Expert System Based VLSI Design Environment. Proc. ISCAS 85, pp. 893ff.

7.9 Horneber, E.-H.; Feldmann, U.: Timing Simulation and Mixed-Mode Simulation of MOS Integrated Circuits. Siemens Research and Development Reports 11 (1982), pp. 12–21.

7.10 De Man, H.; Reynaert, P.; Bolsens, I.: Guided Mixed-Mode Circuit-Timing Simulation. Proc. Europ. Conf. Circuit Theory and Design, pp. 429–432. Stuttgart, Sept. 1983.

7.11 Engl, W. L.; Dirks, H. K.: Functional Device Simulation by Merging Numerical Building Blocks. Proc. of NASECODE II, pp. 34–62. Dublin, 1981.

7.12 Le Faou, C.: Hierarchical Multilevel Mixed-Mode Simulation in CASCADE. IMAG/ARTEMIS Res. Rep. No. 513, Grenoble, Mars 1985.

7.13 Dirks, H. K.; Eickhoff, K.-M.: Numerical Models and Table Models for MOS Circuit Analysis. Proc. of NASECODE IV Conf., Dublin, 1985.

7.14 Selberherr, S.: Analysis and Simulation of Semiconductor Devices. Wien: Springer 1984.

7.15 Miller, J. J. H. (ed.): New Problems and Solutions for Device and Process Modeling. Dublin: Boole Press 1985.

7.16 Selberherr, S.: Numerical Modeling of MOS Devices: Methods and Problems. pp. 122–137 in [7.15].

7.17 Shigyo, N.; Onga, S.; Dang, R.: A Three-Dimensional MOS Device Simulator. pp. 138–149 in [7.15].

7.18 Fukuma, M.: Recent Activities in Process and Device Modeling in NEC. pp. 24–34 in [7.15].

7.19 Prendergast, J.: An Integrated Approach to Modeling. Proc. of NASECODE IV Conf., Dublin, 1985.

Subject Index

MIX
Papier aus verantwortungsvollen Quellen
Paper from responsible sources
FSC® C105338

If you have any concerns about our products,
you can contact us on
ProductSafety@springernature.com

In case Publisher is established outside the EU,
the EU authorized representative is:
Springer Nature Customer Service Center GmbH
Europaplatz 3, 69115 Heidelberg, Germany

Printed by Libri Plureos GmbH
in Hamburg, Germany